POTATO: PRODUCTION, PROCESSING, AND PRODUCTS

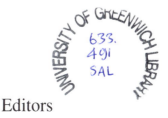

Editors

D. K. Salunkhe, Ph.D.
Professor Emeritus
Department of Nutrition and Food Sciences
Utah State University
Logan, Utah

S. S. Kadam, Ph.D.
Professor and Head
Department of Food Science and Technology
Mahatma Phule Agricultural University
Rahuri, Maharashtra, India

S. J. Jadhav, Ph.D.
Professor and Head
Department of Chemistry and By-Products
Vasantdada Sugar Institute
Manjri (BK), Pune
Maharashtra, India

CRC Press
Boca Raton   Ann Arbor   Boston

Library of Congress Cataloging-in-Publication Data

Salunkhe, D. K.
   Potato : production, processing, and products / D.K. Salunkhe,
  S.S. Kadam, S.J. Jadhav.
     p.   cm.
   Includes bibliographical references and index.
   ISBN 0-8493-6876-6
   1. Potato products. I. Kadam, S. S. II. Jadhav, S. J.
  (Sadashiv Jotiram), 1944-   . III. Title.
  TP444.P6S25 1991
  664′.80521--dc20

90-43555
CIP

This book represents information obtained from authentic and highly regarded sources. Reprinted material is quoted with permission, and sources are indicated. A wide variety of references are listed. Every reasonable effort has been made to give reliable data and information, but the authors and the publisher cannot assume responsibility for the validity of all materials or for the consequences of their use.

Direct all inquiries to CRC Press, Inc., 2000 Corporate Blvd., N.W., Boca Raton, Florida, 33431.

International Standard Book Number 0-8493-6876-6

Library of Congress Card Number 90-43555
Printed in the United States

PREFACE

The production of a successful commercial crop depends on the development of suitable varieties in addition to agronomic and climatic conditions. While high yields are important to the farmer, the crop must be of sufficiently high quality to meet the demands of the consumer market and the requirements of the food industry for processing. Such quality factors include color, appearance, texture, and flavor together with nutritional quality and wholesomeness. Quality of potato tubers for fresh and processing utilization depends upon genetic, climatic, biotic, chemic, and edaphic factors: genetic—varietal characteristics as determined by its hereditary make-up; climatic—precipitation, temperature, and duration and intensity of sunshine; biotic—animals and competition with other plants and the crop itself in relation to its environment; chemic—applications of chemicals such as herbicides, fungicides, pesticides, growth regulators, hormones, and other chemicals; edaphic—physical, chemical, and biological properties of the soil in influencing the capacity to supply the crop with necessary nutrients and water. The successful production of potatoes of high quality is fundamentally influenced by the individual as well as cumulative effects of these five factors. Though these are well-recognized factors, their effects on quality need further exploration and definition. The particular factors such as selection of variety, date of planting, fertilizers, herbicides, pesticides, fungicides, vine-killing treatments, soil structure and texture, irrigation, location, and their cumulative effects on fresh and processed products need periodic definitions. Likewise, storage temperature, duration and subsequent conditioning, specific gravity of the tubers, and several other factors influence the final quality in color, flavor, texture, and nutritive value of the fresh and processed products.

The potato is a plant that is perhaps the largest producer of proteins, carbohydrates, thiamine, ascorbic acid, and minerals on a per acre dry-weight basis than any other cultivated crop. Since 1960, potato production and the consumption of potato products have been significantly increased in developed and developing nations. Due to research and development there are now convenient and instant processed products from potato tubers, such as potato chips, French-fried potatoes, potato flakes, hash-brown potatoes, potato granules, diced potatoes, potato flour, potato starch, canned potatoes, prepeeled potatoes, potato pancake mixes, potato nuts, potato puffs, potato salad, chemicals such as solanine, chaconine, alcohol, lactic acid, and many more products.

Increased concern for a healthier diet has resulted in a larger consumption of fresh produce and processed products in the world over the past decade. The greater urbanization has placed enormous demands on the food industry for a greater variety of processed vegetables, particularly potato products, in the food market. This has led to the development of new processing technologies that effectively preserve the quality attributes associated with the original fresh produce. The future of potato processing and its convenience products is bright, particularly for potato chips, restructured potato chips, frozen French fries, and dehydrated potato granules and flakes.

This book brings out up-to-date information relative to potato production, processing, and utilization and will be useful to researchers, teachers, undergraduate and graduate students, potato growers, shippers, processors, and product development personnel in the potato processing industry worldwide.

D. K. Salunkhe
S. S. Kadam
S. J. Jadhav

FOREWORD

During the past thirty years, potato production and utilization have changed considerably around the world. No longer is the potato crop suitable only for cool climates. The potato has moved down the hills into warm climates and into rice paddies during the dry season. In fact, the potato fits very well into cereal-based farming systems; as the countries of Asia become self-sufficient in rice, the potato is becoming a priority alternative crop.

Only a small portion of the genetic wealth which exists in potato germplasm collections has been utilized up to now in breeding programs. New breeding techniques are rapidly expanding the ability to identify and utilize previously unexploited sources of resistance. Advances in biotechnology are also providing tools that can help in constructing varieties with the specific resistance components required. Thus, limits to climates where potatoes can be grown and potentials for yield are changing rapidly.

Production of quality tuber seed is no longer confined to remote, cool-season production areas. New seed technology is permitting countries in the tropics to establish their own seed programs independent of expensive importations. The use of true potato seed instead of tubers is providing a way for good, clean, planting material to reach small growers in tropical climates.

In the developing world the growth of the fast-food industry in recent years has markedly changed the way potatoes are consumed, and has called for changes in the varieties produced and in the method of storage. The rapid expansion in potato processing which took place some time ago in the developed world is now taking place in third-world countries. While potato production has stabilized in the developed world, in the developing world it is increasing faster than that of any other major world food crop. Many developing countries are only now starting to rely on the potato as a major food source. This trend can be expected to continue as scarcity of land resources requires a change from extensive production of certain food crops to intensive production with priority consideration given to calories and high-quality protein per unit area and per unit of time.

As the potato has moved into warm climates, the problems of post-harvest storage have become magnified and the need for processing increased. Maintaining an adequate storage environment is difficult in warm climates and storage losses increase rapidly. Thus, there is the need for new innovations in potato processing in order to quickly convert the potato into nonperishable, culturally acceptable, and highly nutritional food products.

I am pleased to see that in this book the recent research that has permitted the rapid expansion of potato production into nontraditional climates has been concentrated. Policy makers, researchers, and teachers will need a continuing flow of the latest information available, because in the decisions for changes to be made, sustainability of production, storage, processing, environmental enhancement, and input requirements will be as important, if not more so, than production per se.

Richard L. Sawyer
Director General
International Potato Center
Lima, Peru

THE EDITORS

D. K. Salunkhe is Professor of Nutrition and Food Sciences at Utah State University, Logan, Utah. Under his guidance, 80 postgraduate students received their M.Sc. or Ph.D. degrees. He has authored about 400 scientific papers, book chapters, and reviews. Some of his articles received recognition and awards as outstanding articles in biological journals.

He was Alexander Humboldt Senior Fellow and Guest Professor at the Institut fur Strahlentechnologie der Bundesforschungsanstalt fur Lebensmittelfrischaltung, West Germany. He was Guest Lecturer at the Technological Institute, Moscow, U.S.S.R., and an exchange scientist to Czechoslovakia, Romania, and Bulgaria on behalf of the National Academy of Sciences, National Research Council, as well as Advisor to U.S. Army Food Research Laboratories and many food storage, processing, and consumer organizations.

Professor Salunkhe was Sigma Xi President, Utah State University Chapter; Fellow of Utah Academy of Sciences, Arts and Letters; Fellow of the Institute of Food Technologists; and Danforth Foundation Faculty Associate. He delivered the Utah State University 50th Faculty Honor Lecture, ''Food, Nutrition, and Health: Problems and Prospects'' on the basis of his creative activities in research and graduate teaching.

Professor Salunkhe was Vice-Chancellor of Marathwada Agricultural University, Parbhani, and Mahatma Phule Agricultural University, Rahuri, in India during 1975 to 1976 and 1980 to 1986, respectively.

Professor Salunkhe was a member of the editorial boards of the *Journal of Food Biochemistry, International Journal of Plant Foods for Human Nutrition, Journal of Food Science,* and *Journal of Food Quality.*

He has authored or co-authored two textbooks: *Storage, Processing and Nutritional Quality of Fruits and Vegetables* (CRC Press, 1974) and *Postharvest Biology and Handling of Fruits and Vegetables,* with Professor N. F. Haard (AVI Publishing Company, 1975).

Professor Salunkhe has authored the following reference books: *Postharvest Biotechnology of Fruits* (CRC Press, 1984), Vol. 1 and 2; *Postharvest Biotechnology of Vegetables* (CRC Press, 1984), Vol. 1 and 2; *Postharvest Biotechnology of Oilseeds* (CRC Press, 1986) with Dr. B. B. Desai; *Postharvest Biotechnology of Cereals* (CRC Press, 1985); *Postharvest Biotechnology of Food Legumes* (CRC Press, 1985) with Dr. J. K. Chavan and Dr. S. S. Kadam; and *Modern Toxicology* (Metropolitan, 1984), Vol. 1, 2, and 3 with Dr. P. K. Gupta. He has edited the following books: *Nutritional and Processing Quality of Sorghum* (Oxford and IBH, 1984) with Dr. J. K. Chavan and S. J. Jadhav; *Quality of Wheat and Wheat Products* (Metropolitan, 1986) with Dr. S. S. Kadam and Dr. A. Austin; and *Legume-Based Fermented Foods* (CRC Press, 1986) with Dr. N. R. Reddy and Dr. M. D. Pierson; *Vegetable and Flower Seed Production* (Agricole Publishing Academy, 1987) with Dr. B. B. Desai and Dr. N. R. Bhat; *Aflatoxins in Foods and Feeds* (Metropolitan, 1987); and *Postharvest Biotechnology of Sugar Crops* (CRC Press, 1988) with Dr. B. B. Desai.

S. S. Kadam obtained his B.Sc. (Agriculture) degree from Maharashtra Agricultural University in 1969. In 1971 and 1975, he received his M.Sc. and Ph.D. degrees in Biochemistry from the Indian Agricultural Research Institute, New Delhi. He was a Postdoctoral Fellow at Forschungstelle Vennesland der Max-Planck Gesselschaft, Berlin, Germany in 1977 and 1978. Dr. Kadam has authored over 80 scientific papers and reviews and recently authored three books with Professor Salunkhe. He worked as Biochemist and Associate Professor of Biochemistry in the College of Agricultural Technology, Marathwada Agricultural University, Parbhani from 1975 until 1981. Dr. Kadam was Visiting Scientist to CIMMYT, Mexico in the year 1982. He worked as Professor of Food Science and Technology at Mahatma Phule Agricultural University, Rahuri from 1981 to 1984. He was Visiting Professor in the Department of Agricultural Biochemistry and Nutrition, University of Newcastle-upon-Tyne, England, in the year 1985 to 1986. Dr. Kadam was Visiting Scientist to the U.S. in conjunction with a P. L. 480 research grant in 1988. He worked as Professor and Head, Department of Biochemistry, Mahatma Phule Agricultural University, Rahuri from 1984 to 1987. Currently, he is working as Professor and Head, Department of Food Science and Technology, Mahatma Phule Agricultural University, Rahuri, India.

Dr. S. J. Jadhav received his B.Sc. (Chemistry) at Shivaji University, Kolhapur and M.Sc. (Organic Chemistry) at Poona University, Poona, India, in 1965 and 1967, respectively. He completed his Ph.D. degree in Nutrition and Food Science Technology and was a Post-Doctoral Fellow at Utah State University, Logan, Utah in 1969 to 1973. He worked as Research Associate in charge of the Alberta Potato Commission's Research Unit at the University of Alberta in 1974 to 1975 and Food Scientist at I & S Produce Ltd., Edmonton, Alberta, Canada in 1976 to 1980. He became Head and Professor of Food Technology at the Marathwada Agricultural University, Parbhani, India, in 1981 to 1982. He has authored over 100 research papers and acted as co-editor of a book on *Nutritional and Processing Quality of Sorghum* (Oxford and IBH, 1984) with Professor D. K. Salunkhe. He is presently working as Chief Scientist, Professor and Head, Department of Sugar Chemistry and By-Products, Vasantdada Sugar Institute at Pune, India.

CONTRIBUTORS

R. N. Adsule, Ph.D.
Department of Biochemistry
Mahatma Phule Agricultural University
Rahuri, Ahmednagar
Maharashtra, India

J. K. Chavan, Ph.D.
Department of Biochemistry
Mahatma Phule Agricultural University
Rahuri, Ahmednagar
Maharashtra, India

U. T. Desai, Ph.D.
Department of Horticulture
Mahatma Phule Agricultural University
Rahuri, Ahmednagar
Maharashtra, India

S. S. Dhumal, Ph.D.
Department of Plants, Soils,
    and Biometeorology
Utah State University
Logan, Utah

S. J. Jadhav, Ph.D.
Vasantdada Sugar Institute
Manjri (BK), Pune
Maharashtra, India

N. D. Jambhale, Ph.D.
Department of Botany
Mahatma Phule Agricultural University
Rahuri, Ahmednagar
Maharashtra, India

S. S. Kadam, Ph.D.
Department of Food Science
    and Technology
Mahatma Phule Agricultural University
Rahuri, Ahmednagar
Maharashtra, India

A. Kumar, Ph.D.
Vasantdada Sugar Institute
Manjri (BK), Pune
Maharashtra, India

G. Mazza, Ph.D.
Agriculture Canada
    Research Station
Morden, Manitoba, Canada

R. B. Natu, Ph.D.
Vasantdada Sugar Institute
Manjri (BK), Pune
Maharashtra, India

S. P. Phadnis, Ph.D.
Vasantdada Sugar Institute
Manjri (BK), Pune
Maharashtra, India

D. K. Salunkhe, Ph.D.
Department of Nutrition and Food Sciences
Utah State University
Logan, Utah

D. M. Sawant, Ph.D.
Department of Plant Pathology
Mahatma Phule Agricultural University
Rahuri, Ahmednagar
Maharashtra, India

B. N. Wankier
H. J. Heinz Company
Ore-Ida Foods Incorporated
Ontario, Oregon

TABLE OF CONTENTS

Chapter 1

INTRODUCTION

D. K. Salunkhe and S. S. Kadam

The potato (*Solanum tuberosum* L.), with an annual production of nearly 300 million metric tons, is one of the major food crops grown in a wide variety of soils and climatic conditions.[1] It is the most important dicotyledonous source of human food. It ranks as the fourth major food crop of the world, exceeded only by wheat, rice, and maize.[2] The dry matter production of potatoes per unit area exceeds that of wheat, barley and maize (Table 1). Yields of protein per unit of land exceed those of wheat, rice, and maize by factors of 2.02, 1.33, and 1.20, respectively. Because of increasing yield per unit area of land, total potato production has been increasing in both developed and developing countries in the past 20 years. In addition, the rate of production in developing countries has increased significantly more than that of developed countries (Figure 1). Similarly, potato production rate has increased as compared to other tuber and root crops (Figure 2). The production figures of potatoes in major producing countries are presented in Table 2. Per capita availability of potato production is highest in Europe, and especially in Eastern Europe, where a large share of total production is fed to livestock (Table 3). Most foreign trade in potatoes takes place within Europe. The countries of Western Europe, the U.S., and Japan have the highest potato yields in the world. The lowest yielding countries are in tropical Africa, Asia, and the Andean region of South America.[3]

The potato is thought to have originated in South America. Archeologists and historians are able to trace the potato back to at least A.D. 200, at which time it was being cultivated in mountainous areas of Peru. For many centuries, the potato served as the main food source of the people of Peru. Apparently, sufficient potatoes were dried to provide for periods when potatoes were not available between successive crops or when shortage developed because of frost or other unfavorable growing conditions. When the Spanish explorers arrived in the New World, potatoes were widely grown throughout South and Central America. It appears likely that the potato was derived from some wild species occurring in South America, perhaps in the Andes of Peru and Bolivia.[4] Following the conquest of Peru, the Spaniards introduced potatoes to their own country, spreading this crop to many European countries including Italy, Belgium, Germany, France, Switzerland, and the Netherlands by the end of the 16th century. However, the potato was grown only as a garden crop. A considerable time had to elapse before it was considered fit for human consumption, probably because it was not mentioned in the Bible. According to Nash,[5] in Russia potatoes were referred to as "Devil's apples", while in France they were considered fit only for animals and poor people, and Sweden was particularly slow in adopting them.

Potatoes were probably introduced to Britain before the close of the 16th century. A well-known agricultural writer, Donaldson of Scotland, in his book on agriculture published in 1697, entitled *Husbandary Anatomized,* advocated the potato crop for men with large families. Barring Ireland, potatoes were possibly not grown as a field crop in any of the Western European countries until the latter half of the 18th century when it became regarded as the most profitable new crop, mainly because it could be used for human consumption, and pigs were found to thrive well on potatoes. Potatoes continue to be widely grown in the 20th century, with a major portion (80% being produced and consumed by Western Europe and the U.S.S.R.,[5] (Figure 3).

The early potato remained as a botanical curiosity until about the mid-18th century, despite the fact that it was a completely accepted cultivated crop in South America. The

TABLE 1
Total Production of Dry Matter (DM) and Total Protein, and
Production of DM and Protein Per Unit Area of Major Staple Food
Crops[2]

Crop	Total DM production (t × 10⁷)	Total protein production (t × 10⁶)	DM production (t/ha)	Protein production (t/ha)
Wheat	27.5	32.9	1.30	0.156
Rice	26.7	23.2	1.97	0.172
Maize	23.5	24.7	2.13	0.224
Barley	11.4	11.6	1.46	0.148
Sorghum/millets	8.2	7.4	0.73	0.066
Potato	6.6	6.0	2.93	0.266
Sweet potato/yam	3.9	2.9	3.82	0.280
Cassava	3.4	0.8	4.92	0.115
Soybean	4.2	16.7	2.62	1.043

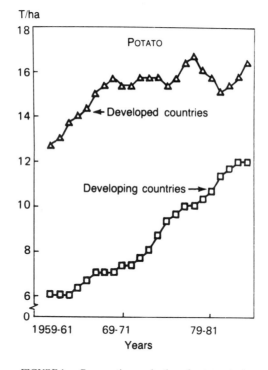

FIGURE 1.   Comparative production of potatoes in developed and developing countries from 1961 to 1985.

crop was probably introduced into Europe on two occasions, firstly into Spain in about 1570, and secondly into England a little later, in 1590. After these introductions the potato spread into nearly every part of the world. From Spain it moved to continental Europe and parts of Asia, and from England it spread to Ireland, Scotland, Wales, and parts of northern Europe, and one or more of the British sources spread to the British overseas colonies, including what was later to become the U.S.[4]

Today, the potato is cultivated in Peru as intensively as in Inca times before the advent of Spanish explorers. In the Lima market it is reported that one may see extraordinary and unusual types of potatoes (Figure 4). In Peru, the natives, through centuries of cultivation,

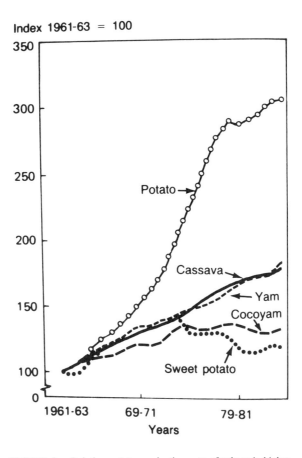

Index 1961-63 = 100

FIGURE 2. Relative potato production rate of tubers is higher than other tuber and root crops in developing countries.

TABLE 2
Major Potato Producing Countries of the World[3]

Country	Production × 1000 metric tons				
	1969—71	1978	1979	1980	1985
U.S.S.R.	93,739	86,124	90,956	66,900	73,000
Poland	45,013	46,648	49,572	26,400	36,546
U.S.	14,483	16,616	15,535	13,653	18,331
China	11,029	12,529	12,536	12,537	45,528
Germany (East)	10,432	10,777	12,243	8,568	11,500
India	4,482	8,135	10,125	8,306	12,642
France	8,569	7,467	7,450	7,485	7,814
Germany (West)	15,804	10,510	8,716	6,694	8,704
U.K.	7,359	7,330	6,485	6,327	6,850

have developed varieties that are scarcely recognizable as potatoes. Some are golden yellow, others are red, purple, pink, pale lilac, and blue; spotted and striped; round and oblong; crenated and cylindrical; and are smooth skinned or warty as toads. These varieties are presently used throughout the world by plant breeders and geneticists to develop new cultivars for specific uses. Spanish and English explorers recognized the value of the potato at an

TABLE 3
Per Capita Availability and Consumption of Potato in the Top Ten Countries in 1980/82[12]

Country	Availability per capita (kg/year)	Human consumption (%)	Consumption per capita (kg/yr)
Poland	938	12.9	121
Germany (East)	628	26.9	169
Netherlands	358	24.7	88
U.S.S.R.	279	34.9	97
Ireland	265	35.8	95
Czechoslovakia	221	38.2	84
Romania	205	46.1	95
Austria	165	27.5	45
Belgium-Luxembourg	155	84.7	131
Germany (West)	149	40.3	60

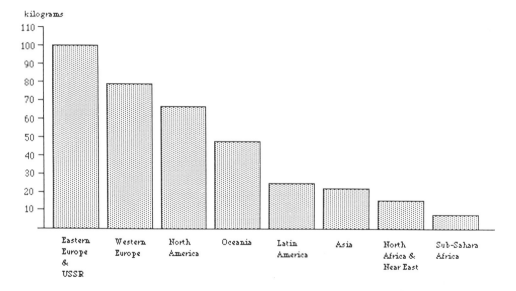

FIGURE 3.   Per capita potato consumption by region, 1979/81. (From FAO, Food Balance Sheets, Rome, 1984. With permission.)

early date and obtained them from natives for use in provisioning their ships. Then, it was not many years before the potato was introduced into European countries. Potatoes were regarded as a luxury in Spain. However, in Italy, potatoes were grown for cattle feed. In 1719, a group of Irish settlers brought potatoes with them to New Hampshire. Being so familiar with potatoes, they did much to promote its cultivation and acceptance in New England. Hence, potatoes are often termed as "Irish" Potatoes. In recent years the potato has become a very popular crop in many African and Asian countries.

In Europe, experiments with many types of dried potatoes were made with the desire to increase the usefulness of tubers as provisions for ships. Dried potatoes were ground to make potato flour, which was stored for years. During World War I and II, a number of dehydrated potato products were manufactured for military use. Thus, potato processing and potato products became popular and acceptable throughout Europe and the southeastern countries. Potato starch is a potato product that is utilized in significant quantities in Europe, the U.S., Canada, and Asia.

FIGURE 4.   Variation in shape, size, and color in market samples of potatoes in Lima, Peru market.

Potatoes are an excellent source of carbohydrates and contain significant amounts of phosphorus, potassium, calcium, and vitamins, especially vitamin C.[6] Their protein content of over 10% on a dry weight basis brings them relatively close to the 11% protein in wheat flour. Because of the relatively high nutritive value and lysine, methionine, cystine, and cysteine contents of potato protein, it is a valuable supplement to cereal proteins.[7] Potatoes thus serve as a significant source of proteins (10 to 15% of total protein requirements), a major source of vitamin C, and an important source of energy. In addition to ascorbic acid, potatoes also provide minerals, principally iron, and vitamins such as thiamin, nicotinic acid, riboflavin, and pro-vitamin A (β carotene).[8]

Potatoes are generally processed before consumption. Many times the produce is stored and then used for processing. Thus storage and processing are frequently needed to prevent seasonal gluts and to increase the availability of potatoes to consumers throughout the year (Figure 5). Several changes occur in nutritional composition of the potato during storage and processing. Significant losses in quantity and quality occur during storage if potatoes are not stored properly. In order to meet the demand of growing population in the developed countries, efforts are required to improve storage facilities which will help to avoid losses during storage.

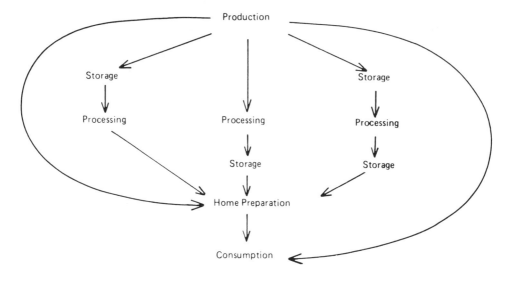

FIGURE 5.   Possible routes of production to consumption of potatoes.

Quality of potato tubers for processing depends upon genetic, climatic, biotic, chemic, and edaphic factors: genetic—varietal characteristics as determined by its hereditary make up; climatic—precipitation, temperature, and the duration and intensity of sunshine; biotic—animals and competition with other plants and the crop itself in relation to its environment; chemic—applications of chemicals such as herbicides, fungicides, pesticides, growth regulators, hormones, and other chemicals; edaphic—physical, chemical, and biological properties of the soil influencing the capacity to supply the crop with necessary nutrients and water. The successful production of potatoes of high quality is fundamentally influenced by the individual as well as the cumulative effects of these five factors. Though these are well-recognized factors, their effects on quality need further exploration and definition. The particular factors such as selection of variety, date of planting, fertilizers, herbicides and pesticides, fungicides, vine-killing treatments, soil structure and texture, irrigations, locations, and their cumulative effects on processed products need periodic definitions. Likewise, storage temperature, duration and subsequent conditioning, specific gravity of the tubers, and several other factors influence the final quality—color, flavor, texture, and nutritive value of the processed products.

In the last 30 years, potato cultivation has moved into warm, tropical rice-fields in the dry season (October to January). The potato crop fits into cereal-based agriculture; as the countries of Asia become self-sufficient in cereals, the potato is becoming an excellent alternative food crop, both for processing and fresh consumption after cooking.

Since 1960, potato production and consumption of potato products have been significantly increased due to research and development in convenience and instant processed products from potato tubers such as potato chips, French-fried potatoes, potato flakes, hash-brown potatoes, potato granules, diced potatoes, potato flour, potato starch, canned potatoes, prepeeled potatoes, dehydrated instant mashed potatoes, restructured potato chips, potato soup, shoestring potatoes, potato pancake mixes, potato nuts, potato puffs, potato salad, and chemicals such as solanine, chaconine, alcohol, lactic acid, and many more products.[9,10] Per capita consumption of processed potatoes increased from 1.9 lb in 1940 to 80 lb in 1978 when processing used about 48% of the total crop produced in the U.S.[10] Frozen products account for more than one half of all potatoes that are processed.[10] Similarly, dehydrated products and chips are the other major products of the potato-processing industry. This situation in the U.K. and European countries is very similar.[11] In 1965, out of a total yearly supply of 100 kg per head, 6.6% was sold in processed form including 4.8% as crisps. In 1978, out of a total of 103 kg per head, 21.6% was sold in processed form (including 7.1% crisps, 7.8% frozen and par-fried products, 5.1% as canned and dehydrated products, and 2% prepeeled).[11] The future of potato processing and instant and convenience products is bright, particularly for frozen products, potato chips, restructured potato chips, and mashed potato granules.

This book presents up-to-date information relative to potato production, processing, and products and will be useful to researchers, teachers, professors, undergraduate and graduate students, potato growers, processors, and product-development personnel in the potato cultivation and processing industry.

REFERENCES

1. **Salunkhe, D. K. and Desai, B. B.,** *Postharvest Biotechnology of Vegetables,* CRC Press, Boca Raton, FL, 1985, 238.
2. **Hooker, W. J., Ed.,** *Compendium of Potato Diseases,* American Phytopathological Society, St. Paul, MN, 1986.
3. **Anon.,** F.A.O. Production Year Books, Food and Agriculture Organization, Rome, 1985.
4. **Hawkes, J. G.,** History of the potato, in *The Potato Crops: The Scientific Basis for Improvement,* Harris, P. M., Ed., Chapman and Hall, New York, 1978.
5. **Nash, M. J.,** *Crop Conservation and Storage in Cool Temperature Climates,* Pergamon Press, Oxford, 1978, 171.
6. **Woolfe, J. A.,** *The Potato in the Human Diet,* Cambridge University Press, 1987, 1.
7. **Ugent, D.,** The potato, *Science,* 170, 1161, 1971.
8. **Kay, D. E.,** *Root Crops, Trop. Prod. Inst. Crop and Product Digest No. 2,* Tropical Products Institute, London, 1973, 100.
9. **Champson, H. C. and Kelley, W. C.,** The potato, in *Vegetable Crops,* 5th ed., McGraw-Hill, New York, 1957, 374.
10. **Talburt, W. F. and Smith, O.,** *Potato Processing,* Van Nostrand Reinhold, New York, 1987, 8.
11. **Churchill, D.,** Frozen foods, *Financial Times (London),* p. 37, July 20, 1979.
12. **Horton, D. E. and Fano, H.,** *Potato Atlas,* International Potato Centre, Lima, Peru, 1985.

Chapter 2

STRUCTURE, NUTRITIONAL COMPOSITION, AND QUALITY

S. S. Kadam, S. S. Dhumal, and N. D. Jambhale

TABLE OF CONTENTS

I. INTRODUCTION

Potato tubers have been used for thousands of years to maintain and support the growth and health of human beings. In recent years these are used for preparation of various products. The nutritional composition and quality of tubers are influenced by genetics as well as environmental factors.[1-4] The different components of potato tubers vary in nutrient composition. This chapter presents information on structure and nutrient composition of potato tubers used for domestic consumption and commercial processing.

II. BOTANY

Potato (*Solanum tuberosum* L.) belongs to the family Solanaceae,[1] which also includes tomato, tobacco, pepper, eggplant, petunia, black nightshade, belladonna, and others. Out of several hundred species of *Solanum,* only the potato (*S. tuberosum*) and a few others bear tubers. Salaman[5] postulated that the modern potato was derived from hybrids of *S. tuberosum* and *S. andigenum.* Some varieties with resistance to late blight have been evolved from crosses with *S. demissum,* but for all practical purposes, the English potato or Irish potato is considered to be *Solanum tuberosum.*[6]

The potato is a herbaceous dicotyledon (Figure 1), sometimes regarded as a perennial because of its ability to reproduce vegetatively, although it is cultivated as an annual crop. The tubers are modified, thickened, underground stems, their size, shape, and color varying according to the cultivar. On the surface of the tubers are the eyes from which arise the growing buds. According to some authorities, as many as 20 species of cultivated potatoes have been recognized, but the widely accepted Dodd's classification includes a cultivated species *S. tuberosum,* plus five horticultural groups, and two hybrid cultivars, viz., *S. juzepazukii* and *S. curtilobum.*[7] Ugent[8] reported the botanical origin of the potato and described how it first became domesticated.

A. STEM

The above-ground stem is herbaceous and erect in its early stages of development. Later, it becomes spreading and prostrate. It attains a height of 2 to 5 ft or more. Several axillary branches usually are produced. The stems are round to sub-triangular or quadrangular in cross section and are of a green or purplish color. In many varieties, ridges or wings form on the angular margins and are especially prominent in young stems. The stems may be green or may contain an anthocyanin to give them a purple or reddish color. Most varieties have an upright and erect stem until flowering. The stolons are slender, underground lateral stems arising from buds on the underground portion of the stem. There is a considerable variation in the number, length, and diameter of stolons produced by different potato varieties grown in different climatic conditions. Initially, one stolon is produced at each node, but others may emerge later. The tubers are the greatly enlarged tips of stolons, although some stolons may not form tubers. The size, shape, and color of tubers may vary greatly.[8,9] The flesh color may be white, yellow, pink, red, or blue, the white types being accepted.

B. LEAVES

Leaves are alternate in a counterclockwise spiral. Petioles are semicircular in cross section, convex on the lower side, and slightly concave on the upper side. The petiole base is flattened and extends around about one third of the stem at the node. Wing-like margins of the petiole extend downward on the stem through one or two internodes. The first leaves arising from the seed piece usually are simple. Later formed leaves are compound, irregularly odd pinnate with petioled leaflets.

Although the number of leaflets varies with the variety, there are usually three or four

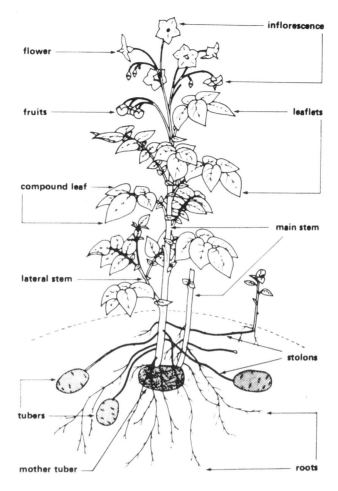

FIGURE 1. The potato plant leaves, stems, and tubers. (Courtesy of
International Potato Center, Lima, Peru, 1986. With permission.)

pairs of large oval leaflets with entire or serrate margins in addition to the terminal one.
Many times, smaller, secondary leaflets grow between the primary leaflets.[6]

C. ROOTS

Potatoes have a fibrous root system. Plants grown from seed develop a slender tap root
from which lateral branches arise to form the fibrous system.[7] Plants arising from tubers
also have a fibrous system consisting of adventitious roots arising in groups of three just
above the nodes of the underground stem. Although the greater portion of the root system
is in the surface foot of soil, they may penetrate as deep as 5 ft.

D. FLOWERS

The flower cluster is determinate, although it appears to be lateral due to the series of
superposed branches which resemble a simple axis. The corolla is five lobed and white,
yellow, blue, purple, or striped according to the variety. The calyx is tabular and lobed.
The five stamens are borne on the corolla tube and converge around the pistil. Anthers
dehisce by two terminal pores. The pistil consists of two carpels which form a two-loculed
ovary with a single style and stigma.[6]

E. TUBERS

The potato tuber is the underground stem (Figure 2A) which resembles the areal stem of the plant. This is formed at the tip of the stolon (rhizome) as a lateral proliferation of storage tissue resulting from rapid cell division and enlargement. The longitudinal section of a potato tuber showing principal structure components is presented in Figure 2B. The integument (periderm), early in its development, is only a few cells thick, but later becomes massive with a long micropyle. The mature integument consists of three layers: (1) an external region of a single layer of epidermal cells; (2) an intermediate area consisting of an external and an internal region; and (3) an internal layer of cells adjoining the endosperm (Figure 3). The principal areas in the mature tuber from the exterior inward are the periderm, the cortex, the vascular cylinder perimedullary zone, and the central pith (Figure 4). The periderm is six to ten cell layers thick, acting as a protective area over the surface of the tuber. Small, lenticel-like structures occur over the surface of the tuber. These develop in the tissue under the stomata and are initiated in the young tuber when it still has an epidermis. Periderm thickness varies considerably between different varieties. Cultural conditions also, however, influence the thickness of the periderm, rendering this characteristic too variable to be used for variety identification. Underlying the periderm, a narrow layer of parenchyma tissue is present. Vascular storage parenchyma high in starch content lies within a shell of cortex. The size of parenchyma cells changes with advancement in maturity of the tubers (Figure 5). The pith or water core is located at the center of the tuber, which consists of cells containing starch, less than that found in the vascular area and the innermost part of the cortex. The external features of the stem show eyes, which are rudimentary scale leaves, or leaf scar with axillary buds. Wound healing develops under cut, bruised, or torn surfaces. Suberin forms within 3 to 5 d in walls of living cells under the wound. A corn cambium layer developing under the suberized cells gives rise to a wound periderm. The tuber surface permits or excludes the entrance of pathogen, regulates the rate of gas exchange or water loss, and protects against mechanical damage.

The composition of the tuber varies with the cultivar and growing conditions. The tuber contains 63 to 87% water; 13 to 30% carbohydrates; 0.7 to 4.6% protein; 0.02 to 0.96% fat; and 0.44% ash. In addition, sugars, nonstarchy polysaccharides, enzymes, ascorbic acid and other vitamins, phenolic substances, and nucleic acids are also present. Carbohydrates are stored within storage parenchyma cells of pith and cortex in the form of starch granules with characteristic markings. The starch grains from the bud and apical ends exhibit marked differences in size and shape (Figure 6).

Wound healing is more rapid in recently harvested tubers. It is also influenced by temperature. Irradiation by sunlight or ionizing gamma rays impairs the wound-healing process. Infection of pathogen is generally reduced by the rapid development of suberin and periderm under wounds.

III. NUTRITIONAL COMPOSITION

A. COMPOSITION

Although the potato is rich in its carbohydrate content, it provides significant quantities of other nutrients such as proteins, minerals (iron), and vitamins (B-complex and vitamin C).[10-19] Burton[16] reported that 100 g fresh weight of potato provided 2.1 g protein (N × 6.25), 0.3 MJ energy, 25 mg vitamin C, 0.1 mg thiamine, 0.02 mg riboflavin, 0.5 mg nicotinic acid, and 1.0 mg of iron. A report of the National Food Survey Committee, London, stated that on an average in the U.K., potatoes provided 4 to 4.5% of the daily energy and protein intake and over 25% of the daily requirement of ascorbic acid.[17]

The nutritional or chemical composition of the potato tubers varies with variety, storage, growing season, soil type, preharvest-nutrition, and method of analysis used by the inves-

The Tuber
Swollen Underground Stem

A. Morphology

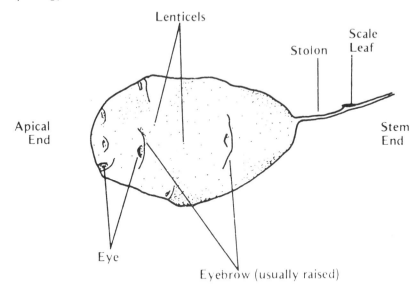

B. Anatomy

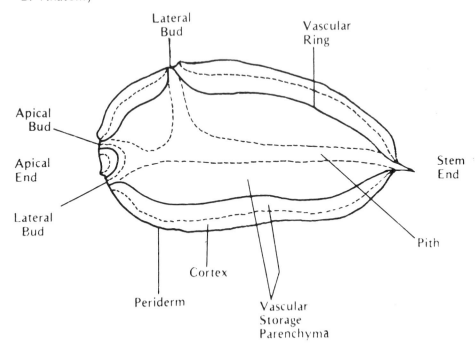

FIGURE 2.   Morphological (A) and anatomical (B) structure of a potato showing principal features. (Courtesy of International Potato Center, Lima, Peru, Bull. 6, 1986. With permission.)

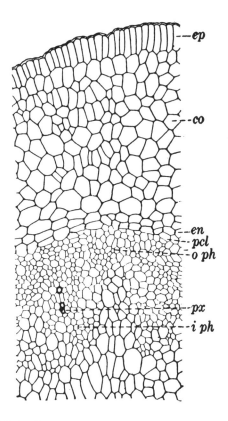

FIGURE 3.   Transection of the tip of a young potato tuber. (ep = Epidermis, co = cortex, en = endodermis, pcl = pericycle, o ph = outer phloem, px = protoxylem, and i ph = inner phloem. (From Hayword, H. E., in *The Structure of Economic Plants,* Macmillan, New York, 1953, 533.)

tigator. Mondy[18] described various factors influencing the nutritional quality of potatoes. According to Mondy,[18] the average nutritional composition of the potato is as follows (values in percent): water, 80; carbohydrates, 18; protein, 2; lipid, 0.1; and minerals, vitamins, etc., less than 0.1. A study carried out at the Animal Nutrition Laboratory, Cornell, in which rats were fed for their whole life on a diet comparable to that eaten by one third of the population of the U.S., indicated that the rats fed with the diet rich in potatoes had the best survival in old age and the greatest mean span of life.[18]

1. Carbohydrates/Energy

Carbohydrates constitute about 80% (range, 63 to 86%) of the total solids (Table 1). They are the constituent of highest concentration other than water, and are comprised largely of starch. Although the potato is an important source of energy, its major disadvantage is the considerable bulk (approximately 3.5 kg) that would have to be consumed to meet daily requirements.[2] One medium-sized potato yields about 100 cal. The contribution of the potato to energy intake, however, varies markedly depending upon whether it is cooked in water or fat (oil). Whereas boiled potatoes provide about 0.3 MJ per 100 g fresh weight, chips (French fries) provide up to 0.6 MJ, providing 5 and 15%, respectively, of the daily requirements at U.K. levels of consumption.[9]

a. Starch

Like other tuber crops, starch is the major component in potato tubers (Table 2). Potato starch contains amylose and amylopectin with 0.093% phosphorus.[20] Much of the starch is

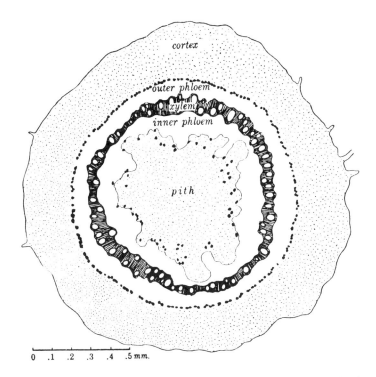

FIGURE 4. Diagrammatic transection of a potato tuber showing proportional amounts of tissue in the main zones. (From Hayword, H. E., in *The Structure of Economic Plants,* Macmillan, New York, 1953, 544.)

present in starch granules. The amylose content increases with the increase in maturity.[21] Fertilization of the crop influences the starch content of tubers.[22] The starch content in potatoes varies according to storage temperatures,[23] diseases of potatoes, and application of chemicals.[24] The amylose in starch ranges from 18.5 to 32.0%. Tubers with small starch granules contain a higher amylose content than those with large granules.[25] Mechanical treatment causes a considerable increase in the degradation of potato starch.[26]

b. Sugars

The sugar content ranges from traces to as high as 10% of the dry weight of the tuber. Sucrose, glucose, and fructose comprise the major sugars of the potato. Traces of ketoheptose, melibiose, melezitose, and raffinose have been detected in potato tubers.[21] Verma[27] reported that mature tubers harvested 161 d after planting contained less reducing sugars and sucrose than tubers harvested 136 d after planting. However, the tubers had more sugars than those harvested after 110 and 117 d. Large amounts of sugars accumulate during low-temperature storage. Sprouting increases sugar content in the potato.[28] The sugar concentration is higher at the center of the tuber than in the outer region. Potatoes stored under a nitrogen atmosphere do not accumulate sugars, but instead, lose starch content.

c. Other Carbohydrates

Potatoes contain cellulose, pectic substances, hemicellulose, and other polysaccharides as nonstarch polysaccharides to the extent of about 0.2 to 3.0%. Cellulose constitutes 10 to 12% of nonstarch polysaccharides of the potato. Pectic substances range from 0.7 to 1.5%. The skin contains about ten times more pectin than does the flesh. The application of auxin stimulates synthesis of pectin at the expense of cellulose.[29] The pectic substances include protopectin, soluble pectin, and pectic acid. The protopectin constitutes about 70% of total

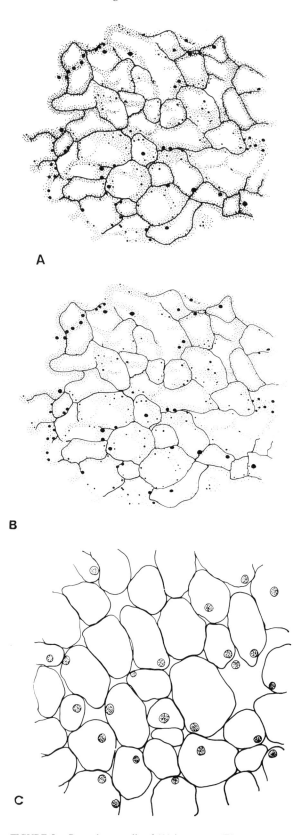

FIGURE 5.   Parenchyma cells of (A) immature, (B) mature, and
(C) overmature potato tubers.

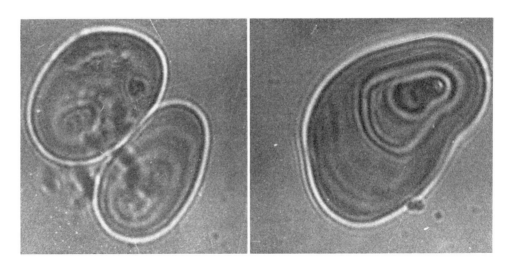

FIGURE 6. Starch grains of potato tubers. The grain on the left is from the basal end, and the grain on the right is from the apical end.

TABLE 1
Average Composition of Fresh Potato Tubers

| Constituent | Average values (%) | | Range (%) |
	(Ref. 20)	(Ref. 18)	(Ref. 2)
Water	77.5	80	63—86
Dry matter	22.5	20	13—36
Carbohydrates	19.4	16.9	13—30
Proteins	2.0	2.0	0.7—4.6
Lipids	0.1	1.0	0.02—0.96
Ash	1.0	1.0	0.44—1.9

TABLE 2
Chemical Composition of Potato Tubers
(Dry Matter Basis)[3]

Constituent	Reported range (%)	Average range (%)
Starch	60—80	70
Sucrose	0.25—15	0.5—1.0
Reducing sugars	0.25—3.0	0.5—2.0
Total N	1.0—2.0	1.0—2.0
Protein N	0.1—1.0	0.5—1.0
Fat	0.1—1.0	0.3—0.5
Dietary fiber	3—8	6—8
Minerals	4—6	4—6

pectic substance. Storage of the potato increases soluble pectin and decreases protopectin. Soluble pectin contributes about 10% of total pectic substances. The pectic acid fraction constitutes 13.25% of total pectic substances. Hemicellulose contains glucuronic acid, xylose, galacturonic acid, and arabinose. It has been shown that 1% of the total nonstarch polysaccharide of potato is present as hemicellulose.

d. Dietary Fiber

Dietary fiber has received increasing attention due to its significance in humans, particularly its roles in diverticulosis, heart attack, colon cancer, and diabetes. Dietary fiber includes plant polysaccharides and is resistant to hydrolysis by digestive enzymes.[30] The dietary fiber content ranges between 1 to 2% in fresh potatoes.[31-34] Many times, dietary fiber includes starch that is resistant to hydrolysis by enzymes used to remove starch prior to dietary fiber determination. Jones et al.[34] reported that there was little resistant starch in raw potato, but that it formed 20 to 50% by weight of total dietary fiber in cooked potato. However, it is not known whether or not this resistant starch is digested in human digestive tract.[35] The significance of dietary fiber in human nutrition has been recently described.[36-39] Fresh potatoes have a dietary fiber content similar to that of the sweet potato, but somewhat lower than that of other roots and tubers, and much lower than that of cereals and legumes. Unpeeled potatoes contain higher amounts of dietary fiber than peeled raw or boiled potatoes.

e. Nutritive Value

The potato contains a lower average energy content than other tubers and raw cereals. However, a significant variation exists in energy values of commercial varieties of potato.[40,41] When calculated on the basis of a moisture content equivalent to that of dry staples, the energy content of potatoes is similar to that of cereals or legumes.[42] As the potato is a low-energy density food, it is advantageous to include this in the diet of the population of the developed world where obesity is prevalent in a large section of the population.[43] Starch provides most of the energy supplied by the potato. The digestibility of potato starch is low in raw potatoes, but is markedly improved during cooking or processing.

2. Nitrogenous Constituents

The nitrogen content in the potato ranges from 1 to 2% of dry weight. There is an inverse relationship between distribution of starch and nitrogen. The nitrogen content in tubers is influenced by cultivar and environmental conditions under which the potato crop is grown.

a. Proteins

The protein nitrogen in potato tuber is mainly contributed by salt-soluble globulin fraction. It is found to consist of two proteins.[44] Gel electrophoretic studies of potato proteins have revealed several protein components in potato tubers.[45]

b. Amino Acids

The nonprotein fraction which constitutes from one half to two thirds of total nitrogen is present as free amino acid. Asparagine and glutamine are present in approximately equal amounts and together constitute about one half of the total amino acids. Methionine and cystine/cysteine are limiting amino acids in the potato. The amount of lysine in potatoes is similar to that in typical animal protein[46] (Table 3). Chang and Avery[47] found that the nutritive value of potato protein was superior to that of rice. Weight gains and protein efficiency ratios were higher in those rats fed with the potato diet.

c. Nutritive Value

Protein content in the potato is comparable to that of root and tuber samples.[48,49] Potatoes contain a higher concentration of lysine than in cereals.[50,51] However, it contained a lower concentration of sulfur-containing amino acids than the cereals. Thus potatoes can supplement a cereal-based diet. It has been shown that 100 g of potato can supply 7, 6, and 5% of daily energy and 12, 11, and 10% of the daily protein needs of children aged 1 to 2, 2 to 3, and 3 to 5 years, respectively. Based on NDp Cal % data of different foods, the potato is a well-

TABLE 3
Essential Amino Acid Composition of
Potato Tubers[3]

Amino acid	Reported range (mg/g)
Histidine	1—4
Isoleucine	2.7—4.2
Leucine	3.9—6.1
Lysine	4.2—5.7
Met + Cys	1.0—2.9
Phenylalanine + tyrosine	5.8—8.2
Threonine	2.5—3.8
Tryptophan	1.2—1.4
Valine	5.1—6.4

balanced food in terms of protein and energy. Potatoes baked in their skins or roasted or fried in fat can make much greater contributions to intake of energy and protein than when boiled. This is due to the loss of nutrients during cooking. The nutritive value of potato protein varies considerably between lots of the same variety, but little is known about the influence of variety, cultural practices, and climatic or environmental factors on potato quality.

d. Enzymes

The presence of several enzymes such as polyphenol oxidase, peroxidase, catalase, esterase, proteolytic enzymes, invertase, phosphorylase, and ascorbic acid oxidase in potato tubers has been documented. These enzymes influence the processing property of tubers and are involved in sprouting.

3. Lipids

The amount of lipid present in the potato is small, approximately 0.1% on fresh weight basis. Mondy[18] stated that the nutritional importance of lipid in the potato could not be judged solely by its quantity, especially considering its role in membrane structures.

4. Vitamins

Potatoes are an excellent source of ascorbic acid, thiamine, niacin, and pyridoxine and its derivatives.[32] Fresh potatoes may contain 30 mg or more ascorbic acid per 100 g when newly harvested, although values decline when potatoes are stored (Figure 7), cooked, or processed. Potatoes contain higher quantities of ascorbic acid than do carrots, onions, and pumpkins. The vitamin content of potatoes depends upon variety,[52] soil type[53] and nitrogen fertilization,[54] date of harvesting,[55] and phosphorus application.[56] The vitamin content of potatoes during storage is shown in Figure 8.

According to Mondy,[18] of the vitamins included in the recommended daily dietary allowances of the National Research Council, potatoes offer substantial amounts of ascorbic acid, niacin, thiamine, and riboflavin. Potatoes contribute more vitamin C to our food supply than any other major food. The vitamin content of the potato varies markedly with variety, maturity, preharvest mineral nutrition of the crop, soil type, and the storage conditions employed.

According to Rosenberg,[57] vitamin C in the potato tuber is present in both the reduced (ascorbic acid) and oxidized state (dehydroascorbic acid), but the content of the latter is usually low. On cooking, dehydroascorbic acid is readily converted irreversibly to diketo-gulonic acid, which represents a loss of vitamin C (Figure 9). Both temperature and length of period of storage significantly influence the vitamin C content. Carter and Carpenter[58] reported that only 23% of the total niacin in cooked potato was found to be in an available form.

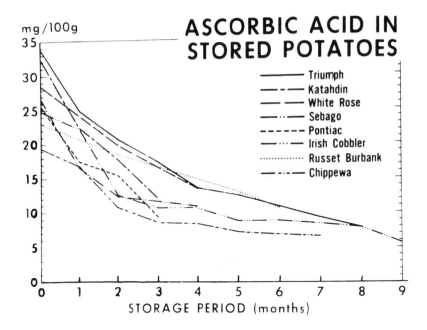

FIGURE 7.   Losses of vitamin C in potatoes during storage.

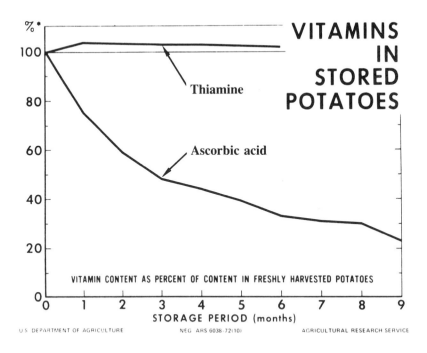

FIGURE 8.   Vitamin contents in stored potatoes.

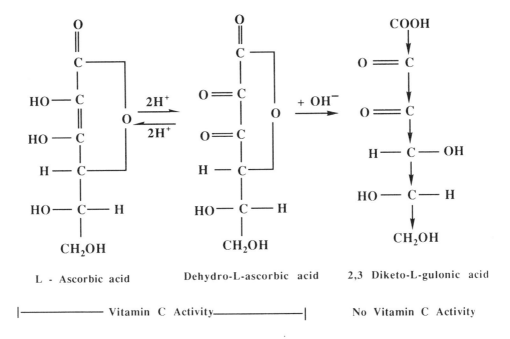

FIGURE 9.   Structural change during oxidation of L-ascorbic acid.

TABLE 4
Mineral and Vitamin Contents
in Potato Tubers (100 g Edible
Portion)

	Content (mg)	
Constituent	(Ref. 18)	(Ref. 3)
Minerals		
Calcium	9	7
Phosphorus	50	53
Iron	0.8	0.6
Vitamins		
Thiamine	0.10	0.09
Riboflavin	0.04	0.03
Niacin	1.5	1.5
Ascorbic acid	20	16

5. Minerals

The potato is a good source of iron and magnesium, and contributes some trace mineral elements lacking in milk. The iron content in the potato is comparable to that in other roots and tubers. A positive correlation was found between ascorbic acid content of potatoes and the amount of iron solubilized from potatoes by gastric juice *in vitro*. The potato appears to have a moderate iron availability superior to that of vegetable foods. Potatoes are a good source of phosphorus[59] (Table 4). A relatively small percentage of the total phosphorus in potatoes occurs in the form of phytic acid. Phytic acid is known to interact with calcium, iron, and zinc in the form of phytate, thus rendering them unavailable for absorption into the body. Approximately 25% of the total phosphorus was found in the phytic acid.[58] A mean of 8.3% of the total phosphorus was found in phytic acid amongst 23 samples of potato grown in India. The lower phytic acid content in the potato may be advantageous in having higher availability of minerals in a meal which includes potatoes.

Potatoes are a poor source of calcium, but a rich source of potassium. Sodium content is also low. Potatoes can therefore be used in diets designed to restrict sodium intake in patients with high blood pressure, where a high potassium:sodium ratio may be of additional benefit. It was found that 97% of zinc was available in a potato-based diet for rats with a phytic acid content of 0.23 mg/100 g, whereas only 23% was available in a corn-based diet with a phytic acid content of 9.93 mg/100 g.[60] Potatoes do not contain excessive fiber and its phytic acid content is low, so zinc availability should be high. These are known to contain other minerals such as magnesium, copper, chromium, manganese, selenium, and molybdenum. It is considered to be an excellent source of fluoride.

B. ANTINUTRITIONAL/TOXIC COMPOUNDS

1. Glycoalkaloids

The steroidal glycoalkaloid present in potato tubers has been shown to be a mixture of α-solanine and α-chaconine (Figure 10) in potato tubers. Earlier reviews[61-63] have shown that the glycoalkaloids present in normal tuber range from 0.01 to 0.1% on a dry weight basis. The peel contains more glycoalkaloids than the flesh. Sprouts contain much more glycoalkaloids than tubers.[61] The formation of the glycoalkaloids is influenced by certain environmental conditions and mechanical injury (Chapter 4). The glycoalkaloids of the potato are not destroyed by cooking, baking, or frying. Jadhav and Salunkhe[62] have reviewed the literature on the formation, distribution, and control of glycoalkaloids in potato tubers and have evaluated their toxicity. Potatoes containing more than 0.1% glycoalkaloids (dry weight basis) are considered to be unfit for human consumption. The alkaloids are regarded as normal constituents in all solanaceous plants, but in high concentration they are recognized as toxins and teratogens. Glycoalkaloids are cholinesterase inhibitors[62] which can cause headache, nausea, diarrhea, and serious illness if consumed in concentrations greater than 2.5 mg/kg body weight.[63] Concentrations greater than 3 mg/kg body weight can be lethal to humans.[63]

2. Protease Inhibitors

Potato tubers contain protease inhibitors including: trypsin inhibitor, chymotrypsin inhibitor,[64] hemagglutinin activity,[65] and kallikrein inhibitor.[66] The tubers are known to contain high concentrations of protease inhibitors.[67-70] Lau et al.[71] reported enterokinase-inhibiting activity in potato tubers. Since enterokinase initiates the cascade reaction which activates the digestive proteinases in animals, it is expected that its inhibitor would also be a potent inhibitor of digestive proteolysis. The amino acid composition of different protease inhibitors in the potato is very much similar.[72] These inhibitors are implicated in the tuber defense mechanism against pest attack.[67] Chymotrypsin inhibitor functions as a storage protein during potato plant development.[72] The trypsin and chymotrypsin inhibitors are implicated in field resistance to late blight.[73]

Ryan and Hass[68] reported that boiling, microwave heating, or baking the tubers destroyed most of the protease inhibitor activity, but the carboxypeptidase inhibitor was extremely stable in all three methods of cooking. Huang et al.[74] reported that a significant chymotrypsin inhibitor activity also survived baking and boiling, although trypsin inhibitor activity was completely destroyed.

Livingstone et al.[75] found that potato chymotrypsin inhibitor was responsible for poor nitrogen utilization in pigs. Partial cooking reduced inhibition by one third, but steaming at 100°C for 20 min completely destroyed inhibitor activity. Digestibility of nitrogen in raw potato was 32.8%, while that of a heat-treated diet was 89.8%. The digestibility of nitrogen in partially cooked potatoes was 48% of that of the completely cooked sample. These studies indicated that the low nutritional value of nitrogen from raw potatoes is due to antinutritional factors.

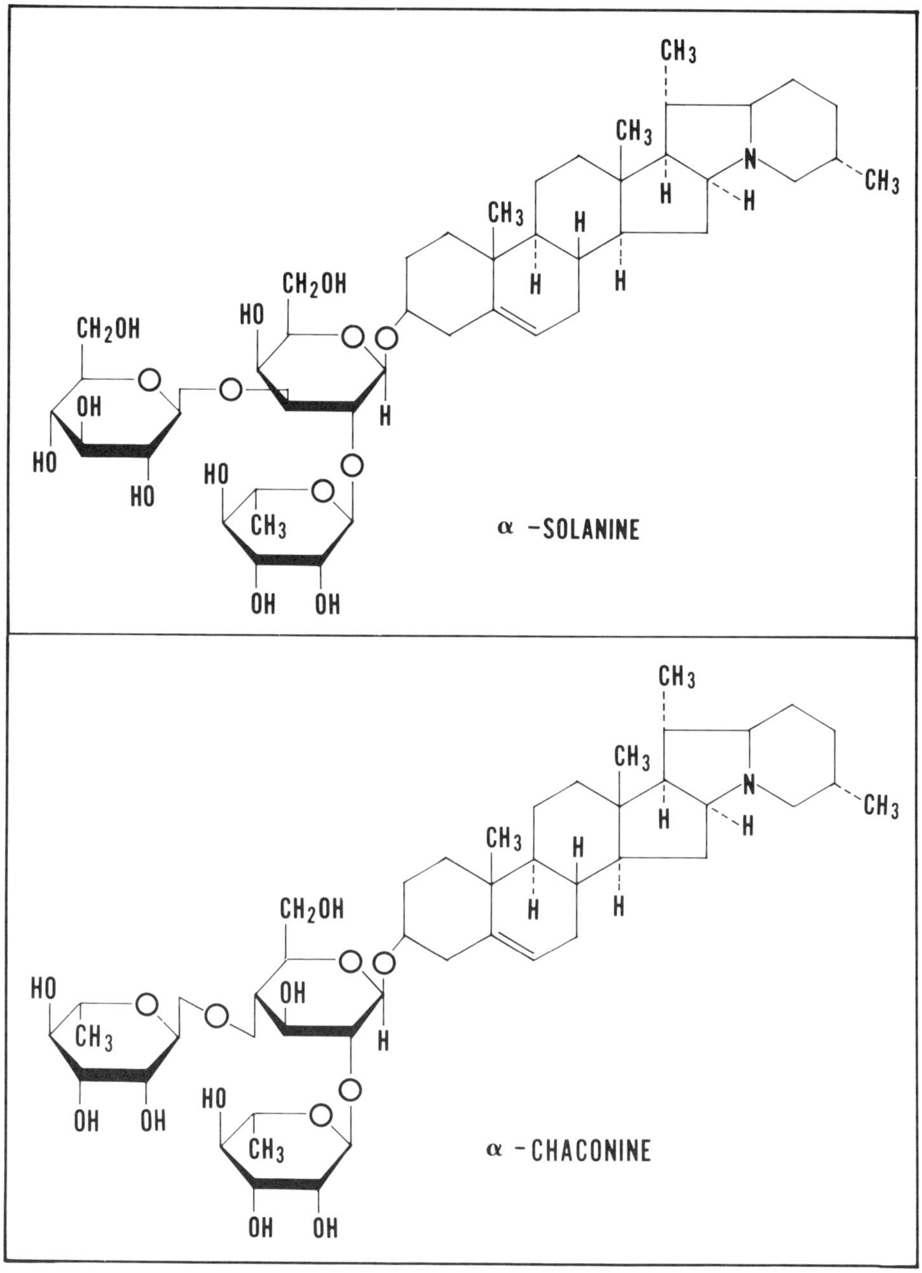

FIGURE 10.   Structure of α-solanine and α-chaconine.

3. Lectins

¡Lectins from potato have been shown to agglutinate erythrocytes of several animals and humans.[76,77] The potato tuber lectin has a saccharide specificity similar to that of wheat germ agglutinins.[77] However, very few studies are available on the nutritional significance of potato lectin. It is anticipated that under certain conditions, potato lectin could also exert a toxic effect.

4. Phenolic Compounds

Many reports[78-81] indicate the presence of phenolic compounds in potato tubers. These

¡are monohydric phenols, coumarins, anthocyanins and flavones, and polyphenols. Tannins are mostly localized in suberized tissue of the potato and impart tan coloration to the skin. Coumarins have been implicated in discoloration of cooked potato. The tyrosine, a monohydric phenol, constitutes 0.1 to 0.3% of the dry weight of the potato, whereas chlorogenic acid constitutes 0.025 to 0.15% of the dry weight of the potato tuber.[78]

Phenolic compounds are located largely in the cortex and peel tissues of the potato.[81] High levels of phenols have been associated with after-cooking discoloration. This discoloration occurs mostly in the stem end of cooked potatoes, and phenols are more concentrated in the stem end than in the bud end.[81] Mondy et al.[82] found a high positive correlation among phenolic content and bitterness and astringency.

Considering the nutritional quality of the tuber, the potato is an ideal food for the year 2000.[18] Zgorska[83] described the factors affecting the quality of table potatoes, giving data on the contents of dry matter, starch, reducing sugars, sucrose, phenolics, citric acid, and tyrosine, together with their relationship with quality characteristics such as darkening of raw and cooked flesh, chip color, and discoloration after tuber damage. Klein et al.[84] reported the nutritional composition and quality of potatoes in terms of total nitrogen, nonprotein nitrogen, total protein, amino acids, minerals, and firmness as influenced by magnesium fertilization. The nutritional composition of the tuber was largely related to the cultivar, the climatic conditions during the growing season, and the storage conditions. Chip color was closely related to the content of reducing sugar and sucrose, while the flesh darkening of the tuber and discoloration after damage were related to the contents of phenolic compounds and tyrosine.

IV. QUALITY OF RAW POTATOES

Several factors such as shape and size of the tuber, depth of the eyes, color of the flesh and skin, extent of surface blemishes due to diseases and pests, and superficial damage or internal damage or a combination of both determine the final tuber quality and the acceptability of the product to the consumer. These quality factors are of prime importance for potatoes destined for processing or for direct table use. Use of poor-quality product can lead to marked wastage and an increase in the cost of preparation.

A. MORPHOLOGICAL CHARACTERISTICS
1. Size of Tubers

The size of tubers required by the consumer varies considerably, and depends upon the ease of handling for household purposes and upon the acceptable levels of peeling loss during processing. Such losses are proportionately greater for small-sized tubers than for large ones. According to Gray and Hughes,[9] tubers within the size range of 4.0 to 7.5 cm are preferred, but smaller tubers with a 2.0- to 4.0-cm diameter are used for canning. The variety of potato and the tuber set per stem partly determine the tuber size.

Plant density or population per unit area influences the size of the potato tubers significantly. Plant density directly affects the tuber size by altering the number of the tuber by each plant, which in turn depends upon stem density, spacial arrangement, variety, and season.[85] Scott and Younger[86] reported that an increase in stem density over the commercial range resulted in a significant reduction in the number of tubers set per stem. Plantation of increasing numbers of seed tubers per unit area increased the number of tubers per plant despite the reduction in the number of tubers per stem (Figure 11A). The increase in stem density by planting more tubers decreased the number of tubers per plant and the number of tubers per stem without affecting the number of stems per seed tuber. High stem densities produce tubers with higher dry-matter contents than similar-sized tubers from low stem densities.[87] The tuber size also affects the dry matter content of the tuber. Planting big size

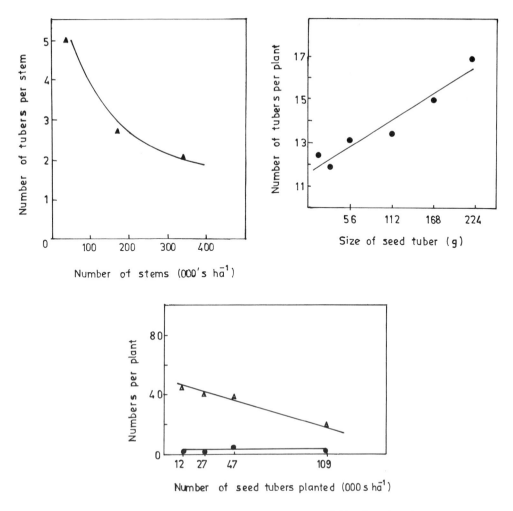

FIGURE 11. Relationship between the number of tubers per stem (top left) and number of stems per unit area; the number of tubers per plant and the size of the seed tuber (top right); and the number of stems per plant and number of seed tubers planted (bottom). (From Scott, R. K. and Younger, A., *Outlook Agric.*, 7, 3, 1972. With permission.)

seed tubers increases the number of tubers per plant (Figure 11B). The stem density affects tuber size distribution and the salability of the crop by influencing the prevalence of tuber greening and growth disorders.[83] The number of stems per plant depends on the number of seed tubers planted (Figure 11C).

2. Shape of Tubers

Uniformity of tuber shape within a population is an important processing characteristic. Potato processing involving peeling and slicing operations has an important bearing on the shape of the tubers. The crop variety and plant spacing mainly influence the tuber shape, close spacing giving a more uniform sample. The tuber shape is also partly controlled by the prevailing climate during growth, especially the temperature. Temperatures in the range of 12 to 20°C have been reported to produce tubers of more even shape than temperatures above or below this range.[87]

Four distinct shapes, viz., round, oval, pointed oval, and kidney shape, have been recognized.[88] Whitehead et al.,[89] however, recognized seven different shapes of potato tubers.

3. Depth of Eyes and Skin Quality

Depth of the eyes is largely a genetically controlled characteristic.[89] Depth of the eyes and depth and appearance of the skin are closely related to the color and appearance of the skin. Depth and appearance of the skin may be important in determining susceptibility to scuffing damage in netted, skinned varieties such as Netted Gem. The thickness of the periderm is usually a varietal feature, but it can be markedly influenced by cultural factors. Yamaguchi et al.[87] reported that high levels of nitrogen fertilizers and deep planting produced a thin skin, and high soil temperatures induced the formation of a rough or scaled skin.

4. Flesh and Skin Color

The flesh color of potatoes, with some notable exceptions, is either white or yellow. The presence of anthocyanins dissolved in the cell sap of the periderm or peripheral cortical cells forms the basis of the skin pigmentation.[9,16] A dozen different types of carotenoids were identified in potato tubers which were shown to have a close relationship with flesh color.[9] Smith[90] reported that six anthocyanins occurring either as the 5-glucoside, 3-rhamnosylglucoside acetylated with *p*-coumaric acid, or as 3-rhamnosylglucoside were found in cultivated potatoes. The aglycones consisted of pelargonidin, cyanidin, peonidin, petunidin, malvidin, and delphinidin pigments.

B. SENSORY CHARACTERISTICS

According to Howard,[91] assessment for more subtle types of flavor involves the use of organoleptic tests, and the results can be difficult to interpret as different evaluators have different sensory preferences. One special difficulty with potatoes is that texture differences tend to confuse flavor differences.[91] Pronounced bad flavors are encountered when wild species are used in breeding programs to introduce disease and pest resistance. One such variety, 'Lenape', bred from wild species *Solanum chacoense,* had under certain conditions poisonous amounts of glycoalkaloids in its tubers.[92] It has therefore been suggested that in choosing parents for such work, the use of wild species should be given the last preference.[91]

C. MECHANICAL DAMAGE/PHYSICAL INJURY

Mechanical damage and physical injuries in the form of bruisings, cuts, splits, and crushing are the major problems of potato growers. According to a survey carried out in the U.K.,[93] about one third of the potatoes showed some form of physical or mechanical damage by the time the tubers reached the store. Gray and Hughes[9] divided mechanical damage broadly into two groups: (1) external damage, including skin damage in the form of scuffing splits and crushing injury; and (2) internal bruising or black spot. Although "scuffing" does not usually include damage to the flesh, this aspect of damage is often considered under splits. Most impact damages occur during harvesting, especially mechanical harvesting and postharvest handling and marketing, as a result of tubers striking moving or stationary parts of the equipment, stones, clods, or one another. Development of pressure spots during prolonged storage leads to internal bruising. All types of mechanical damage and physical injuries are markedly influenced by factors such as variety, stage of maturity, growing season, and storage conditions. Cracking and internal bruising due to impact are generally encountered when potatoes are handled at lower temperatures for prolonged periods.[9]

The susceptibility of potatoes to mechanical damage is evaluated by measuring the rheological properties of the tuber and the external forces exerted upon it, such as the height of fall and the weight of the tuber, the shape of the tuber, the surface area of contact, the speed of moving parts, and the cushioning effect of soil. Selection of tubers with appropriate rheological properties by manipulating variety, maturity, and environmental factors during growth and storage (temperature) therefore forms an important basis of reducing physical damage. Similarly, the external forces imposed upon the tubers during lifting and handling can be reduced by using properly designed mechanical harvesting, lifting, and handling machinery and equipment. The extent of both external damage and internal black spot is closely associated with the tissue strength, elasticity, or firmness, which are in turn determined by the strength of cell walls and intercellular adhesion. Tuber elasticity and firmness are evaluated by measuring the force to achieve a given deformation or deformation at a standard force or the force to deformation ratio. The modulus of elasticity is measured by the resonance frequency technique, and it is influenced by the rigidity of the cell wall, the stiffness of intercellular bonding, and turgor pressure within the cells.[9]

1. External Damage

Although very wide varietal differences are observed in the susceptibility of different cultivars to external damage, susceptibility to damage is also markedly influenced by environmental conditions, particularly the site and the season. The effect of individual environmental factors is often difficult to evaluate because of the interaction of growing conditions and also their diverse effects on the rheological properties and size of the tubers. The effects of those factors on the potential damage behavior of tubers may be masked by the harvesting conditions (soil structure, moisture, temperature, and harvesting equipment used) which are known to greatly influence damage.[92] Several investigators have used rheological tests such as tissue strength, elasticity, and firmness to predict damage.[94] The strength of the tuber skin and the underlying tissue greatly influence splitting and cracking. The increased strength of potato tissue, as measured by penetrometer or force-deformation curves, has been shown to be closely associated with a decrease in field damage.[94] Finney et al.[97] demonstrated that certain rheological properties associated with damage altered towards the end of the growing season and during curing after harvest. They suggested that the tubers would become more susceptible to external damage as they matured shortly before harvesting, but were much less susceptible after curing when the energy they could withstand increased by 30 to 80%.

Field damage is significantly influenced by site, season, fertilizers, and cultural practices.[9] More damage is generally found in tubers harvested from sandy soils than in tubers harvested from clayey soils,[93] and rainfall and irrigation often increase external damage. The cultural conditions which reduce the turgor or firmness of the tissue, such as root cutting and cessation of irrigation well in advance of harvesting, reduce susceptibility of potatoes to mechanical damage.[98] Increased tuber weight, however, has been shown to increase external damage.[96] Environmental factors which increase tuber size thus may also increase mechanical damage. Environment also influences the rheological properties of the tubers and their susceptibility to physical damage. Hunnius et al.[95] reported that nitrogen fertilization reduced the percentage of severely damaged or multiply damaged tubers. These authors suggested that the increased ability of these tubers to withstand damage was due to a reduction in cell size, which is known to occur with use of nitrogen fertilizers.[96]

Wu and Salunkhe[99] reported that the mechanical injuries of potato tubers, such as bruising, cutting, dropping, puncturing, and hammering, greatly stimulated glycoalkaloid synthesis in both peel and flesh of tubers. The extent of glycoalkaloid formation depended on the cultivar, the type of mechanical injury, the storage temperature, and the duration of storage. High-temperature storage stimulated more glycoalkaloid formation than at low temperature. Most of the injury-stimulated glycoalkaloid formation occurred within 15 d after treatments. Mechanical injury caused by cutting of tubers resulted in the highest contents of glycoalkaloids in both the flesh and the peel tissues of the potato. Most of the glycoalkaloid formation occurred within 30 d after damage, although in some cases there were further increases, which were very small. The peels of potatoes contained much higher glycoalkaloid content than did the flesh. Wu and Salunkhe[99] noted that mechanical injury of potatoes not only decreased their quality and caused physiological disorders, but increased the antinutritional factor, the glycoalkaloid content.

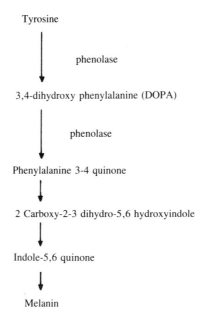

Tyrosine

↓ phenolase

3,4-dihydroxy phenylalanine (DOPA)

↓ phenolase

Phenylalanine 3-4 quinone

↓

2 Carboxy-2-3 dihydro-5,6 hydroxyindole

↓

Indole-5,6 quinone

↓

Melanin

FIGURE 12.   Synthesis of melanin from tyrosine.

2. Internal Bruising (Blackspot)

Potato cultivars vary widely in their susceptibility to internal bruising or blackspot.[100] Blackspot is more prevalent at the stem end of the tuber and is typically a blue-grey (or some times brown) spherical zone in the vascular region of the tissue. It usually appears 1 to 3 d after impact damage. The skin often shows no visible signs of damage. The pigments responsible for blackspot are produced by the oxidation of phenolic substances such as tyrosine and chlorogenic acid by phenolases. During the course of the reaction brown, red, and finally black (melanin) pigments are formed as the oxidation products of tyrosine (Figure 12) and chlorogenic acid. The first two reactions are catalyzed by potato phenolase; the others are, or can be nonenzymatic. Under stress conditions, owing to breakage of cells or membranes and the disorganization of cellular components, enzymes and substrates react to cause enzymatic browning. It was demonstrated that the intensity of blackspot in the impacted tubers of different varieties was similar, whether the blackspot was allowed to develop naturally over 3 d or rapidly (2 h) using oxygen under pressure. The results suggested that the time needed to develop blackspot in potato was dependent upon the availability of oxygen.[9,111]

Different intrinsic factors determining the susceptibility of a tuber to blackspot can be divided into two broad categories:[9] (1) the potential of the tissue to produce colored oxidation products, which in turn depends upon the presence of sufficient enzymes, substrates, and oxygen; and (2) the susceptibility of the tissue to damage. Whereas the former is determined by the biochemistry of the tuber, the latter is influenced by the rheological properties of the tuber and externally applied stress. Therefore, the biochemical properties of the tuber alone do not explain the blackspot. It is well established that the oxidation of tyrosine by enzyme phenolase produces melanin, and this reaction is significantly influenced by variety and environmental and storage conditions. The relative role and rheological properties of the potato skin and the underlying tissue involved in the total damage have not been established completely. It has been suggested that blackspots occur in potatoes with strong skin, but weak underlying tissue. Firmness, turgor pressure, and specific gravity are also closely associated with the development of blackspot in potatoes.[9]

According to Gray and Hughes,[9] two groups of factors are important in determining differences in internal tissue damage among varieties: (1) the amount of deformation for a given force, depending upon the stress in the walls (turgor pressure); and (2) the amount of deformation the tissue can withstand before damage to membranes and walls occur. Both these factors may in turn be influenced by cell size. Increased cell size in certain instances has been shown to increase susceptibility of tissues to internal damage and bruising.[101]

The development of blackspot within the varieties is closely associated with the specific gravity of the tubers;[102-104] among varieties, the relationship has been inconsistent. It is also not known whether the effect of specific gravity is direct or if it reflects changes in the water content with its concomitant effect on turgor.[105] Sawyer and Collin[106] reported that in stored potatoes, specific gravity and firmness, as measured by a durometer, are very closely related within the varieties. A similar relationship was observed between water loss and tuber firmness as measured by compression. According to Gray and Hughes,[9] increasing levels of applied potassium or nitrogen reduce the specific gravity of tubers, but the effects of these two fertilizers on blackspot are quite different. While potassium generally reduces blackspot, nitrogen may have the opposite effect. It has been shown that both the potential of the tissue to produce colored oxidation products and the susceptibility of the tissue to damage were lower in tubers from plants fed with high levels of potassium. Despite these observations, it was noted that tuber firmness and the amount of dynamic deformation on impact were similar in tubers from all potassium treatments. This may be due to differences in the properties of cell walls, cell size, and specific gravity, which make the tissue resistant to damage. The increase in susceptibility to blackspot owing to high levels of nitrogen may be due to an interaction of a number of factors, including increased tuber size and differences in tuber turgor.[105]

Low temperatures during lifting,[106] handling, and storage[107] increase blackspot, possibly through their effects on the rheological properties of the cells. The injury may be reduced by raising the temperature of susceptible tubers before handling[108,109] (Figure 13).

Pressure spots developed during extended storage may also influence the amount of blackspot occurring in potatoes when they are unloaded in stores. The pressure areas on tubers (softened, flattened, or indented areas occurring as a result of the pressure of adjacent tubers) make them more susceptible to the development of blackspot. Gray and Hughes[9] stated that careful handling of tubers at suitable temperatures and the production of tubers which are not excessively large were the key factors in controlling all forms of damage. The manipulation of environmental conditions which affect tuber firmness may help in the control of damage. However, it may be difficult to obtain the precise degree of firmness required to obtain the minimal amount of damage,[110] because too-firm tubers will crack and internal bruising may occur in too-soft tubers.

Hughes and Grant[112] recently described the factors influencing mechanical damage which caused considerable wastage through direct rejection of damaged potatoes and increased water loss and microbial spoilage during storage. According to these authors, damage (cracks, splits, blackspot) is influenced by impact conditions during all stages of handling and the properties of tubers, which can be reduced by improved design and operation of harvesting and handling machinery, and avoiding harvesting and grading tubers at low temperatures. Hughes and Grant[112] showed involvement of different tuber properties in controlling the various forms of impact damage (Figure 14). Although all these properties and damage

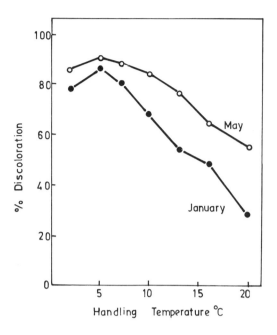

FIGURE 13.   Effects of the length of storage period and the temperature of handling on blackspot of the potato (cultivar Eigenheimer). (From Ophuis, B. G., Hensen, J. C., and Kroesbergen, E., *Eur. Potato J.*, 1, 48, 1958. With permission.)

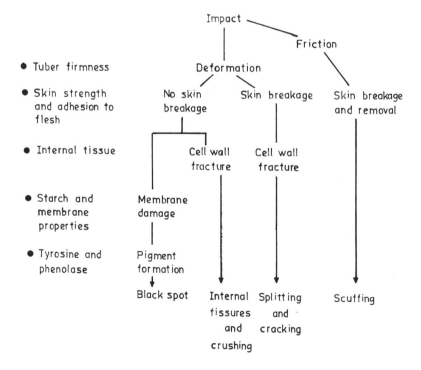

FIGURE 14.   Properties of potato tubers involved in controlling various forms of impact damage. (From Hughes, J. C. and Grant, A., *J. Sci. Food Agric.*, 32, 99, 1981. With permission.)

susceptibility of the tubers are genetically controlled, they are markedly influenced by environmental conditions during both growth and storage. Climatic conditions during growth and duration and type of storage affect the water status of tubers and the form and amount of damage.[113,114] Generally, turgid tubers are more prone to cell-wall fracture. In varieties like 'Pentland Crown', this may result in splitting or cracking, whereas in "Record" it may just result in localized damage below the skin (fissures or crushed zones) without apparent skin breakage. Flaccid tubers, on the contrary, are not brittle, but are rather prone to blackspot (membrane damage), especially when tubers have a high content of starch and tyrosine as in 'Record' potatoes.[111]

REFERENCES

1. **Hooker, W. J.,** The potato, in *Compendium of Potato Diseases,* Hooker, W. J., Ed., America Phytopathological Society, St. Paul, MN, 1986, 3.
2. **Talburt, W. F. and Smith, O.,** *Potato Processing,* AVI/Van Nostrand Reinhold, New York, 1987.
3. **Woolfe, J. A.,** *The Potato in the Human Diet,* Cambridge University Press, London, 1987.
4. **Burton, W. G.,** Postharvest behaviour and storage of potatoes, in *Applied Biology,* Vol. 2, Coaker, T. H., Ed., Academic Press, New York, 1978.
5. **Salaman, R. N.,** *The History and Social Influence of the Potato,* Cambridge University Press, New York, 1949.
6. **Thompson, H. C. and Kelley, W. C.,** *The Potato in Vegetable Crops,* 5th ed., McGraw-Hill, New York, 1957, 372.
7. **Kay, D. E.,** *Root Crops: Crop and Product Digest No. 2.,* Tropical Products Institute, London, 1973, 100.
8. **Ugent, D.,** The potato, *Science,* 170, 1161, 1970.
9. **Gray, D. and Hughes, J. C.,** Tuber quality, in *The Potato Crop: The Scientific Basis for Improvement,* Harris, P. M., Ed., Chapman and Hall, New York, 1978, 504.
10. **Hayword, H. E.,** *The Structure of Economic Plants,* Lubrecht & Cramer, Monticello, NY, 1967.
11. **Augustin, J.,** Variations in the nutritional composition of fresh potatoes, *J. Food Sci.,* 40, 1295, 1975.
12. **McCay, C. M., McCay, J. B., and Smith, O.,** The nutritive value of potatoes, in *Potato Processing,* Talburt, W. F. and Smith, O., Eds., AVI Publishing, Westport, CT, 1975.
13. **Mondy, N. I., Koch, R. L., and Chandra, S.,** Influence of nitrogen fertilization on potato discoloration in relation to chemical composition. II. Phenols and ascorbic acid, *J. Agric. Food Chem.,* 27, 418, 1979.
14. **Watt, B. K. and Merill, A. L.,** Composition of Foods: Raw, Processed, Prepared, Agriculture Handbook No. 8, U.S. Department of Agriculture, Washington, D.C., 1964.
15. **Howard, H. W.,** *Genetics of the Potato Solanum tuberosum,* Logos Press Ltd., London, 1970.
16. **Burton, W. G.,** The potato, in *A Survey of its History and of the Factors Influencing its Yield, Nutritive Value, Quality and Storage,* Veenman, H. and Zonen, N. V., Eds., European Association of Potato Research, Wageningen, 1966.
17. **Anon.,** Household food consumption and expenditure, Annu. Rep. National Food Survey Committee, Her Majesty's Stationery Office, London, 1972.
18. **Mondy, N. I.,** Factors affecting the nutritional quality of potatoes, in *Proc. Int. Congress "Research for the Potato in the year 2000",* Hooker, W. J., Ed., International Potato Center (CIP), 1983, 136.
19. **Woodward, C. F. and Talley, E. A.,** Review of nitrogenous constituents of the potato. Nutritive value of the essential amino acids, *Am. Potato J.,* 30, 205, 1953.
20. **Banks, W. and Greenwood, C.,** The starch of the tuber and shoots of the sprouting potato, *Biochem. J.,* 73, 237, 1959.
21. **Schwimmer, S., Vevenue, A., Weston, W., and Potter, A.,** Survey of major and minor sugar and starch components of white potato, *J. Agric. Food Chem.,* 2, 1284, 1954.
22. **Petrov, A. A.,** Effects of forms of nitrogen fertilizers on the level of starch and ascorbic acid on potato tubers, in *Nauch Rab. Molodykn Veh Chuvash Sel Skokhoz. Inst.,* 1, 65, 1970.

23. **Samotus, B. and Palasinski, M.,** Transformation of carbohydrate in potato tubers transferred from low to high temperature during storage, *Zesz. Nauk. Wyzsz. Szk. Roln. Krakowie Roln.,* 10, 81, 1964.

24. **Fisher, A. M. and Physhtaleva, O. T.,** Effect of chlorocholine chloride on the yield and food qualities of potato, *Tr. Nauchno-Issled. Inst. Kraev. Patol. Alma-Ata,* 26, 183, 1974.

25. **Janicki, J., Szebiotko, K., Grzeskowiak, Z., Piasecki, M., and Piorunski, N.,** Suitability of Polish varieties and strains of potatoes for processing, *Hodowla Rosl. Aklim. Nasienn.,* 11, 455, 1967.

26. **Potavina, V. C. and Falunina, Z. F.,** Reactivity of amylases towards starch, *Prikl. Biokhim. Mikrobiol.,* 2, 210, 1966.

27. **Verma, S. C., Joshi, K. C., and Sharma, T. R.,** Some observations on the quality of potato varieties grown in India, *Abstr. Conf. Pap. Trienne Conf. Eur. Assoc. Potato Res.,* 6, 162, 1975.

28. **Mondy, N. I. and Ponnampalam, R.,** Effect of sprout inhibitor isoprophyl N (3 chlorophenyl carbamate) on total glycoalkaloid content of potatoes, *J. Food Sci.,* 50, 258, 1985.

29. **Buffel, K. and Carlier, A.,** The cell wall composition of hydrated potato tissue and action of auxins, *Agriculture (Louvain),* 4, 157, 1956.

30. **Trowell, H., Southgate, D. A. T., Wolever, T. M. S., Leeds, A. R., Gassull, M. A., and Jenkins, D. J. A.,** Dietary fiber redefined, *Lancet,* 1, 967, 1976.

31. **Paul, A. A. and Southgate, D. A. T.,** *McCance and Widdowson's The Composition of Foods,* 4th ed., MRC Spec. Rep. 297, Her Majesty's Stationery Office, London.

32. **Finglas, P. M. and Faulks, R. M.,** A new look at potatoes, *Nutr. Food Sci.,* 92, 12, 1985.

33. **Wills, R. B. H., Lim, J. S. K., and Greenfield, H.,** Variation in nutrient composition of Australian retail potatoes over a 12 month period, *J. Sci. Food Agric.,* 35, 1012, 1984.

34. **Jones, G. P., Briggs, D. R., Wahlquist, M. L., and Flentje, L. M.,** Dietary fiber content of Australian foods. I. Potatoes, *Food Technol. Aust.,* 37, 81, 1985.

35. **Dreher, M. L., Breedon, C., and Orr, P. H.,** Percent starch hydrolysis and dietary fiber content of chipped and baked potatoes, *Nutr. Rep. Int.,* 28, 687, 1983.

36. **Sandberg, A. S.,** *Dietary Fiber: Determination and Physiological Effects,* Dept. of Clinical Nutrition and Department of Surgery II., University of Göteborg, Sweden 1982.

37. **Reistad, R.,** Content and composition of non-starch polysaccharides in some Norwegian plant foods, *Food Chem.,* 12, 45, 1983.

38. **Theandner, O.,** Advances in the chemical characterization and analytical determination of dietary fiber components, in *Dietary Fiber,* Birch, G. G. and Parker, K. J., Eds., Applied Science Publishers, New York, 1983.

39. **Englyst, H., Wiggins, H. S., and Cummings, J. H.,** Determination of the non-starch polysaccharides in plant foods by gas chromatography of constituent sugars as alditol acetates, *Analyst,* 107, 307, 1982.

40. **Toma, R. B., Orr, P. H., D'Appolonia, B., Dintzis, F. R., and Takekhia, M. M.,** Physical and chemical properties of potato peel as a source of dietary fiber in bread, *J. Food Sci.,* 44, 1403, 1979.

41. **Toma, R. B., Augustin, J., Shaw, R. L., True, R. H., and Hogan, J. M.,** Proximate composition of freshly harvested and stored potatoes, *Solanum tuberosum* L., *J. Food Sci.,* 43, 1702, 1978.

42. **Doughty, J.,** Water, the hidden ingredient, *Appropr. Technol.,* 8, 11, 1982.

43. **WHO,** Energy and Protein Requirements. Rep. Joint FAO/WHO UNV. Expert Consultation, WHO Tech. Rep. Ser. No. 724, World Health Organization, Geneva, 1985.

44. **Groot, E., Jansen, L., Kentie, A., Oosterhuis, H., and Trap, H. A.,** A new protein in potato, *Biochim. Biophys. Acta,* 1, 410, 1947.

45. **Loeschcke, V. and Stegeman, H.,** Potato varieties and their electropherogram characteristics, *Eur. Potato J.,* 9, 111, 1966.

46. **Hughes, J. C.,** Chemistry and after cooking discoloration in potato, *Natl. Inst. Agric. Bot. J.,* 9, 235, 1962.

47. **Chang, Y. O. and Avery, E. E.,** Nutritive value of potato protein, *J. Am. Diet. Assoc.,* 55, 565, 1969.

48. **Mondy, N. I. and Munshi, C. B.,** Chemical composition of potato as affected by herbicide: enzymatic discoloration, phenols, and ascorbic acid content, *J. Food Sci.,* 53, 475, 1988.

49. **Cameron, M. and Hofvander, Y.,** Manual on Feeding Infants and Young Children, 2nd ed., Protein calorie, Advisory Group, United Nations, 1976.

50. **Kaldy, M. S.,** Protein yield of various crops as related to protein value, *Econ. Bot.,* 26, 142, 1972.

51. **Chick, H. and Slack, E. B.,** Distribution and nutritive value of nitrogenous substances in the potato, *Biochem. J.,* 45, 211, 1951.

52. **Swaminathan, K. and Pushkarnath,** Nutritive value of Indian potato varieties, *Indian Potato J.,* 4, 76, 1962.

53. **Leichsenring, J. M.,** *Factors Influencing the Nutritive Value of Potatoes,* Minnesota Tech. Bull. No. 96, University of Minnesota, Agricultural Experiment Station, Minneapolis, MN, 1951.

54. **Augustin, J., Toma, R. B., True, R. H., Shaw, R. L., Teitzel, C., Johnson, S. R., and Orr, P.,** Composition of raw and cooked potato peel and flesh: proximate and vitamin composition, *J. Food Sci.,* 44, 805, 1979.

55. **Augustin, J., Swanson, B. G., Teitzel, C., Johnson, S. R., Pometto, S. F., Artz, W. E., Huang, C. P., and Schomaker, C.,** Changes in nutrient composition during commercial processing of frozen potato products, *J. Food Sci.,* 44, 807, 1979.

56. **Klein, L. B., Chandra, S., and Mondy, N. I.,** The effect of phosphorus fertilization on the chemical quality of Katahdin potatoes, *Am. Potato J.,* 57, 259, 1980.

57. **Rosenberg, H. R.,** in *Chemistry and Physiology of the Vitamins,* Interscience, New York, 1942.

58. **Carter, E. G. A. and Carpenter, K. J.,** The availability of niacin in foods, *Fed. Proc. Abstrs. Part I,* CA, 1536, 1980, 557.

59. **Quick, W. A. and Li, P. H.,** Phosphorus balance in potato tubers, *Potato Res.,* 19, 305, 1976.

60. **Maga, J. A.,** Potato glycoalkaloids, *Crit. Rev. Food Sci. Nutr.,* 12, 371, 1980.

61. **Gull, D. D. and Isenberg, F. M.,** Chlorophyll and solanine content and distribution in four varieties of potato tubers, *Proc. Am. Soc. Hortic. Sci.,* 75, 545, 1960.

62. **Jadhav, S. J. and Salunkhe, D. K.,** Formation and control of chlorophyll and glycoalkaloids in tubers of *Solanum tuberosum* L. and evaluation of glycoalkaloid toxicity, *Adv. Food Res.,* 21, 307, 1975.

63. **Morris, S. C. and Lee, T. H.,** The toxicity and teratogenicity of Solanaceae glycoalkaloids particularly those of the potato *(Solanum tuberosum)*: a review, *Food Technol. Aust.,* 36, 118, 1984.

64. **Balls, A. and Ryan, C.,** Concerning a crystalline chymotrypsin inhibitor from potatoes and its binding capacity for the enzyme, *J. Biol. Chem.,* 238, 2976, 1963.

65. **Marinkovich, V.,** Purification and characterization of the hemagglutinin present in potatoes, *J. Immunol.,* 93, 732, 1964.

66. **Werle, E., Appel, W., and Happ, W.,** The kallikrein inactivation of potatoes and its differentiation from proteinase inhibitor, *Fermentforschung,* 10, 127, 1959.

67. **Santarius, K. and Belitz, H. D.,** Protease activity in potato plants, *Planta,* 141, 145, 1978.

68. **Ryan, C. A. and Hass, G. M.,** Structural, evolutionary and nutritional properties of proteinase inhibitors from potatoes, in *Antinutrients and Natural Toxicants in Foods,* Ory, R. L., Ed., Food and Nutrition Press, Westport, CT, 1981.

69. **Hass, G. M., Derr, J. E., Makus, D. J., and Ryan, C. A.,** Purification and characterization of the carboxypeptidase inhibitors from potatoes, *Plant Physiol.,* 64, 1022, 1979.

70. **Richardson, M.,** The proteinase inhibitors of plants and micro-organisms, *Phytochemistry,* 16, 159, 1977.

71. **Lau, A., Ako, H., and Werner-Washburne, M.,** Survey of plants for enterokinase inhibitors, *Biochem. Biophys. Res. Commun.,* 92, 1243, 1980.

72. **Ryan, C. A., Kuo, T., Pearce, G., and Kunkel, R.,** Variability in the concentration of 3 heat stable proteinase inhibitor proteins in potato tubers, *Am. Potato J.,* 53, 443, 1976.

73. **Peng, J. H.,** Increase activity of proteinase inhibitors in disease resistance to *Phytophthora infestans, Diss. Abstr. Int. B,* 26, 24, 1975.

74. **Huang, D. Y., Swanson, B. G., and Ryan, C. A.,** Stability of proteinase inhibitors in potato tubers during cooking, *J. Food Sci.,* 46, 287, 1981.

75. **Livingstone, R. M., Baird, B. A., Atkinson, T., and Crofts, R. M. J.,** The effect of either raw or boiled liquid extract from potato *(Solanum tuberosum)* on digestibility of diet based on barley in pigs, *J. Sci. Food Agric.,* 31, 695, 1980.

76. **Goldstein, I. J. and Hynes, C. E.,** The lectins: carbohydrate binding proteins of plants and animals, in *Advances in Carbohydrate Chemistry and Biochemistry,* Tipson, R. S. and Horton, D., Eds., Academic Press, New York, 1978.

77. **Kilpatric, D. C.,** Isolation of a lectin from pericarp of potato *(Solanum tuberosum)* fruits, *Biochem. J.,* 191, 273, 1980.

78. **Clark, R., Kuc, J., Henze, R., and Quackenbush, F.,** The nature of and fungitoxicity of an amino addition product of chlorogenic acid, *Phytopathology,* 49, 594, 1959.

79. **Zucker, M.,** Influence of light on synthesis of protein and of chlorogenic acid in potato tuber tissues, *Plant Physiol.,* 38, 575, 1963.

80. **Craft, C. and Audia, W.,** Phenolic substances and barrier formation in potato, *Bot. Gaz. (Chicago),* 123, 211, 1962.

81. **Reeve, R. M., Hautala, E., and Weaver, M. L.,** Anatomy and compositional variation within potatoes. II. Phenolics, enzymes and other minor components, *Am. Potato J.,* 46, 347, 1969.

82. **Mondy, N. I., Metcalf, C., and Plaisted, R. L.,** Potato flavor as related to chemical composition. I. Polyphenols and ascorbic acid, *J. Food Sci.,* 41, 459, 1971.

83. **Zgorska, K.,** Factors affecting the quality of table potatoes Ziemniak, 183-206, 1979; quoted from *Field Crops Abstr.,* 35 (6), 548, 1982 (No. 5431).
84. **Klein, L. B., Chandra, S., and Mondy, N. I.,** Effect of magnesium fertilization on the quality of potatoes: total N, non-protein N, protein, amino acids, minerals and firmness, *J. Agric. Food Chem.,* 30, 754, 1982.
85. **Allen, E. J.,** Plant density, in *The Potato Crop: The Scientific Basis for Improvement,* Harris, P. M., Ed., Chapman and Hall, London, 1978, 278.
86. **Scott, R. K. and Younger, A.,** Potato agronomy in a changing industry, *Outlook Agric.,* 7, 3, 1972.
87. **Yamaguchi, M., Timm, H., and Spurr, A. R.,** Effects of soil temperature on growth and nutrition of potato plants and tuberization, composition and periderm structure of tubers, *Proc. Am. Soc. Hortic. Sci.,* 84, 412, 1964.
88. **Wurr, D. C. E. and Allen, E. J.,** Some effects of planting density and variety on the relationship between tuber size and tuber dry matter percentage in potatoes, *J. Agric. Sci. (Cambridge),* 82, 277, 1974.
89. **Whitehead, T., McIntosh, T. P., and Findlay, W. M.,** The potato, in *Health and Disease* 3rd ed., Oliver and Boyd, Edinburgh, 1953.
90. **Smith, O.,** Chemical composition of the potato, in *Potatoes: Production, Storing, Processing,* Smith, O., Ed., AVI Publishing, Westport, CT, 1968, 59.
91. **Howard, H. W.,** The production of new varieties, in *The Potato Crop: The Scientific Basis for Improvement,* Harris, P. M., Ed., Chapman and Hall, New York, 1978, 607.
92. **Anon.,** *Potatoes: Consumer Buying Behavior and Attitude During 1972—1973,* Potato Marketing Board, London, 1974.
93. **Anon.,** *Report on a National Damage Survey, 1973,* Potato Marketing Board, London, 1974.
94. **Robertson, T. M.,** Prediction of susceptibility of bruising (Abstr.), in *Proc. Fourth Triennial Conf., European Association of Potato Research,* London, 1970.
95. **Hunnius, W., Buchthaler, G., and Munzert, M.,** The influence of nitrogen on the ability of the potato tuber to withstand combined harvesting, *Potato Res.,* 15, 54, 1972.
96. **Blight, D. P. and Hamilton, A. T.,** Varietal susceptibility to damage in potatoes, *Potato Res.,* 17, 261, 1974.
97. **Finney, E. E., Hall, C. W., and Thompson, N. R.,** Influence of variety and time upon the resistance of potatoes to mechanical damage, *Am. J. Bot.,* 41, 178, 1964.
98. **Finney, E. E. and Findlen, H.,** Influence of preharvest treatments upon turgor of 'Katahdin' potatoes, *Am. Potato J.,* 44, 383, 1967.
99. **Wu, M. T. and Salunkhe, D. K.,** Changes in glycoalkaloid content following mechanical injuries in potato tubers, *J. Am. Soc. Hortic. Sci.,* 101, 229, 1976.
100. **Hudson, D. E.,** Relationship of cell size, intercellular space and specific gravity to bruise depth in potatoes, *Am. Potato J.,* 52, 9, 1975.
101. **Mondy, N. I., Klein, B. P., and Smith, L. A.,** The effect of maturity and storage on phenolic content, enzymatic activity and discoloration of potatoes, *Food Res.,* 25, 693, 1960.
102. **Lujan, L. and Smith, O.,** Potato quality. XXV. Specific gravity and after-cooking darkening of 'Katahdin' potatoes as influenced by fertilizers, *Am. Potato J.,* 41, 274, 1964.
103. **Hughes, J. C.,** Chemistry of after-cooking discoloration in potatoes, *Natl. Inst. Agric. Bot. J.,* 9, 235, 1962.
104. **Kunkel, R. and Gardner, W. H.,** Potato tuber hydration and its effect on black spot of "Russet Burbank" potatoes in the Columbia Basin of Washington, *Am. Potato J.,* 42, 109, 1965.
105. **Kunkel, R. and Gardner, W. H.,** Blackspot of "Russet Burbank" potatoes, *Proc. Am. Soc. Hortic. Sci.,* 73, 436, 1959.
106. **Sawyer, R. L. and Collin, G. H.,** Black spot of potatoes, *Am. Potato J.,* 37, 115, 1960.
107. **Johnston, E. F. and Wilson, J. B.,** Effect of soil temperatures at harvest on the bruise resistance of potatoes, *Am. Potato J.,* 46, 75, 1969.
108. **Wiant, J. S., Findlen, H., and Kaufman, J.,** Effect of temperature on black spot in "Long Island" and "Red River Valley" potatoes, *Am. Potato J.,* 28, 753, 1951.
109. **Ophuis, B. G., Hesen, J. C., and Krosbergen, E.,** The influence of the temperature during handling on the occurrence of blue discolorations inside potato tubers, *Eur. Potato J.,* 1, 48, 1958.
110. **Thornton, R. E., Smittle, D. A., and Peterson, C. L.,** Reducing potato harvester damage, in *Proc. 12th Annu. Washington Potato Conf. Trade Fair,* 1973.
111. **Duncan, H. J.,** Rapid bruise development in potatoes with oxygen under pressure, *Potato Res.,* 16, 306, 1973.
112. **Hughes, J. C. and Grant, A.,** Factors influencing damage, in *Agriculture Group Symp.: Factors Influencing Storage Characteristics and Cooking Quality of Potatoes (Abstr.), J. Sci. Food Agric.,* 32, 99, 1981.

113. **Harris, J. R.,** Potato processing industry—what the market requires, in *Agriculture Group Symp.: Factors Influencing Storage Characteristics and Cooking Quality of Potatoes (Abstr.), J. Sci. Food Agric.,* 32, 104, 1981.
114. **French, W. M.,** Varietal factors influencing the quality of potatoes, in *Agriculture Group Symp.: Factors Influencing Storage Characteristics and Cooking Quality of Potatoes (Abstr.), J. Sci. Food Agric.,* 32, 144, 1981.

Chapter 3

PRODUCTION

D. M. Sawant, S. S. Dhumal, and S. S. Kadam

TABLE OF CONTENTS

I. INTRODUCTION

Several cultural factors and environmental conditions influence the production of potatoes. These include date of planting, soil type, soil pH, moisture, season, location, mineral nutrition, weed control, control of insects and diseases, temperature during growing season, and time and method of harvesting.[1,2,4] In addition to these, the choice of cultivar has a significant contribution. These factors vary from region to region, and a better understanding of these in relation to local conditions can help in developing appropriate methods of cultivation to obtain high yields.

II. FACTORS INFLUENCING PRODUCTION OF POTATOES

A. CULTIVARS

Potato varieties greatly differ in time of maturity, appearance, yield, quality, and resistance to diseases and pests. Old potato varieties, grown for more than 50 years, originated as chance seedlings. They have more deep eyes, tend to be rough and irregular in shape, and have little resistance to diseases or insect pests. The U.S. Department of Agriculture and various state experiment stations introduced ''new'' varieties since 1925. They have few eyes, smooth, uniform shape, and higher yields than old varieties, and resistance to some diseases and pests. Thompson and Kelly[3] described the following commercial varieties of potato grown in the U.S.:

Katahdin—The first new heading potato variety to be introduced, it is late maturing, produces oval to round tubers with smooth white skin and shallow eyes, is resistant to mild mosaic, net necrosis, and brown rot, and is immune to warts.

Red Pontiac—The most popular red-skinned variety, this potato is late maturing and high yielding, has oblong to round tubers with smooth red skin and shallow eyes, and is fairly drought resistant.

Russet Burbank or Netted Gem—A very old variety introduced by Luther Burbank, produced mainly in Idaho and Montana, this variety produces long tubers with russeted skin and many eyes of medium depth. It has excellent cooking quality and is slightly resistant to scab, but is very susceptible to verticillium wilt.

Irish Cobbler—An old variety of wide adaptation, Irish Cobblers are early maturing, produce white smooth-skinned tubers, are round with blunt ends, often deeply notched, and have many deep eyes.

White Rose—An old, late-maturing variety that produces long, white-skinned tubers with many eyes, it is a leading variety in California with outstanding yields.

Bliss Triumph—An early-maturing, red-skinned variety grown for a long time, this potato yields round, thick tubers with many deep eyes, but is susceptible to most diseases, especially to hopper burn.

Kennebec—A new, high-yield, late-maturing variety that is widely adapted, producing large oblong tubers with a few shallow eyes and smooth white skin, the Kennebec potato is resistant to common strains of late blight, mild mosaic, and net necrosis, and possesses good cooking and processing qualities.

Cherokee—This is a new, medium-early variety, replacing Irish Cobbler in some areas. The tubers are medium in size, round with blunt ends, and white skinned with a few eyes of medium depth. The variety is resistant to common strains of late blight, common scab, mild mosaic, and net necrosis.

Chippewa—A new, midseason variety producing large, elliptical tubers with a white skin, few shallow eyes, and a speckled appearance, it is resistant to mild mosaic, net necrosis, brown rot, and late blight and has good cooking quality.

"Russet Rural", "Red McClure", "Green Mountain", "Superior", "Norchip Norgold Russett", and "Norland" are some other popular old varieties grown in the U.S. and Canada, but have limited adaptability. Tiechelaar[4] described "Belleisle", a new variety originated by the Canada Department of Agriculture Research Station at Fredericton, NB. It has high-specific gravity, oval-shape tubers that are bruise resistant and resistant to common scab, black wart, and *Fusarium sambucinum*.

Both old *desi* potato varieties, such as Phulwa, Darjeeling Red Round, Craig's Defiance, Great Scot, and Satha, as well as the improved varieties mainly developed at the Central Potato Research Institute, Simla, such as Hybrid 9, 45, 208, 209, 2236, Kufri Red, Kufri Safed, Kufri Kuber, Kufri Kisan, Kufri Kundan, Kufri Kumar, Kufri Jeewan, and Kufri Chandramukhi, are grown in different parts of India.[5,6] Pushkarnath[5] and Yawalkar[6] have given the following description of some important commercial potato varieties of India:

Phulwa—These tubers are white, round, and small to medium in size with yellow flesh. They are waxy and firm, have good cooking and keeping qualities, are very late maturing and are susceptible to all diseases.

Kufri Safed—A new, improved, disease-free clone of Phulwa, Kufri Safeds are high yielding (25% higher than Phulwa).

Darjeeling Red Round—These are red, medium-sized tubers with yellow flesh and waxy, firm texture and good cooking quality, yielding high under plains, maturing earlier than Phulwa, and susceptible to all important diseases and viruses.

Kufri Red—This variety is a selection from Darjeeling Red Round (DRR), and is superior in yield and quality as compared to DRR.

Up-to-Date—This is an early, high-yielding variety, producing big, oval-shaped white tubers and white flesh. It cooks readily and breaks if cooked for a long time, and is popular in North India.

Craig's Defiance—An English potato variety improved in 1936, this is high yielding with wide adaptability to both hills and plains, is early maturing, has good keeping quality, and is resistant to physical damage. It is susceptible to early blight and to virus Y, but immune to other viruses A, B, C, and X.

Kufri Kuber—An early cultivar that is suited to plains and is quick growing, producing uniform, medium-sized tubers (originally designated as O.N. 2236).

Kufri Kundan—This variety is a hybrid cross between Ekishirazy and Katahdin, with an assured yield. Its tubers are white, attractive, and uniform in size. It withstands water-logging and is moderately resistant to late blight, but is susceptible to other diseases and viruses.

Choudhury[7] and Singh and Poal[8] described the following newly evolved varieties of potato at the Central Potato Research Institute at Simla, India.

Kufri Kisan—These are late maturing, high yielding in plains, tubers medium in size with a smooth surface. They have deep eyes, white skin, round shape, and a good-keeping quality.

Kufri Neela—A late maturing variety with round, white tubers, the Kufri Neela is a substitute for "Great Scot", high yielding (70% higher than Great Scot), and resistant to late blight.

Kufri Sindhuri—This is a midseason potato with round, light-red, medium-sized tubers, a substitute for Kufri Red and other later varieties, and is moderately immune to viruses.

Kufri Chandramukhi—Earlier maturing and higher yielding than Up-to-Date, these are attractive, oval-shaped white tubers suited to plains. Although susceptible to late blight, this variety escapes the disease due to its early habit.

Kufri Khasi-Garo—A variety that is early maturing, resistant to late blight, and moderately resistant to early blight and viruses.

Kufri Naveen—These are resistant to late blight and wart and moderately resistant to *Cercospora* leaf spot and early blight.

Kufri Chamatkar—An early bulking variety (110 to 120 d) that is of uniform size, producing smooth tubers suited to plains in all seasons.

Kufri Neelamani—Suited to the Nilgiri Hills in South India, this potato yields 100% higher than Great Scot.

Kufri Sheetman—These are frost resistant, suited to northern and eastern parts of India.

Kufri Jyoti—Large, oval, flattened, white-skinned tubers with dull white flesh, these are widely adaptable, fertilizer responsive, resistant to late blight and warts, and are moderately resistant to *Cercospora* leaf spot.

Kufri Alankar—A very early maturing (75 d), photo-insensitive variety, high yielding, suited to plains and for multiple cropping; it produces large tubers that are oval-shaped with white flesh and fleet eyes.

Kufri Jeewan—These are late maturing, high yielding, resistant to late blight, warts, and Cercospora leaf spot.

Kufri Moti—This potato is resistant to late blight, with oval, white, round tubers. It is immune to one race of late blight and yields 100% higher than Great Scot.

Kufri Lavkar—Early maturing, suited to the plains of central India, this variety produces round, white tubers with firm, light cream flesh.

Kufri Dewa—These tubers are oval to round with white to yellow flesh, resistant to visible mosaic and leaf roll, but highly susceptible to late blight.

B. CLIMATE

Climate plays one of the important roles in influencing potato yields. The climatic factors that influence potato yields are temperature, light intensity and photoperiod, rainfall, and length of growing season.

The potato is classified as a cool-season crop and thrives most in moderate temperatures during the growing season. The potato production zones by climate in developing countries are shown in Table 1. Best growth is achieved when the mean July temperature is about 70°F (20°C).[125] Temperature influences the rate of absorption of plant nutrients and their translocation within the plant, the rate of photosynthesis, and the rate of respiration, and also hastens early growth. Temperatures during growing determine the length of the growing season. At low temperatures, the rate of respiration is less than the rate of photosynthesis, resulting in more accumulation of carbohydrates in the tubers and an increase the specific gravity of the tubers.

The influence of light on growth and yield of potatoes is dependent upon intensity, quality, and day length. Effects of the photoperiod on the growth and development of potatoes have been studied extensively.[126-128] During short days, the light-demanding growth stage

TABLE 1
Potato Production Zones by Climate in Developing Countries

Climate	Estimate of total production (%)	Growing season	Observations
Hot summer, cool to cold winter, unreliable precipitation	5	Cooler months	Irrigation
Hot, dry year around	6	Cooler months	Irrigation
Hot, wet summer; hot, dry winter	2	Dry season	Irrigation
Tropical rain forest	1—2	Rainy season	Humid
Warm to hot, wet summer	35	Winter	Low land Asia
Hot, dry summer; cool, wet winter	2	Various	Dry winter irrigation
Cool to warm, wet summer; cool, wet winter	10	Summer	Temperate
Cool to warm, wet summer; cool to cold, dry winter	12	Summer	Temperate
Highlands (<1500 m)	21	Various	Complex production

Courtesy of the *Annual Report of International Potato Center*, Lima, Peru, 1988.

of the potato is completed faster than on a long day. During the stage of tuber formation, potato plants prefer long days. At this stage the plant devotes a large portion of its carbohydrates resulting from photosynthesis to the growth of tubers and less to vegetative growth. A long photoperiod increases the maturity period of potato plants, whereas with a short period they tend to mature earlier.

Adequate moisture is required for steady growth and maximum yield of potatoes. In areas that receive low rainfall, additional irrigation water should be distributed so as to provide adequate moisture for the growth of the plants.

C. SOIL

The soil factors that influence tuber growth are structure, water-holding capacity, aeration, temperature, drainage, and the nutrient-supplying capacity of the soil. Potatoes can be grown in all types of soil except saline and alkaline soils. Sandy loam and loamy soils rich in organic matter are most suitable for potato cultivation. In general, loam soil produces high-specific gravity tubers, probably because they have optimum soil moisture and uniform temperature and structure relationships for potato production than the heavy-textured soils. Sandy soils are conducive to rapid growth, provided sufficient nutrients and moisture are present, resulting in early maturity of the crop. Soils that are too sandy and especially those low in organic matter usually are not desirable for high yields, as a result of their low water-holding capacity. To obtain high yields, the soil should be loose and friable with good drainage and aeration. Such soils are easy to till and do not provide conditions favorable for the development of blight rot of tubers. Potatoes do not grow well on heavy-textured or undrained soils. Potatoes grown on some fertile clay soils may produce good yields, but not on sticky soils that hinder digging and marketing of potatoes due to the soil adhering to them.

The sensitivity of potato plants to soil aeration has been well demonstrated.[129] Potato plants respond well in soils that have 50% or more of pore or air space.

The optimum soil pH for potatoes is about 5.0 to 5.5. Acidic soil conditions tend to limit potato scab disease, which is favored by alkalinity. Successful production of potatoes can be achieved on soils with neutral or alkaline pH. The effects of soil pH on yield of potatoes are probably associated with the availability and uptake of plant nutrients.

Potatoes should be cultivated in well-prepared, pulverized, and moist soil. Soils that lack desirable physical and chemical characteristics are not suitable for potato cultivation.

TABLE 2
Influence of Fertilizer Application (4-14-8) on Yields
of Three Cultivars in a New Red Latosol of Brasilia
D. F., 1981

Fertilizer t/ha	Aracy		Achat		Bintje	
	1980	1981	1980	1981	1980	1981
0	12.1	12.2	9.9	3.4	11.7	4.8
2	48.5	24.5	32.7	14.6	25.8	9.6
4	46.9	30.0	41.2	18.9	31.4	15.6
6	50.1	29.9	42.4	23.8	33.6	14.6
8	46.8	32.8	45.4	21.9	29.6	14.6
Tukey 5%	14.24	5.71	16.96	3.66	7.21	4.79

From Lindbergue, C. and Campos, T., Report of the Centro, Nacional de Pesquisa de Hortalicas Embrapa, Brasilia, D. F., Brazil, 1981. With permission.

The moisture content of the soil is an important factor that determines the yield of potatoes. Potatoes require a uniform supply of moisture throughout the growing season for maximum yields. The most critical period concerning water requirements is at the beginning of tuberization.[130] Excess soil moisture may, however, produce low dry matter yields. Fluctuation in soil moisture tends to promote unequal growth in vines and tubers, resulting in lower yields. The soil moisture should be maintained at about 60% of field capacity for optimum yields. To obtain potatoes high in dry matter, water is withheld late in the season to help them become mature at harvest. The specific gravity of the tubers is reduced when they are supplied with excessive irrigation late in the season.

D. FERTILIZER

Potatoes require large quantities of mineral nutrients for maximum growth, particularly large amounts of nitrogen and potassium. The kind and amount of fertilizers required for maximum yield vary depending upon soil type, soil fertility, climate, crop rotation, variety, length of growing season, and moisture supply. The ratio of nitrogen, phosphorus, and potash is important. The most widely used and recommended ratios are 1-2-1, 1-2-2, 2-3-3, and 1-1-1. The results of studies on the effect of different levels of fertilizer on yield of potatoes are presented in Table 2. In many areas, fertilizers are applied before or at the time of planting. In other areas, in addition to applied fertilizers at the time of planting, a side-dressing application is made during the growing season. In addition, nitrogen and micronutrients such as magnesium, copper, and zinc are applied in certain areas. Green manuring improves organic-matter status of the soil.

Consideration should be given to soil pH when selecting the source of nitrogen, phosphorus, and potassium fertilizers. High yield responses in potatoes in most areas results from application of nitrogen. Nitrogen enhances extensive vine growth, and thus tends to increase the growing season, which in turn influences maturity at harvest. A solid or liquid source of nitrogen can be employed. Phosphorus application to potatoes has some effect on dry matter. At the early stage of growth the need for phosphorus is critical. High-yielding potato cultivars require large amounts of potassium for growth and development. Potassium removal by the potato crop is high, but the response to potassium application is less than that of nitrogen and phosphorus. Potassium sulfate as a potassium fertilizer is often recommended for use on potatoes, because it is thought that it causes less of a problem with tuber quality than with potassium chloride in terms of specific gravity. In addition to the reduction in specific gravity with high rates of potassium application, further reduction has

TABLE 3
Bacterial Diseases of Potatoes

Disease	Causal organism
Bacterial soft rot	*Erwinia carotovora* var. *carotovora* (Jones) Dye and *E. carotovora* var. *atroseptica* (Van Hall) Dye
Brown rot	*Pseudomonas solanacearum* E. F. Smith
Ring rot	*Corynebacterium sepedonicum* (Spieck and Kott) Skapt and Burkh
Pink eye	*Pseudomonas fluorescens* Migula
Common scab	*Streptomyces scabies* (Thaxter) Waksman and Henrici (Syn. *Actinomyces scabies* (Thaxter) Gussow)

been shown with the use of potassium chloride over the use of K_2SO_4.[121] Micronutrients do affect potato quality and yield when present in deficient quantities.

E. WEED CONTROL

In many potato growing areas, it is a practice to control weeds early in the season with the application of certain chemicals to the soil. These chemicals usually are applied as sprays to the soil before potatoes emerge, but after weeds are in the seedling stage. Intercultural operations like hoeing, weeding, and earthing up help in checking the weeds. Linuron, simazine, and patoran are more effective for weed control in the spring crop because of their persistent effects.[9,10] The application of dinobo compounds such as dinitro-*o*-secondary butyl phenol and dinitro-*o*-secondary amyl phenols are available under a trade name. Premerge® and sinox-PE® are effective for pre-emergence weed control.[11-15]

F. DISEASES OF POTATOES

A potato disease is an interaction between a host (the potato) and a pathogen (bacteria, fungus, virus, mycoplasma, nematode, or adverse environment) that impairs productivity or usefulness of the crop. The host-pathogen interaction is influenced by environment acting on either the potato or the pathogen or on both, and is determined by the genetic capabilities of (1) the potato, in being either susceptible or resistant, and (2) the pathogen, in being pathogenic or nonpathogenic. Disease or adverse environment may severely affect production or the quality of potatoes. Various diseases caused by bacteria, fungi, viruses, mycoplasma, and nematodes were compiled by the American Phytopathological Society and control measures have been suggested.[16]

Attack by microorganisms probably results in the most serious gross losses caused during potato production, seed tuber production, and postharvest potatoes. The diseases may cause losses quantitatively and qualitatively. Quantitative pathogenic losses result from the frequently rapid and extensive breakdown of host tissues as in the case of blights, pink rot, dry rot, bacterial soft rots, and in some, virus diseases. The pattern of attack is often an initial infection by a specific pathogen, followed by an attack of broad-spectrum secondary organisms, which are only weakly pathogenic or saprophytic on the dead tissues remaining from the primary infection. These secondary invaders may be aggressive and can have an important role in postharvest diseases, frequently surviving to multiply and exaggerate the initial damage by the primary pathogens. Qualitative pathogenic losses are typically the result of diseases such as common scab, powdery scab, black scurf, and silver scurf, or deforming diseases such as wart. These diseases, although inducing little or no tuber rotting, affect the appearance of the potato and thereby influence the market value. Another group of diseases, namely skin spot and *Rhizoctonia* scurf, invade and kill potato eyes and hence are of great importance on seed potatoes.

The list of bacterial, fungal, and viral diseases is, respectively, given in Tables 3, 4 and 5. The diseases described here merely represent a sample of the more prevalent and

TABLE 4
Fungal Diseases of Potatoes

Disease	Causal organism
Powdery scab	*Spongospora subterranea* (Wallr.) Lagerh. f. sp. *subterranea* Tomlinson
Wart	*Synchytrium endobioticum* (Schilb.) Perc.
Skin spot	*Polyscytalum pustulans* (Owen and Wakef.) M. B. Ellis (Syn. *Oospora pustulans* Owen and Wakef.)
Leak	*Pythium ultimum* Trow, *P. debaryanum* Hesse and other *Pythium* spp.
Pink rot	*Phytophthora erythroseptica* Pethybr.
Late blight	*P. infestans* (Mont.) de Bary
Powdery mildew	*Erysiphe cichoracearum* DC. ex. Merat
Early blight	*Alternaria solani* Sorauer (Syn. *Macrosporium solani* Ellis and Martin)
Septoria leaf spot	*Septoria lycopersici* Speg.
Cercospora leaf blotches	*Mycovellosiella (Cercospora) concors* (Casp.) Deighton *Cercospora solani-tuberosi* Thirumalachar.
Phoma leaf spot	*Phoma andina* Turkensteen
Gray mold	*Botrytis cinerea* Pers.
White mold	*Sclerotinia sclerotiorum* (Lib.) de Bary (Syn. *Whetzelinia sclerotiorum* [Lib.] *S. minor* Jagger Korf. and Dumont) *S. intermedia* Ramsey
Stem rot	*Sclerotium rolfsii* Sacc.
Rosellinia black rot	*Rosellinia* spp.
Rhizoctonia canker	*Rhizoctonia solani* Kuhn *Thanatephorus cucumeris* (Frank.) Donk.
Violet root rot	*Helicobasidium purpureum* (Tul.) Pat. (Syn. *Rhizoctonia crocorum* [Pers.] DC.)
Silver scurf	*Helminthosporium solani* Dur. and Mont (Syn. *Spondylocladium atrovirens* Harz.)
Black dot	*Colletotrichum atramentarium* (Berk. and Br.) Taub. (Syn. *Colletotrichum coccodes* [Wallr.] Hughes)
Charcoal rot	*Macrophomina phaseoli* (Maubl.) Ashby (Syn. *M. phaseolina* [Tassi] Goid., *Sclerotium batalicola* Taub.)
Gangrene	*Phoma exigua* Desm.
Fusarium dry rots	*Fusarium solani* (Mart.) App. and Wr. emend Synd. and Hans. and *F. roseum* (Lk.) Snyd. and Hans.
Fusarium wilt	*F. eumartii* Carp. (Syn. *F. solani* f. sp. *eumartii* [Carp.] Snyd. and Hans.) *F. oxysporum* Schl. (Syn. *F. oxysporum* Schl. f. sp. *tuberosi* [Wr.] Snyd. and Hans.) *F. avenaceum* (Fr.) Sacc. (Syn. *F. roseum* [Lk.] Snyd. and Hans.) *F. solani* (Mart.) App. and Wr. (Syn. *F. solani* f. sp. *eumartii* [Carp.] Snyd. and Hans.)
Verticillium wilt	*Verticillium albo-atrum* Reinke and Berth.
Thecaphora smut	*Angiosorus solani* (Barrus) Thirum. and O'Brien (Syn. *Thecaphora solani* Barrus)
Common rust	*Puccinia pittieriana* P. Henn.
Deforming rust	*Aecidium cantensis* Arthur.

devastating diseases. The prevalence and incidence of potato diseases vary with the season and the locality in which the crop is grown.

1. Bacterial Diseases
a. Black Leg and Bacterial Soft Rot

Soft rot caused by different species of soft-rot coliform bacteria is a common storage and transit disease. These bacteria cause soft rot of tubers in storage and also black leg and wilt of plants in the field. Black leg and bacterial soft rot are principally caused by two varieties of the same species of bacterium, *Erwinia caratovora*. They are found wherever potatoes are grown.

In the field the typical ''black leg'' is characterized by a striking brown-black or jet-black color of the stem at the soil level (Plate 1A*). The discoloration usually starts from

---

* Plates 1A and B follow page 56.

TABLE 5
Important Virus and Mycoplasma Diseases of Potatoes

Virus/mycoplasma	Mode of transmission	Symptoms
Virus		
Potato leaf roll virus	Insect (aphids)	Potato leaf roll
Potato virus Y	Mechanical/insect (aphids)	Rugose mosaic
Potato virus A	Mechanical/aphids	Mosaic
Potato virus X	Mechanical/tubers/fungi	Mild to severe mosaic
Potato virus M	Mechanical/tuber/insect	Mottle, mosaic, crinkling, and rolling of leaves
Potato virus S	Mechanical/insect	Deepening of veins, rugosity of leaves, when severe necrotic spots
Potato virus T	Tubers/insects/mechanical	Mottling, vein necrosis, chlorotic spots
Andean potato mottle virus	Mechanical/tubers	Mild patchy mottle, leaf deformation, systemic necrosis stunting
Andean potato latent virus	Mechanical/potato flea beetle	Mosaics and/or chlorotic netting
Cucumber mosaic virus	Mechanical/insect	Chlorosis
Potato mop-top virus	Powdery scab fungus/mechanical	Bright yellow marking, blotches, rings, or diagnostic V-shapes in lower leaves; pale V-shaped markings in upper leaves; stunting of stem.
Tobacco rattle virus	Nematode/tuber/mechanical	Mottling of primary stem
Potato yellow dwarf virus	Insect/mechanical	Pith necrosis
Alfalfa mosaic virus	Mechanical/insect	Tuber necrosis
Potato aucuba mosaic virus	Mechanical/insect	Mosaic and top necrosis
Potato yellow vein virus	Not identified	Yellow of veins/interveinal yellowing
Potato spindle tuber virus	Mechanical/pollen and true seed	Necrotic spot and surface cracking
Mycoplasma		
Aster yellows and Stolbur	Insect	Rolling and purple pigmentation of leaves, formation of gummy tubers
Witches' broom	Insect	Formation of many auxiliary and basal branching having upright habit of growth, rolling of leaflets, and chlorotic discoloration in apical leaves

the old seed tubers. The cortical tissues may shrivel and rot. The plants may achieve normal height, but usually they remain dwarfed and stunted. The foliage turns yellow, with a slight metallic luster, and soon wilts and dies. Curling of leaves similar to that caused by potato leaf curl virus may also be found. Sometimes, young seedlings arising from diseased seed tubers are destroyed before or soon after emergence. When infection occurs late and tubers have developed, they carry bacterium to the storage godowns. Stems, petioles, and leaves may also become infected through wounds such as petiole scars or hail or wind damage. Infection may progress up or down the stems or petioles, thus producing typical black leg symptoms on plants that do not show infection from infected seed pieces. In wet weather, the decay is soft and slimy and may spread to most of the plant. Under dry conditions, infected tissues become dry and shriveled and are generally restricted to the underground portion of the stem. Tubers produced by the infected plants may show symptoms ranging from slight vascular discoloration at the stolon end to soft rot of the entire tuber (Plate 1B*).[17]

In soft rot of tubers, which may occur in the field if the soil is moist and the temperature is high or during transit and storage, the tubers are transformed partly or totally, slowly or quickly, into a soft, decayed, pulpy mass. When a soft rot tuber is cut open a colorless, putrid mass turns pinky red on exposure to air, rapidly becoming brownish red to brown-black. Infection of tubers in storage or in the soil before harvest occurs through lenticels and wounds or through the stolon end of the tuber via the infected mother plant. Lesions associated with lenticels appear as slightly sunken, tan to brown, circular, water-soaked areas, approximately 0.3 to 0.6 mm in diameter. Lesions associated with wounds are irregular in shape, sunken, and usually dark brown. Soft-rotted tissues are wet, cream to tan, with a soft, slightly granular consistency. Brown to black pigments develop near the margins of the lesions.[18,19]

Control

1. Use only healthy tubers as seed material.
2. Avoid using cut potatoes as seed.
3. Plant seed tubers and cut seed tubers in well-drained soil.
4. Avoid excessive irrigation to prevent an anaerobic soil condition which favors seed piece decay and subsequent stem invasion.
5. Treat the seed tubers with effective antibiotics[20,21] or approved fungicides well before planting to reduce infection by *Fusarium* sp. and other pathogens that predispose to bacterial invasion.
6. Plant shallow in heavy soil rather than in light soil.
7. Do not plant too early or too late.
8. Give best tillage to the field, but avoid injuring young plants.
9. Avoid injury to tubers during harvesting, transit, and storage.
10. Remove plant debris and tuber material from the field after harvest and destroy them.
11. Pick out the diseased and injured tubers before packing or storing the tubers.
12. Wash the tubers with chlorinated water before storage.
13. Keep the stores dry, well ventilated, and cool.
14. Frequently clean and disinfect seed cutting and handling equipment, at least between different seed lots.
15. Remove infected plants as soon as they appear.

b. Brown Rot

This disease has a wide host range, including potato, tomato, tobacco, and other Solanaceae. It is principally a disease of the southern states and of other warmer climates. It limits growing of potatoes and other susceptible crops in parts of Asia, Africa, and South and Central America. The first plant symptom is a wilting of leaves during the heat of the day with recovery at night, but eventually the plants fail to recover, and they die. The vascular bundles are stained brown, giving the stems a streaked appearance and the foliage a bronze tint.[22] A white, slimy mass of bacteria will ooze from the vascular bundles which are broken or cut. The vascular area of infected tubers becomes brown. A valuable diagnostic sign of brown rot is glistening beads of a gray to brown slimy ooze on the infected xylem in the stem cross sections.[22] A bacterial exudate oozes from the eyes and stolon end where it becomes mixed with soil. The soil and exudate may dry and adhere to the surface of infected tubers. Underground stems, stolons, and roots of plants with initial foliage symptoms show few advanced symptoms of infection. Grayish-brown discoloration indicates well-established infection.

Control

1. Plant disease-free seed potatoes in uninfected soil.
2. Disinfect the cutting knife.
3. Crop rotation may reduce the severity of disease if solanaceous crops are avoided and seed hosts are controlled. Green Mountain, Sebago, and Katahdin are moderately resistant to this disease.[23]

c. Ring Rot

Plant symptoms begin with wilting of leaves and stems after midseason.[24] Lower leaves slightly roll at the margins and become pale green and show usually the first wilting. As wilting progresses, pale yellowish areas develop between veins. Two important diagnostic features are the wilting of stems and leaves and a milky-white exudate that can be squeezed from the vascular rings of tubers and stems when cross sectioned at their base. In tubers, the disease sometimes may not be seen before harvest, and symptoms appear later in storage.[25] Squeezing the tubers, particularly those from storage, expels creamy, cheeselike ribbons of odorless bacterial ooze, which leaves a distinct separation of the tissues adjacent to the ring. Secondary invaders (usually soft-rot bacteria) cause further tissue breakdown in advanced diseased stages, displaying ring rot symptoms. Pressure developed by this breakdown can cause external swelling, ragged cracks, and reddish-brown discoloration near the eyes.

This disease can be controlled by using disease-free seed; disposing of all potatoes from the farm when the disease is noticed; disinfecting warehouses, crates, and all other machinery to be used for handling, planting, harvesting, and grading of potatoes; using new bags for clean seed; and planting certified seeds.

d. Common Scab

Common scab has no foliar or above-ground symptoms. Tuber symptoms vary considerably with the variety and severity of infection. Shallow or surface scab is characterized by raised, corky areas of variable size which are brown or light tan in color. This type of scab is common in thick-skinned varieties such as Russet Burbank. A deep or pitted scab is characterized by roughly circular to irregularly pitted or sunken lesions of corky tissues, varying in diameter from about 1 to 5 mm. This type of scab is typical of thin-skinned white or red varieties.[26] Lesions may also occur on stolons and roots.

Control

1. Plant only scab-free or disinfected seed potatoes in scab-free soil.
2. Adjust or maintain the soil pH at 5.0 to 5.3 in areas where soil is naturally acidic.
3. Avoid the use of fresh manure, wood ashes, or too much lime, which increase soil pH and lower the soil Ca-P ratio.[27] If lime is needed, use dolomite lime and apply it following potatoes in rotation.
4. Treat soil with sulfur and acid-forming fertilizers to increase solid acidity, and pentachloronitrobenzene, urea, formaldehyde and other soil fumigants.
5. Apply seed treatment with organic mercury to avoid introduction of inoculum to new areas.
6. Use mancozeb (8%) dust as a tuber seed treatment to effectively control acid scab.

2. Fungal Diseases
a. Late Blight

Symptoms of the disease may appear on the tops of the plants at any time during development. On leaflets, the first signs of leaf lesions are formed, having typically small, brownish- to purplish-black, irregularly shaped spots. Under favorable environmental conditions they rapidly grow to large, necrotic lesions that may kill the entire leaflet and spread via the petioles to the stem, eventually killing the entire plant (Plates 2A and B*). A pale, green-yellow halo is present outside the area of leaf necrosis.[28,29] Under moist conditions, on the lower side of the leaf, a white mildew appears on the surface of the lesion in the region where the pale green tissues join. The mildew consists of the sporangiophores and sporangia of the pathogen. When the blight is advancing rapidly in a potato field a mildly pungent characteristic odor is given off. This odor actually results from rapid breakdown of potato leaf tissues. Potato tubers are often infected while in the field, and they are subjected to infection at harvest and sometimes in storage. The effect under field conditions is secondary or primary. The early attack of the disease resulting in death of the tops will reduce the size and number of tubers formed, but the primary invasions of the tubers through soil cause rotting, a dry or wet rot resulting according to the amount of moisture and the temperature prevailing at the time.[30] The first sign is a brown to purple discoloration of the skin, followed by a brownish dry rot which extends to about 1.5 cm below the surface (Plate 2C*). When the disease proceeds without complications, tuber symptoms are quite distinctive and the entire affected tuber may be rapidly decayed. In storage containers, the wet-rot phase of the disease is more common unless ideally cool and dry conditions are maintained, in which tuber lesions develop slowly and may become slightly sunken after several months. Secondary organisms (bacteria and fungi) often follow infection by *Phytophthora infestans*, resulting in a partial or complete breakdown of tubers and complicating diagnosis. In a moist atmosphere, white tufts of mycelium and sporangiophores of the fungus appear on the surface of the tuber.

Control

1. Plant healthy seed potatoes.
2. Destroy or eliminate dump piles.
3. Spray or dust with an effective fungicide.[31-33]
4. Kill potato tops before harvest or spray them with copper sulfate to prevent tuber infection during harvest.
5. Control powdery scab to reduce susceptibility of tubers to late blight infection.
6. Use resistant varieties where possible.
7. Prevent rot in storage by removing infected tubers before storage and maintaining adequate circulation and temperature as cool as is incompatible with other considerations.

b. Early Blight

The disease is of common occurrence wherever potatoes are grown. In India, it is the most common and destructive disease of this crop. It appears 3 to 4 weeks after the crop is sown, much earlier than the late blight. The late blight epidemics are common in cooler areas, while early blight occurs in cool as well as in warm areas. The disease also occurs on tomato, peppers, eggplant, and related wild hosts.

Initial infection is most frequent on lower, older leaves (Plate 3A**). Symptoms of the disease first become visible as small (1 to 2 mm) dry spots, dry and papery in texture, later becoming brown-black and circular to ovoid as they expand. Advanced lesions have angular margins because of limitations by leaf veins. In the necrotic tissues concentric ridges develop to produce a "target-board" or "bull's-eye" appearance, the most characteristic appearance of the disease. There is usually a narrow chlorotic zone around the spots which fades into

---

*   Plates 2A—C follow page 56.
** Plates 3A—C follow page 56.

normal green, and increases with the increase in size of the spots. There may be only two spots or a large number of them may appear, occupying a major portion of the leaf surface. As new lesions develop and older ones expand, the entire leaf becomes chlorotic, later necrotic, and desiccates, but usually does not abscise. Damage to leaves is considerably in excess of the tissue actually destroyed by lesions, suggesting that toxins—alternaric acid— cause leaf death some distance from the site of infection. In dry weather, the spots become hard and the leaves curl. In humid weather the affected areas coalesce and big rotting patches may appear. In severe attacks leaves shrivel and fall down. Potato stems may also be affected, showing brown to black necrotic lesions on the skin (Plate 3B). These lesions may lead to collapse of the branches or the entire above-ground portion of the plant. Tuber lesions are dark, sunken, circular to irregular in shape, and often are surrounded by a raised border of a purplish to gun-metal color (Plate 3C). Tissue in advanced decay is often water-soaked and yellow to greenish yellow, and finally causes rotting of the tubers.[34] Damage to the crop may be considerable as the peak period of attack coincides with the period of tuber formation and causes formation of undersized tubers and a reduced number of tubers. Early blight tuber lesions are not as prone to invasion by secondary organisms as are many other tuber rots.

Control

1. Keep plants healthy and vigorous by providing proper nutrition and water, by controlling insects such as flea beatles, and by controlling other diseases.
2. Spray or dust protectant fungicides such as Difolatan, Maneb or Zineb, fentin hydrochloride, and chlorothalonil, which could effectively control the early blight on foliage.[35,36]
3. Fungicide applications scheduled by spore trapping[37] or other methods,[38] so as to coincide with secondary spread of the disease, are most effective. Early season applications of fungicides before secondary inoculum is produced often have little or no effect on the spread of the disease.
4. Crop rotation and field sanitation are essential to effectively check disease.
5. Permit tubers to mature in the ground before digging and avoid bruising in handling.
6. Avoid disturbing seed tubers until ready to plant.
7. Cultivars with a level of field resistance are available, but no cultivars are immune.

c. Powdery Scab

The first symptom on potato tubers consists of faintly brown areas, about the size of a pinhead, which enlarge in 6 to 8 d to about $^1/_2$ cm. In the meantime, the brownish color disappears and the diseased tissue becomes jelly-like, which contains the mature yellow-brown spore balls (cystosori) of the fungus. The lesion is usually surrounded by the raised, turned edges of the burst periderm. In very wet soils, wound periderm does not develop and the lesion expands in depth and width, forming hollowed-out areas of very large warts. This is a cankerous form of powdery scab. In storage, powdery scab may lead to a dry rot or to more warts or cankers. If infected tissues have not burst through the periderm, infection and necrosis may spread laterally, producing one or two necrotic rings surrounding the original infection. Under humid conditions, the periderm ruptures and warts become larger and secondary warts may develop. Powdery scab lesions may serve as infection courts for late blight and a number of wound pathogens. Infection on roots and stolons parallel to that of tubers may occur as small necrotic spots developing into milky-white galls varying in diameter from 1 to 10 mm or more. Galls on roots may become so severe that young plants wilt and die. As galls mature, they turn dark brown and gradually break down, liberating powdery masses of cystosori into the soil.[39]

Control

1. Crop rotations of 3 to 10 years have been recommended, depending on climate and soil conditions.
2. Plant disease-free seeds.
3. Crop in porous and well-drained soils and avoid planting on soils known to be infested.
4. Do not use manure from animals fed with infected tubers.
5. Soaking infected seed in a solution of formaldehyde or mercuric chloride reduces seed-borne inoculum.
6. Resistant cultivars are recommended, but no immune varieties are known.

d. Black Wart

Symptoms of the disease appear on all underground parts other than roots. Buds on stems, stolons, and tubers are the centers of infection and abnormal growth activity leading to wart formation. The warts vary from very small protuberances to intricately branched systems. A typical wart is roughly spherical, soft, and pulpy and can be cross-sectioned more readily than tubers. Morphologically, it consists of distorted, proliferated branches and leaves grown together into a mass of hyperplastic tissue.[40] The underground wart is usually similar in color to a normal stolon or tuber. Above-ground galls are green to brown, becoming black at maturity and later decaying. Occasionally, galls appear on the upper stem, leaf, or flower. Below-ground galls form at stem bases, stolon tips, and tuber eyes. Tubers may be disfigured or completely replaced by galls. Subterranean galls are white to brown, becoming black after decay.

Control

1. Worldwide control of spread is being attempted through quarantine legislation.
2. No chemical control is known that is not also injurious to soil and crops.
3. The herbicide dinoseb reduces infection to some extent.
4. Resistant cultivars have been developed in Europe and North America.

e. Black Scurf (Rhizoctonia Canker)

When seed tubers sprout, the growing tips of emerging plants are most severely damaged, killing underground sprouts and delaying emergence, especially in cold, wet soils. This results in poor, uneven stands of weak plants and subsequent yield reduction.[41] Emerging potato sprouts may also be infected with the cankers on the developing stem, often causing girdling and stem collapse. Partial or complete girdling may cause stunting of plant tops, cortical necrosis of woody stems, purple pigmentation of leaves, upward leaf roll, and most severe chlorosis at the top of the plant. Reddish-brown lesions on stolons cause stolon pruning or tuber malformation. Roots are also pruned, resulting in a sparse root system. Black or dark-brown sclerotia develop on the surfaces of mature tubers. Sclerotia may be flat and superficial or large, irregular lumps resembling soil that will not wash off.[42] Other tuber symptoms may include cracking, malformation, pitting, and stem-end necrosis.

Control

1. When the nature of soil permits, shallow planting of potato seed tubers reduces apical shoot injury and stem cankers. In practice, seed pieces are dropped in the usual trench with minimum covering, and the soil may be worked to the plants in succeeding cultivation.
2. Seed treatment is not effective in heavily infested soils. Use disease-free seed combined

with seed treatment such as the systemic fungicide[43] (benomyl, thiabendazole, or carboxin) or, where acceptable, organic mercury.[44]

3. Soil treatments of benomyl or pentachloronitrobenzene reduce soil-borne inoculum, but the returns may not be economical.

f. Verticillium Wilt

The disease apparently occurs wherever potatoes are grown, although it may be confused with other diseases that cause early maturity.

Symptoms of infected plants include pale green or yellow leaves that die prematurely, and these leaves are described as "early drying" or having "early maturity". During the growing season, plants may lose their turgor and wilt, especially on hot, sunny days. Single stems or leaves on one side of the stem may wilt first. The vascular tissue of stems becomes light brown. Tubers from infected plants usually develop a light-brown discoloration in the vascular ring, and severe vascular discoloration may extend over halfway through the tubers. A pinkish or tan discoloration may develop around the eyes or as irregular blotches on the surface of affected tubers.[45]

Control

1. Rotate potatoes with cereals, grasses, or legumes. Avoid rotation with highly susceptible solanaceous crops such as eggplant and most tomato cultivars.
2. Do not plant susceptible cultivars.
3. Control weeds.
4. Seed tubers contaminated with infested soil should be disinfected before planting. Liquid seed treatments[46] are more effective than dusts. Organic mercuries are very effective, but their use is generally prohibited. Systemic fungicides (benomyl or thiophanate-methyl) and nonsystemics (mancozeb, Captan, or metiram) applied to seed tubers are reported effective.[47]
5. Several soil treatments show promise; sodium methyldithiocarbamate, benomyl,[48] and systemic insecticides[49] (aldicarb, acephate) have delayed symptoms and increased some yields.
6. Several nematodes increase the incidence and severity of the disease. Soil fumigation with nematocides alone (trichloro-nitromethane or 1,3-dichloropropene and related compounds) or with chemicals that control both fungi and nematodes is effective.

g. Fusarium Wilts

These diseases are widespread and most severe where potatoes are grown at relatively high temperatures or when seasons are hot and dry.

Several *Fusarium* spp. cause essentially similar symptoms. Tubers exhibit surface blemishes and decay, including the stem, and browning and decay at the stolon attachment, and vascular discoloration that severely impairs market quality, because such tubers cannot be removed during grading. On vines, symptoms include cortical decay of roots and lower stems; vascular discoloration or rot in the lower stems; wilting; chlorosis, yellowing, or bronzing of the foliage; rosetting and purpling of the aerial parts; aerial tubers in the leaf axis; and premature death of the plant. Additional symptoms vary with the pathogen involved[50] and the environment.[51]

Control

1. Tubers infected with Fusarium wilt should not be used for seed.
2. Avoid contamination of clean fields by inoculum transfer through infested soil or diseased tubers and plant refuse.

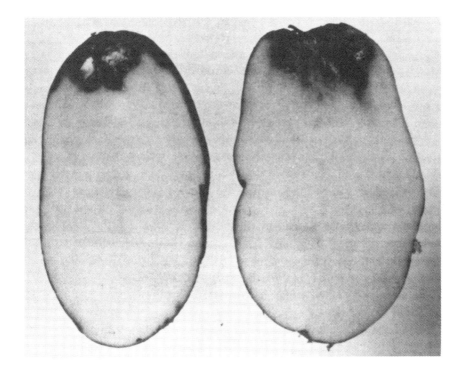

FIGURE 1.   Fusarium rot of potatoes develops during storage, and turns to a dark-brown color subsequently decaying the potato tuber.

h. Fusarium Dry Rot

Fusarium dry rots affect tubers in storage and planted seed tubers. Tuber lesions at wounds are visible as small brown areas after 1 month of storage and later enlarge. The periderm over the lesion sinks and wrinkles in concentric rings as the tissues dry out. Fungus pustules containing mycelia and spores may emerge from the dead periderm. The rotted tubers shrivel and become mummified.[52]

Whole-tuber seed become infected through wounds during storage or preparation for planting, and the infection is carried into the field. Brown to black flecks appear on the cut surface in about 1 week and pits form in 2 weeks. With numerous cut surface infections, lesions coalesce; seed pieces rot from the surface inward and the eyes are destroyed as decay progresses. *Fusarium* spp. alone in association with *Erwinia* spp. partially or completely destroy the seed pieces in the field, resulting in extreme variability in plant size.

Control

1.   For storage and seed purposes, harvest tubers from dead vines.
2.   Use all precautions with machinery and equipment to prevent wounding during harvest and storage.
3.   Provide high humidity and good ventilation early in storage to facilitate wound healing and provide aeration during storage.
4.   Seed tubers may be treated with a fungicide, dust, or liquid spray before storage.
5.   Warm seed tubers from cold storage at 20 to 25°C for a week before planting or cutting pieces.
6.   Spray or dip seed tubers with fungicide suspensions or treat seed pieces with 7 to 8% fungicidal dusts.[53]

7.   Handle treated seed with noncontaminated containers and equipment.
8.   Plant seed immediately after cutting in soils sufficiently warm and moist to promote prompt sprout growth and good wound healing.

i. Charcoal Rot

The fungus is worldwide in its distribution, but is only economically important in warm regions where soil temperatures exceed 28°C.

Under hot conditions, the pathogen can attack potato stems and cause yellowing and a sudden wilt. Stem infection is not important. Tuber attack, which is more important and occurs before harvest and in storage, causes loss of the entire crop. Early symptoms develop around the eyes, near the lenticels, and at the stolon attachment as light gray, water-soaked cavities entirely filled with black mycelium and sclerotia rapidly invade the tubers. When cut the tubers exhibit semiwatery, flabby breakdown, with color changing from yellowish to pinkish to brown and finally to black.[54] Wet rot may later develop from secondary invaders.

Control

1.   Harvest early before soil temperatures go high.
2.   Avoid bruising and wounding of tubers in harvest and postharvest handling.
3.   Field irrigation may be useful to prevent excessive soil temperature.
4.   Harvest the crop immediately after plants have matured and do not harvest during periods in which soil temperatures exceed 28°C.
5.   Do not store tubers at high temperatures.
6.   Do not use seeds originating from areas where the disease is prevalent.

3. Viral and Mycoplasmal Diseases

Viral and mycoplasmal diseases, although not lethal, reduce plant vigor and yield potential of both commercial crop as well as seed tubers. The list of important viral and mycoplasmal diseases along with the mode of transmission and important symptoms is given in Table 5.

a. Interveinal Mosaic (Leaf Rolling Mosaic)

The disease causes slight dwarfing, diffuse mottling, mosaic, crinkling, and rolling of the upper leaves; leaflet deformation and twisting and necrosis of petioles and stems;[55] and necrosis of the phloem in tubers.[56] These symptoms are usually quite mild and are often masked in hot weather.

Control

1.   Use disease-free, virus-tested seed tubers.
2.   Control aphid populations.
3.   Rogue infested plants when first found in the field.
4.   Disease-free stock of some varieties can be obtained by clonal selection and serological tests. Saco is resistant to leaf roll virus.[57]

b. Mild Mosaic

Mild mosaic is characterized by chlorotic mottling of the foliage of susceptible varieties. The mottled areas vary in both size and intensity and lie both on and between the veins. A slight crinkling usually accompanies mottling, and the margins of the leaflets may become wavy. Plants may be slightly dwarfed and tubers may be smaller than normal.[58] The disease may decrease yield of infected potatoes up to 40%. Warm, sunny growing conditions tend to mask the symptoms. Some highly susceptible European varieties exhibit a top necrosis. Some varieties such as British Queen develop a tuber necrosis also.

Control

1.   Plant disease-free seed potatoes.
2.   Use resistant varieties such as Katahdin, Chippewa, Sebago, Kennebec, Houma, and Mohawk.
3.   Plant early and rogue diseased plants.

c. Yellow Dwarf

The most common symptoms[59] are yellowing of the foliage and dwarfing of the plant, as the name implies. Other symptoms include rolling or rugosity of the leaves with downward curving of the longitudinal axis of the leaflet, and pith necrosis of the stems following shortly after foliage chlorosis. Death of the apex results, followed by premature death of the entire plant. Tubers are few, small, deformed, and fail to germinate. A dark-brown necrosis occurs throughout the tuber, with only the xylem elements unaffected.

Control

This disease has been successfully controlled by planting certified seeds, which are produced in areas where the virus is not found, in isolated areas away from clover, by roguing, by control of insects, and by planting tolerant varieties having field tolerance— namely Chippewa, Katahdin, Russet Burbank, and Sebago.

4. Nonpathogenic Diseases

Besides pathogenic disease, the diseases are also incited by genetic abnormalities, adverse environmental conditions, and nutritional imbalances, which are listed in Tables 6 and 7. The symptoms and possible control measures of these diseases are recently described.[60]

5. Disease Resistance

Potatoes suffer from many diseases and pests, and extensive research work has been done to evolve a variety with high yield, good quality, and resistance to all pathogens. The breeding programs for resistance to pathogens usually consider, first, the importance of pathogens; second, the type and degree of resistance available; third, the inheritance of resistance, and fourth, tests for resistance.[93] Most breeding programs therefore begin with a source of resistance which can be hybridized with the standard varieties. It is also important to know whether or not resistance is dominant to susceptibility and how many genes are involved. According to Howard,[93] nonrace-specific resistance (horizontal resistance, "field" resistance, etc.) is due to polygenes and is therefore a difficult type of resistance to handle.

Several types of viral resistance have been distinguished: tolerance, "infection resistance"; hypersensitivity (usually giving "field immunity"), and extreme resistance (or immunity). Tolerance of a virus, i.e., showing no obvious symptoms and little reduction in yield, may be a dangerous type of resistance in potatoes because of difficulties caused in seed-certification schemes and because a tolerant variety may act as a reservoir of virus from which nontolerant varieties can be infected. "Infection resistance" is the type of resistance in which only a small percentage of plants are infected in the field, and is often due to polygenes. Hypersensitivity and immunity, on the other hand, are due to single dominant genes, which are widespread and found in *tuberosum* and *Andigena* potatoes. There are several important viruses to which neither hypersensitivity nor immunity have been found. Most breeding programs include sources of resistance to virus X, virus Y, virus A, leaf roll, and other viruses such as virus S, mop top, tobacco rattle, and spindle tuber.

TABLE 6
Important Nonpathogenic Diseases of Potatoes*

Cause	Symptoms	Ref.
I. Genetic abnormalities		
Weldings	Low growth, close bushy habit, numerous thin stems, reduced number of leaflets, large rounded terminal leaflets, absence of flowers, increased number of stolons with numerous small tubers	60
Feathery wildings	More thin stems and many more small tubers; top leaflets small, narrow, pointed	60
Giant hill	Plant with greater height and stronger, more vigorous vines, and coarser and thicker leaflets; tubers sprout late; plants mature late, reduce yields	60
Tall types	Intermediate between giant hill and normal plants, more vigorous and 2 to 3 weeks later in maturity	60
II. Adverse environment		
Oxygen deficit	Abnormal development of stolons and tubers	61
Low-temperature tuber injury	Diffuse patches of brownish-black metallic discoloration, net necrosis of phloem tissue, blackish patches or blotches near the vascular ring, brownish-red to black discoloration in the central part of the tuber	62
Low-temperature foliage injury	Rapid wilting and collapse of leaves; diffused chlorotic areas, mottling, and necrotic streaks on young leaves	63
Black heart (inadequate oxygen supply) (Plate 8)	Black to brown discoloration in the central portion of the tuber	64
High-temperature field injury	Girdling and rotting of stems; defoliation of leaves	65
Internal heat necrosis	Light-tan, dark-yellowish, to reddish-brown necrosis towards the center of large tubers	66
Secondary growth and jelly end rot (high field temperatures and drought) (Plate 9)	Stolon ends of tubers shrivel, collapse into a wet, jelly-like substance; second growth in the form of germination	67,68
Hollow heart (excessively rapid tuber enlargement)	One cavity with lens or star shaped and angular at the corners formed near the center of the tuber; tissues water soaked with brown necrotic patch	69
Growth cracks (Plate 10)	Growth cracks are developed from internal pressure, virus infection, mechanically produced cracks, and harvest cracks	70
Black spots (bruising injury)	Blue-gray to black discolored lesions develop beneath the tuber skin without periderm, later flatten; spheriodal blue-gray patches in the vascular region	71
Tuber greening and sunscald (exposure to sunlight)	Slight exposure to light for some time in field or after harvest, tuber tissue turns green with purple pigmentation; exposure to intense sunlight results in formation of white-skinned restricted areas covering a sunken necrotic area	72
Internal sprouting	These sprouts are in eyes with tightly cultured multiple rosette, branched or unbranched, and penetrate the tubers directly above; sprouts from tubers with deep eyes penetrate into the side of the eye depression	73
Coiled sprout (over matured seed)	Underground sprouts lose their negative geotrophic habit and coil tightly with curve portion of the swollen stem, with light-brown lesions on the stem inside the coil	74
Secondary tubers or knobs (Plate 11A and B)	Tubers sprout either in storage or in the field, producing new tubers directly without forming a normal plant	75
Hair sprout (hot or dry conditions in late growing season, viruses)	Tubers with hair or spindle sprout germinate early and produce normal sprouts and hair sprouts from different eyes	76

* Plates 8—11B follow page 56.

TABLE 6 (continued)
Important Nonpathogenic Diseases of Potatoes

Cause	Symptoms	Ref.
Nonvirus leaf roll (disturbance in carbohydrate translocation, genetic)	Leaves become leathery, roll upward; top roll affects the apical leaves of the plant	77
Hail injury	Hail tears and perforates leaves, causes defoliation and reduces yields	78
Wind injury	Discolored to brown tissues with glistening appearance extending through leaves and finally tattering at the edges	79
Lightning injury (severe thunderstorms)	Stems collapse and tops of plants irreversibly wilt without effect on leaves; tubers have brown to black skin necrosis with cracking	80
Air pollution; photochemical oxidants (ozone)	Stippling of upper leaf surface by darkly pigmented, chlorotic, bronzed spots within 24 h following heavy exposure; advanced necrosis and chlorosis within 10 to 14 d	81
Air pollution; sulfur oxides	Light-tan to white interveinal necrotic areas accompanied by yield reduction	82
Chemical injury	Necrosis at the stolen attachment and vascular discoloration of the stem end, burning of interveinal tissues; in storage, netting of tuber surfaces and dehydration due to foliage application of maleic hydrazide; abnormal sprouting with other compounds	83
Stem-end browning	Internal brown discoloration of tuber tissue near the stem end or stolon attachment, confined to 12-mm section at the stolon end of the tuber, in deeper penetration to the xylem of the vascular ring	84

Three important bacterial diseases of the potato are black leg (*Erwinia phytophthora* Appel or *E. caratovora* Jones); ring rot (*Corynebacterium sepedonicium* [Spiedk and Kotth] Skapt and Burkh); and brown rot or bacterial wilt (*Pseudomonas solanacearum* E. F. Smith). Small differences in the susceptibility of potato tubers of different varieties to black leg and ring rot have excluded these diseases from the breeding programs, because there are no known good sources of resistance to these diseases. Black leg is controlled by producing disease-free stocks from stem cutting (black leg), and ring rot is controlled by isolating disease-free tubers to produce new "seed" stocks. *Solanum phureja*, a diploid clone of the cultivated species, has shown significant resistance to a serious pathogen of warm soil, viz., *P. solanacearum*. Common scab, caused by *Streptomyces scabies* (Thaxt) Waksm. and Henrici, reduces tuber quality and consumer acceptance. A significant varietal difference in the susceptibility to common scab has made it possible to breed a potato variety for resistance to scab.

The potato is affected by several fungi, rust *(Puccinia pitteriana)*, and smut *(Thecaphora solani)*, the later two in Colombia and Peru, respectively. Amongst fungal diseases, the most important are wart disease *(Synchytrium endobioticum)*, "late blight" *(Phytophthora infestans)*, storage rots, viz., dry rot *(Fusarium solani* var. *coeruleum)*, gangrene *(Phoma exigua* var. *foveata)*, pink rot *(Phytophthora erythroseptica)*, black scurf *(Corticum solani)*, skin spot *(Oospora pustulans)*, and verticillium wilt *(Verticillium alboatrum)*. The most common potato pests include cyst nematodes, *Globodera (Heterodera)* spp., viz., *G. rostochionsis* and *G. pallida*, root-knot nematodes *(Meloidogyne* sp.), false root-knot nematodes *(Nacobbus* sp.) and migratory nematodes *(Trichodorus* sp.), aphids *(Myzus persicae* and *Macrosiphum euphorbiae)*, and Colorado beetle *(Leptinotarsa decemlineata)*. Resistance to all these different diseases and pests is being sought under various climatic conditions and regions of the countries. Howard[94] reported a method for screening potatoes for resistance to *Globodera* sp. under laboratory conditions.

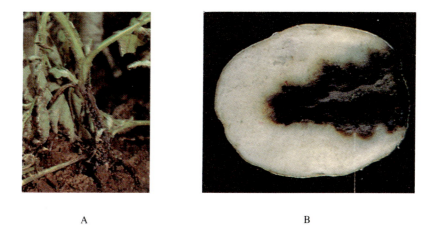

A   B

Plate 1.   Black leg: (A) aerial infection of the above-ground portion of a stem; (B) tuber infection.

A   B

C

Plate 2.   Late blight, *Phytophthora infestans* (Mont.): (A) causing death of leaves; (B) black necrotic lesions on the stem; (C) tuber lesions.

Note: All plates are from *Potato Field Manual*, J. R. Simplot Company, Pocatello, ID, 1989. With permission.

A   B

C

Plate 3.   Early blight, *Alternaria solani:* (A) foliar infection; (B) stem lesions; (C) tuber infection.

A   B

C

Plate 4.   Colorado potato beetle defoliates plants. (A) Adult and eggs; (B) larva; (C) damage.

Plate 5.   Potato aphids, *Macrosiphum euphorbiae*, act as virus vectors.

Plate 6.   Root-knot nematode, *Meloidogyne* sp., affecting the tuber.

Plate 7.   Golden nematode, *Globedera rostochiensis;* galls present on roots.

Plate 8.   Physiological disorder: black heart.

Note: All plates are from *Potato Field Manual,* J. R. Simplot Company, Pocatello, ID, 1989. With permission.

A   B

Plate 9.   Physiological disorder: jelly end rot—(A) pointed ends; (B) translucent end.

Plate 10.   Physiological disorder: growth cracks.

A   B

Plate 11.   Physiological disorder: (A) knobs; (B) knobs after storage.

Plate 12. Nitrogen deficiency symptom.

Plate 13. Phosphorus deficiency symptom.

Plate 14. Potassium deficiency symptom.

Plate 15. Calcium deficiency symptom.

Plate 16. Sulfur deficiency symptom.

Plate 17. Magnesium deficiency symptom.

Note: All plates are from *Potato Field Manual*, J. R. Simplot Company, Pocatello, ID, 1989. With permission.

Plate 18. Zinc deficiency symptom.   Plate 19. Boron deficiency symptom.   Plate 20. Manganese deficiency symptom.

Plate 21. Iron deficiency symptom.   Plate 22. Copper deficiency symptom.

Note: All plates are from *Potato Field Manual*, J. R. Simplot Company, Pocatello, ID, 1989. With permission.

TABLE 7
Nutrient Deficiency Symptoms in Potatoes*

Nutrient	Deficiency symptoms	Ref.
Nitrogen	Deficient plants are chlorotic, slow growing, erect with small erect, pale green leaves (Plate 12); chlorotic and stunted plant growth with loss of lower leaves and reduced yields; nitrogen toxicity causes upward rolling of leaves, poor root development, and reduced yields	85
Phosphorus	Deficiency causes retarded growth of terminals, small, spindly, rigid plant growth; cup-shaped, darker than normal, lusterless leaflets with scorching at the margins (Plate 13); dropping of lower leaves, leaf petiole erect; delayed maturity; reduction in the number and length of roots and stolons; internal rusty-brown necrotic spots in tubers; high phosphorus levels reduce uptake and utilization of zinc or iron	86
Potassium	Deficiency causes browning of old leaves, later necrotic and senescence early (Plate 14), upward rolling leaflet margins from the middle to top of the plant; leaflets are small, cupped, crowded together, bronzed on upper surface, and dark-brown specks on lower; surface coalescing and causing marginal necrosis; poor development of roots, shortening of stolons, and reduced yields; necrotic brown lesions at stolon ends of tuber with necrotic foliage accompanied by drying of tissues and forming hollow spot of 2 mm in diameter surrounded by corky tissue	87
Calcium	Calcium-deficient plants are spindly with small, upward-rolling, crinkled leaflets having chlorotic margins, later becoming necrotic (Plate 15); tubers develop diffuse brown necrosis in the vascular ring near the stolon attachment and later similar fledges in pith; tubers may be extremely small	88
Sulfur	General yellowing with slight upward rolling of leaflets due to sulfur deficiency (Plate 16).	
Magnesium	A pale, light-green color begins at the leaf tips and margins and progresses between veins; leaves are thick, brittle, and roll upward with tissue raised between the veins; necrotic leaves hang on the plant or abscise (Plate 17); roots are stunted	89
Aluminum	Toxicity causes roots to become short and stubby with few branches; plants are small and spindly with branches arising at acute angles	90
Zinc	Deficiency causes upward rolling of young, chlorotic leaves with terminal leaves being vertical; gray-brown to bronze areas (Plate 18), later turning necrotic, develop on leaves near the middle of the plant and later involve all leaves; brownish spots may develop on petioles and stems; stunted plant growth	91
Boron	Deficiency causes death of growing point, sprouting of lateral buds, shortening of inter-nodes, thickening and upward rolling of leaves, assuming busy appearance of plants (Plate 19), tubers are small, showing surface cracking at the stolon end and localized brown areas under the skin near the stolon end or brown vascular discoloration	
Manganese	Deficiency causes upward rolling of leaves near the shoot tips with brown necrotic spots along the veins, drying of leaves (Plate 20); interveinal necrosis may precede leaf death; severe reduction in yield; manganese toxicity causes stem streak, stem streak necrosis, and stem break	92
Iron	Deficiency causes yellowing of young leaves and growing tips without necrosis; at later stage, leaves turn yellow with green veins (interveinal chlorosis-Plate 21); low number of tubers are formed with decreased yields.	
Copper	Deficiency is manifested primarily by a pronounced upward cupping and an inward rolling of young leaf blades (Plate 22).	

G. INSECT PESTS AND NEMATODES OF POTATO
1. Insect Pests

Many insect pests (Table 8) damage the potato. It is obviously difficult to estimate yield losses caused by pests. However, the pest damage to potatoes in field and in storage is a major constraint for potato production. Potato pests cause damage in a number of ways. Some, such as the Colorado beetle (Plates 4A, B, and C**) and the lady beetle, defoliate the plant, whereas aphids remove sap, may inject toxic saliva, act as virus vectors[95,97] (Plate 5**), and reduce tuber yields. Others, such as wireworms, eat tubers, considerably decreasing

* Plates 12—22 follow page 56.
** Plates 4 and 5 follow page 56.

TABLE 8
Insect Pests of Potatoes

Common name	Scientific name	Damage caused
Major pests		
Potato aphid	*Aulacorthum solani* (Kalt.)	Infest foliage; suck sap; virus vectors
Peach-potato aphid	*Myzus persicae* (Sulz.)	Infest foliage; suck sap; virus vectors
Potato tuber moth	*Phthorimaea operculella* (Zeller)	Larvae bore stems and tubers
Epilachna beetles	*Epilachna* spp.	Adult and larvae eat leaves
Colorado beetles	*Leptinotarsa decemlineata*	Adult and larvae eat leaves
Wireworms	*Agariotes* spp.	Larvae in soil eat roots and bore tubers
Minor pests		
Leafhopper	*Amarasca devastans* (Dist.)	Infest foliage; stunt plants
Leafhopper	*Cicadella aureta* (L.)	Infest foliage; stunt plants
Leafhopper	*Typhlocyba jucunda* Herr. Schaff.	Infest foliage; stunt plants
Potato aphid	*Macrosiphum euphorbiae* (Thos.)	Infest foliage; suck sap; virus vectors
Bulb and potato aphid	*Rhopalosiphoninus latysiphon* (Davids.)	Infest foliage; suck sap; virus vectors
Buckthorn-potato aphid	*Aphis nosturtii* (Bomer)	Infest foliage; suck sap; virus vectors
Root mealybug	*Planococcus citri* (Risso)	Infest roots or foliage
Striped mealybug	*Ferrisia virgata* (CK 11.)	Infest foliage
Grape mealybug	*Pseudococcus maritimus* (Ehrh.)	Infest foliage
Green capsid	*Lycocoris pabulinus* (L.)	Sap suckers; toxic saliva
Potato capsid	*Calocoris norvegicus* (Gmel.)	Sap suckers; toxic saliva
Tarnished plant bug	*Lygus rugulipennis* Popp.	Sap suckers; toxic saliva
Green stink bug	*Nezara viridula* (L.)	Sap suckers; toxic saliva
Harlequin bugs	*Bagrada* spp.	Sap suckers; toxic saliva
Swift moths	*Hepialus* spp.	Larvae in soil eat roots and bore tubers
Eggplant fruit borer	*Leucinodes orbonalis* Guem.	Larvae eat foliage
Large yellow underwing	*Noctua pronuba* L.	Larvae are cutworms; live in soil and eat into tubers
Black cutworm	*Agrotis ipsilon* (Hfn.)	Larvae are cutworms; live in soil and eat into tubers
Common cutworm	*A. segetum* (D. and S.)	Larvae are cutworms; live in soil and eat into tubers
Heart and dart	*A. exclamationis* (L.)	Larvae are cutworms; live in soil and eat into tubers
Rosy rustic moth	*Hydraecia micacea* (Esp.)	Larvae bore stems
American boll worm	*Heliothis armigera* (Hb.)	Larvae defoliate
Death's head hawk moth	*Acherontia atropos* (L.)	Larvae defoliate
Leather jackets	*Tipula* spp.	Larvae in soil eat roots
Spotted crane-fly	*Nephrotoma maculata* Meig	Larvae in soil eat roots
Cock chafers	*Melolontha* spp.	Larvae are whitegrubs or chafers; live in soil and eat into tubers
Garden chafer	*Phyllopertha horticola* (L.)	Larvae are whitegrubs or chafers; live in soil and eat into tubers
Summer chafer	*Amphimallon solstitialis* (L.)	Larvae are whitegrubs or chafers; live in soil and eat into tubers
Rose chafer	*Cetonia aurata* (L.)	Larvae are whitegrubs or chafers; live in soil and eat into tubers
Brown chafer	*Serica* spp.	Larvae are whitegrubs or chafers; live in soil and eat into tubers
Tortoise beetles	*Aspidomorpha* spp.	Adult and larvae eat leaves

TABLE 8 (continued)
Insect Pests of Potatoes

Common name	Scientific name	Damage caused
Potato flea beetles	*Phylliodes affinis* (Payk.)	Adults hole leaves
Flea beetles	*Fepitrix* spp.	Adults hole leaves
Wireworms	*Drasterius* spp.	Larvae in soil eat roots and bore tubers
Wireworms	*Lemonium* spp.	Larvae in soil eat roots and bore tubers
Wireworms	*Hypolithus* spp.	Larvae in soil eat roots and bore tubers
Striped blister beetles	*Epicanta* spp.	Adults destroy flowers
Yellow tea mite	*Polyphagotarsonemus* latus (Barks)	Scanty foliage
Red spider mites	*Tetranychus* spp.	Scanty and web foliage

the value of the crop. Several insecticides used to control potato pests include chlorinated hydrocarbon, organophosphorus, carbamates, acaricides, and soil fumigants. The application of these insecticides to control various pests depends upon the availability and relative susceptibility of pests to different insecticides.

2. Integrated Pest Control

In recent years, the use of pesticides for control of insects, nematodes, and other pests has significantly contributed to the increasing production of potatoes. However, prices of synthetic pesticides have increased significantly in recent years. Several pests have developed certain levels of resistance to chemical pesticides.[96] The rate of discovery of new insecticides to counteract the problem of resistance is not keeping pace with the emergence of new strains of resistant insect pests.[98] Moreover, pesticides drastically affect the beneficial insect population. This results in two kinds of phenomena. The population of the target species may quickly recover from pesticide action and rise to a new and higher level. This is called "pest resurgence". Shelton et al.[99] have recently reported the rapid resurgence of the potato tuber worm after insecticide destruction of its natural enemy. Moreover, following pesticide treatments, other nontarget insects may increase to damaging levels, and these new pests may become more important than the primary ones. Due to these reasons, cultivators are compelled to use higher quantities of toxic chemicals, resulting in higher costs and new problems. The insecticide residue can cause environmental contamination and destruction of the ecosystem. Hence, integrated pest control assumes a greater importance in crop production.

Over 100 pests of the potato crop have been described. Out of these few are key pests active over a large area, whereas others may be key pests in more restricted areas. It is therefore important to identify key pests in a particular area and develop appropriate technology for control of these pests. The successful biological control of insect pests can be achieved by analyzing the pest status of each of the injurious pests, developing reliable monitoring techniques, devising schemes for lowering the equilibrium position of key pests, and modifying the pest environments.[98]

3. Nematodes

Nematodes pathogenic to potatoes (Table 9) occur in all climates and cause serious crop losses, but much of this damage is unrecognized or attributed to other causes. These parasites are known to have infected potatoes since 1880 and are globally causing an average 11% loss of potatoes.[100] Recently, 138 species of nematodes, belonging to 44 genera, are reported to be associated with potato culture throughout the world.[100] Among these, root-knot nematodes[101] and cyst nematodes[102] have been recognized as major pests of potatoes. In addition, the rot,[103] lesion,[104] and stubby root[105] nematodes also have been constantly encountered in the potato rhizosphere.

TABLE 9
Nematode Pests of the Cultivated Potato

Scientific and common names	Distribution[a]				Transmission by tubers[b]
	H	C	S	T	
Belonolaimus longicaudatus (sting nematode)	h			t	−
Ditylenchus destructor (potato-rot nematode)				T	+
D. dipsaci (stem and bulb nematode)		c		t	+
Helicotylenchus spp. (spiral nematodes)	h	c	s	t	+
Globodera spp. (round cyst nematodes)					
G. pallida potato cyst nematode (white immature females)		C		T[c]	+
G. rostochiensis potato cyst nematode (golden immature females)		C		T[c]	+
Hexatylus vigissi				t	+
Longidorus maximus (needle nematode)				t[d]	−
Meloidogyne spp. (root-knot nematodes)					
M. acronea				t	+
M. africana	h			t	+
M. arenaria	h	c	s	T	+
M. hapla		C	s	T	+
M. incognita	H	c	S	T	+
M. javanica	H	c	S	T	+
M. thamesi				t	+
Meloinema sp.			s	t	+
Nacobbus aberrans (false root-knot nematode)	h	C		T	+
Neotylenchus abulbosus				t	+
Paratylenchus spp. (pin nematodes)	h	c	s	t	− +
P. spp. (root-lesion nematodes)					
P. andinus		c			+
P. brachyurus			s	t	+
P. crenatus				t	+
P. coffeae	h		s	t	+
P. minyus		c	s	t	+
P. penetrans	h	C	s	T	+
P. pratensis		c	s	t	+
P. scribneri	h	c	s	t	+
P. thornei		c	s	t	+
Rotylenchuhus spp. (reniform nematodes)	h	c		t	+
Rotylenchus spp. (spiral nematodes)	h	c		t	+
Trichodorus spp., *Paratrichodorus* spp. (stubby root nematodes)					
T. allius				t[d]	−
P. christiei		c		t[d]	−
P. pachydermus		c		t[d]	−
T. primitivus		c		t[d]	−
P. teres				t[d]	−
Tylenchorhynchus spp. (stunt nematodes)					
T. claytoni	h	c	s	t	−
T. dubius		c	s	t	−
Xiphinema spp. (dagger nematodes)	h	c	s	t[d]	−

[a] H = hot tropical; C = cool tropical; S = subtropical; T = temperate zone. When capitalized, as shown, the judged relative importance is greater than when lower case letters are used. All attack potatoes and additional members of Solanaceae as well as plants outside Solanaceae.

[b] None are known to be transmitted by true botanical seed.

[c] Limited to Solanaceae.

[d] Plant virus vectors.

From Hooker, W. J., Ed., *Compendium of Potato Diseases,* American Phytopathological Society, St. Paul, MN, 1986, 125. With permission.

Since nematodes attack roots and tubers (Plates 6 and 7*), no diagnostic symptoms appear on above-ground parts of the plant except for unhealthy top growth resulting from poor root systems. Low density in soil causes no top symptoms, but may reduce tuber yields.[106] If the world nematode population increases, soil suitable for potato culture will become more scarce. Consequently, potatoes will be grown more frequently on the best potato land; because monoculture encourages nematode population increase, the nematode damage to potatoes will increase significantly.

Confining nematode populations to areas where they already exist by restricting movement of infected seed tubers and plants[102] may be the most effective way of preventing loss of productive land. Care in purchase of seed and prevention of shipment of infected seed into nematode-free areas cannot be overemphasized. An integrated nematode control approach includes: crop rotation with nonhosts, clean cultivation, use of resistant varieties, adopting a combination of preplant soil treatment[107] and later foliar sprays, and strict followup of plant quarantines.

H. MATURITY

Storage and conditioning of tubers is influenced by tuber maturity. Completely mature potatoes are more desirable for processing than less mature potatoes. Potatoes increase in their specific gravity and yield on complete maturation. The way to obtain complete maturity in potatoes is by early planting, late harvesting, and killing of potato vines rather slowly. Mosher[132] has suggested obtaining more mature tubers by using early-maturing varieties, covering seed pieces low, planting early, using suggested rates of nitrogen, and using low rates of potash fertilizer when possible. Delayed harvesting of potatoes tends to increase the specific gravity and yield of tubers. Mature tubers usually result in higher-quality processed product, as well as in yield of product per acre. Immature tubers are low in specific gravity and low in yields. They are subject to skinning and bruising. Potato tubers are harvested before maturity due to favorable market prices at that time and also to avoid the danger of freezing temperatures which may freeze potato tubers.

I. HARVESTING

The time of harvest can be adjusted to suit market prices. This is a very important operation in potato production. In developed countries, harvesting is done by two-row mechanical diggers or mechanical harvesters. In developing countries, manual harvesting is a common practice. In two-row digger harvesting, potatoes are usually dropped to the ground behind the digger and picked up later by hand. These are picked into containers, sacks, or directly into crates and loaded into trucks and transported to a packing shed or storage. The mechanical harvester digs the potatoes, separates them from soil, vines, and stones, and delivers the tubers into containers or trucks to the storage or packaging shed. The losses in production of potatoes depends upon the maturity stage at harvesting and the method of harvesting.

J. YIELD

Yield of potatoes is related to several other characteristics of the crop, such as the duration of crop growth, the time of maturity and harvest, the plant density, the number of plants, and the weight of individual tubers. Investigations on the frequency of early-maturing seedlings under long-day conditions suggested that the time of potato maturity depends upon a number of genes, and varieties are heterozygous for most of these.[108] According to Moorby,[109] a quantitative understanding of the crop behavior can be used to predict and assess the yield potentialities of unusual genotypes. A parallel approach to conventional growth analysis offers advantages. The crop yield of potatoes was higher in plots planted in mid-June than

---

* Plates 6 and 7 follow page 56.

those planted at the end of May; 110 d after planting, the yield was higher with the earlier planting dates. The yield at 130 d after planting was not significantly different from the yield at 110 d after planting. Burton[110] surveyed the history and various factors in influencing the yield of potatoes.

K. TUBER QUALITY

The quality of potato tubers is related to numerous factors including morphology, structure, and chemical composition, which in turn influence the nutritional, sensory, and processing quality of potato.[108-113] Amongst all the factors affecting tuber quality, the most important ones are the environment during crop growth, the variety, the cultural practices employed, irrigation, fertilization, and the use of other agricultural chemicals, such as pesticides and growth regulators. According to Gray and Hughes,[108] the quality and nutritional value of a tuber at harvest is a product of the effects of varietal, cultural, and environmental factors on the growth of the potato crop. The quality of potato tubers associated with morphology and external appearance include tuber size, shape, depth and appearance of the skin, depth of the eyes, flesh and skin color, and greening. The tuber shape, skin, and flesh color and depth of the eyes are largely influenced by varietal, cultural, and environmental factors.[1]

The size and shape of potato tubers are largely the varietal characteristics,[108] and it has been demonstrated that "long" is dominant to "round" shape. Burton[110] recognized four distinct shapes: round, oval, pointed oval, and kidney shaped. In some varieties of potato, shape is influenced by the cultural and environmental conditions which favor rapid and sustained growth, e.g., wide spacing and ample supply of water and nutrients. It has been reported that high levels of applied nitrogen and irrigation, but low levels of potassium, increase the length of the potato tubers relative to their width.[108]

Tuber size is also determined in part by variety and tuber set per stem, but by far the most significant influencing factor is the plant density or population per unit area. Sanderson and White[111] noted that tuber number and the yield of less than 60-mm-diameter tubers were highest for the spacing of 15 cm within the row. The yield of the tubers having less than 60 cm diameter increased up to 79 d for "Kennebec" and 86 d for "Sabago", and then remained constant. Rioux et al.[112] observed that the rate of increase in tuber size was lower for cultivars "Norland" and "Netted Gem", and for all the planting dates investigated. They also noted that the percentage of tubers with more than 70-mm diameter was four times higher at 110 d than at 90 d after planting.

III. GRADING OF POTATOES

The desirable quality characteristics of potatoes are dictated by the intended use, whereas the acceptability of raw potatoes is determined primarily by size, shape, color, and attractiveness of the tubers; the quality of processed products is evaluated in terms of color, flavor, and texture. High-quality processed products can be made only from good quality raw potatoes. Uniformity of size, shape, and composition is most essential, and in many advanced countries rigid specifications for raw potatoes used for processing have been set.

A. GRADING AND STANDARDIZATION

There are many types of standards which apply to horticultural products in the U.S. Some of these have been drawn up by trade associations and are voluntary, while others are government standards which have been issued by state, county, or municipal authorities, and are often mandatory. The two most important groups of standards are those developed by the Federal Agency and the U.S. Department of Agriculture.[114] According to Schoenemann,[115] the use of grades in marketing of potatoes is not mandatory unless a state or region is operating under a set marketing order or compulsory grading regulations, specifying various grades which may be sold.[116]

B. FEDERAL POTATO GRADE STANDARDS

In the U.S., fresh raw potatoes are marketed under five different federal grade standards.[116] These are graded by a U.S. Department of Agriculture inspector for quality, size, and condition.

1. U.S. Fancy
2. U.S. No. 1
3. U.S. Commercial
4. U.S. No. 2
5. Unclassified

The last one is not actually a grade, but simply indicates that no specific grade has been assigned to it. In addition, the federal grade standards specify certain tuber size standards such as size A, size B, size C, etc., and various other skinning classifications which allow measurement of tuber maturity based on the amount of skin or epidermis missing from the tubers.[116]

The federal potato grading standards also specify grade tolerances, allowing for a limited deviation from the various standards to cope with natural and uncontrolled changes occurring in the produce during transit and to cover a certain degree of allowable human error when grading is done. The U.S. standards for potatoes specify definitions of the terms used in the interpretation of official grade standards issued by the U.S. Department of Agriculture Agricultural Marketing Service.[117]

Schoenemann[115] states that more recently a special set of standards has been made available to aid marketing of potatoes for processing. The U.S. Department of Agriculture now specifies these grades for marketing of raw potatoes used for processing.[118] The needs of the potato processing industry are somewhat different, and thus it uses different grade measurements as compared to the marketing of raw or fresh potatoes. These standards provide for a U.S. No. 1 processing grade and a U.S. No. 2 processing grade. It also provides for a definition of unusable material delivered in a potato lot and standards for different tuber size classification.[115] Testing procedures to determine different quality parameters such as specific gravity, dry matter content, total sugars, glucose content, etc., are also designated. The tests for dry matter, specific gravity, and glucose are of special concern to processors, because the quality of processed potato products depends upon these constituents. Whereas specific gravity determines the textural quality of the processed potatoes, sugars (glucose content) determine the color of the product. The U.S. Department of Agriculture, Agricultural Marketing Service, Washington, D.C., provides information on standards for grading potatoes to meet the various processing grades.

C. ECONOMICS OF GRADING

With a large-scale growth in the processing industry and a shift to grading on a volume basis, use of top grades such as U.S. Fancy Grade[2] is on the reduction ("Stripping" of fancy premium tubers). Grading of the product to a lesser standard like U.S. No. 1 or U.S. No. 2 has often been found to bring higher returns for the total lot or crop. Thus, economically sound grading is being adopted in the U.S. The studies conducted by Perry[119] indicated that it was possible to increase total net returns from a given lot of potatoes through "stripping" of select grades. The commercial profits of this practice depend upon available volume, alternate markets for lesser grades, the premium placed on the fancy grade by retailers, marketing margins taken by wholesalers and retailers, the premium grade-out potential of a particular lot, and the price level needed to provide economic incentive to the growers to produce crops of high quality. Miller[120] showed that price premiums paid for quality-graded potatoes could be substantially higher than the added grading costs involved. According to Schoenemann,[115] the trend toward increased utilization of the potato crop for processing is likely to accelerate the practice of ''stripping'' select grades for fresh market.

D. LIMITS OF TOLERANCE

All U.S. grades of potatoes have defined tolerances for defects and size limits. These tolerance limits provide protection against defective changes occurring within a given lot during the marketing processes. It is rather impossible to grade potatoes so critically as to eliminate all chance of human error while sorting them. The development of defects in transit can be significant.

E. SIZE GRADING

In addition to the physical grading of potatoes within broad limits for size and the elimination of tubers with various defects, modern grade standards for potatoes also provide for marketing of the produce in various tuber size classes.[115] Schoenemann[115] states that the sizing of potatoes has become a more widespread practice in recent years, which has led to the development of special machines to sort out potato tubers into various sizes within fairly narrow limits. Most sizing machines size-grade tubers on the principle of the physical dimensions of the tuber or on the basis of weight. The newer electronic potato-sizing machines are elaborate devices that sort potatoes accurately and rapidly. These advances in sizing machines have led to marketing of potatoes in special count packs, enabling consumers especially to buy containers of potatoes with all tubers of practically identical weight and shape. Such uniformity in tuber size is of special value to the hotel and restaurant trade in the preparation and serving of baked and cooked potatoes.[115]

F. SPECIFIC GRAVITY GRADING

Grade standards of potatoes now in use are mostly based on factors affecting the external or internal quality of potatoes. These are discussed in detail under Section III of this chapter. Smith and Nash[121] attempted to devise a method of grading using certain culinary properties of potatoes. It is well established that the dry matter content or specific gravity of the tuber significantly influences the cooking and processing qualities of potatoes. The possibility of utilizing specific gravity and its application for grading in the potato chip industry was indicated.[122] Schoenemann[115] suggested that potatoes segregated for high dry matter content could be utilized more effectively for the processing purpose.

G. CONSUMER PREFERENCES

Consumer wants and desires involve several factors of choice, such as preference for a variety, color, tuber size, shape, type and size of package, and cooking quality, and reaction to various types of defects. Consumer preferences are influenced by price differences and variations in cooking use, color, quality, and time of year. Most consumers appear to prefer a medium-sized potato, and the reasons given for this choice are[123] (1) the right size for judging portions, (2) easy to peel and handle, (3) easily adapted to several cooking methods, and (4) less waste in peeling. Whereas potatoes varying from 7 to 10 oz (198 to 283 g) are preferred for baking purposes, most high-class restaurants consider 10 to 13 oz (283 to 368 g) to be the best grade.[124] Increasing qualities of potatoes are being graded to within specific tuber-size limits and packed for both the retail and institutional trade in the developed countries.

REFERENCES

1. **Kay, D. E.,** Root crops, in *Tropical Product Institute—Crop and Product Digest No. 2,* Tropical Products Institute, London, 1973, 100.
2. **Ugent, D.,** The potato, *Science,* 170, 1161, 1970.
3. **Thompson, H. C. and Kelly, W. C.,** The potato, in *Vegetable Crops,* 5th ed., McGraw-Hill, New York, 1957, 372.
4. **Tiechelaar, E. C.,** New vegetable varieties list XXI. Garden seed research committee, *HortScience,* 15, 565, 1980.
5. **Pushkarnath,** *Potato in India,* Indian Council of Agricultural Research, New Delhi, 1969.
6. **Yawalkar, K. S.,** *Vegetable Crops of India,* 2nd ed., Agri-Horticultural Publishing House, Nagpur, India, 1980.
7. **Choudhury, B.,** *Vegetables,* National Book Trust, New Delhi, 1976, 214.
8. **Singh, S. P. and Poal, K. R.,** Important potato diseases, *Seeds Farms,* 4(8), 74, 1978.
9. **Koshore, H. and Upadhya, M. D.,** Potato in India: present status and problems, in *Recent Technology in Potato Improvement and Production,* Nagaich, B. B., Ed., Central Potato Research Institute, Simla, India, 1977, 1.
10. **Pushkarnath,** Potato varieties in India, *Exp. Agric.,* 6, 181, 1970.
11. **Swaminathan, M. S.,** The origin of the early European potato—evidence from Indian varieties, *Indian J. Genet. Plant Breed.,* 18, 8, 1958.
12. **Lindbergue, C. and Campos, T.,** Internal Report, Centro National de Pesquisqde Hortalicas EMBRAPA, Brasilia, D. P., Brazil, 1981.
13. **Ahmad, K. U. and Rashid, A.,** Effect of planting and harvesting dates on the production of potato, in *Proc. Third Workshop of Potato Research Workers,* BARI, Dacca, Bangladesh, 1981, 73.
14. **Bogdanova, L. S.,** Effectiveness of using herbicides during the cultivation of potatoes, *Nauchn. Tr. Sev.-Zapadn. Nauchno-Issled. Inst. Sel'sk. Khoz.,* 20, 141, 1971.
15. **Leszozynski, W., Lisinske, G., Sobkowicz, G., and Rola, J.,** Effect of herbicides used in potato farming for the control of dicotyledon weeds on the chemical composition of potato tubers, *Zesz. Nauk, Wyzszg. Szk. Roln. Wroclawiu, Roln.,* 29, 163, 1972.
16. **Hooker, W. J.,** *Compendium of Potato Diseases,* American Phytopathological Society, St. Paul, MN, 1986, 125.
17. **Kotila, J. E. and Coons, G. H.,** Investigations on the black leg disease of potato, *Mich. Agric. Exp. Stn. Tech. Bull.,* 67, 29, 1925.
18. **Mugniery, D.,** A method for screening potatoes for resistance to Globodera sp. under laboratory conditions, in *Proc. Int. Congr. Research for the Potato in the Year 2000,* Hooker, W. J., Ed., International Potato Center (CIP), Lima, Peru, 1983.
19. **Jones, D. R. and Dowson, W. J.,** On the bacteria responsible for soft rot in stored potatoes, and the reaction of the tuber to invasion by *Bacterium carotovorum* (Jones) Lehmann and Neumann, *Ann. Appl. Biol.,* 37, 563, 1950.
20. **Bonde, R.,** Antibiotic treatment of seed potatoes in relation to seed piece decay, black leg, plant growth and yield rate, *Plant Dis. Rep.,* 39, 342, 1955.
21. **Robinson, D. B. and Hurst, R. R.,** Control of potato black leg with antibiotics, *Am. Potato J.,* 33, 56, 1956.
22. **Eddins, A. H.,** Brown rot of Irish potatoes and its control, *Fla. Agric. Exp. Stn. Bull.,* 299, 44, 1936.
23. **Nidsen, L. W. and Haynes, F. L., Jr.,** Resistance in *Solanum tuberosum* to *Pseudomonas solanacearum, Am. Potato J.,* 37, 260, 1960.
24. **Spieckermann, A. and Kotthoff, P.,** Under-suchungen iiber die Kartottelpflanze und ihre krankheiten. I. Die Bakterienrigfaule der Kartoffelpflanze, *Landwirtsch. Jahrb.,* 46, 659, 1914.
25. **Raeder, J. M.,** Ring rot of potatoes, *Am. Potato J.,* 26, 126, 1949.
26. **Hooker, W. J. and Page, O. T.,** Relation of potato tuber growth and skin maturity to infection by common scab, *Streptomyces scabies, Am. Potato J.,* 37, 414, 1960.
27. **Davis, J. R., McDole, R. E., and Callihan, R. H.** Fertilizer effects on common scab of potato and the relation of calcium and phosphate-phosphorus, *Phytopathology,* 66, 1236, 1976.
28. **Dastur, J. F.,** The potato blight in India, *Mem. Dep. Agric. India (Bot. Ser.),* 7, 1, 1915.
29. **Dastur, J. F.,** *Phytophthora* spp. on potatoes in Simla hills, *Indian Phytopathol.,* 1, 19, 1948.
30. **Kaung, Z.,** Activity of *Phytophthora infestans* in soil in relation to tuber infection, *Trans. Br. Mycol. Soc.,* 45, 205, 1962.
31. **Bonde, R. and Schultz, E. S.,** The control of potato late blight tuber rot, *Am. Potato J.,* 22, 163, 1945.
32. **Choudhari, H. C.,** Spray tests for control of potato blight in the hills of West Bengal, *Am. Potato J.,* 31, 263, 1954.
33. **Roy, A. K. and Das, N. D.,** Controlling late blight of potato with some fungicides, *Indian Phytopathol.,* 21, 232, 1968.

34. **Folsom, D. and Bonde, R.,** *Alternaria solani* as a cause of tuber rot in potatoes, *Phytopathology,* 15, 282, 1925.
35. **Mathur, R. S., Singh, B. K., and Nagrhoti, M. S.,** Control of early blight of potato with fungicides, *Indian Phytopathol.,* 24, 58, 1971.
36. **Douglas, O. R. and Groskopp, M. D.,** Control of early blight in eastern and south central Idaho, *Am. Potato J.,* 51, 361, 1974.
37. **Harrison, M. D., Livingstone, C. H., and Oshima, N.,** Control of potato early blight in Colorado. II. Spore traps as a guide for initiating applications of fungicides, *Am. Potato J.,* 42, 333, 1965.
38. **Harrison, M. D., Livingstone, C. H., and Oshima, N.,** Control of potato early blight in Colorado. I. Fungicidal spray schedules in relation to the epidemiology of the disease, *Am. Potato J.,* 42, 319, 1965.
39. **Kole, A. P.,** A contribution to the knowledge of *Spongospora subterranea* (Wallr.) Langerh, the cause of powdery scab of potatoes, *Tijdschr. Plantenziekten,* 60, 1, 1954.
40. **Karling, J. S.,** *Synchytrium,* Academic Press, New York, 1964, 470.
41. **Hide, G. A., Hirst, J. M., and Stedman, O. J.,** Effect of black scurf *(Rhizoctonia solani)* on potatoes, *Am. Appl. Biol.,* 74, 139, 1973.
42. **James, W. C. and McKerizie, A. R.,** The effect of tuber-borne sclerotia of *Rhizoctonia solani* Kuhn on the potato crop, *Am. Potato J.,* 49, 296, 1972.
43. **Biehn, W. L.,** Evaluation of seed and soil treatments for control of Rhizoctonia scurf and Verticillium wilt of potato, *Plant Dis. Rep.,* 53, 425, 1969.
44. **Van Emden, J. H.,** Control of *Rhizoctonia solani* Kuhn in potatoes by disinfection of seed tubers and by chemical treatment of soil, *Eur. Potato J.,* 1, 52, 1950.
45. **Robinson, D. B., Larson, R. H., and Walker, J. C.,** Verticillium wilt of potato in relation to symptoms, epidemiology and variability of the pathogen, *Wis. Agric. Exp. Stn. Res. Bull.,* 202, 49, 1957.
46. **Ayers, G. W.,** Potato seed treatment for the control of *Verticillium* wilt and *Fusarium* seed piece decay, *Can. Plant Dis. Surv.,* 54, 74, 1974.
47. **Robinson, D. B. and Ayers, G. W.,** The control of *Verticillium* wilt of potatoes by seed treatment, *Can. J. Agric. Sci.,* 33, 147, 1973.
48. **Biehn, W. L.,** Control of *Verticillium* wilt of potato by soil treatment with benomyl, *Plant Dis. Rep.,* 54, 171, 1970.
49. **Powelson, R. L. and Carter, G. E.,** Efficacy of soil fumigants for control of *Verticillium* wilt of potatoes, *Am. Potato J.,* 50, 162, 1973.
50. **McLean, J. G. and Walker, J. C.,** A comparison of *Fusarium avenacearum, F. oxysporum* and *F. solani* var. *eumartii* in relation to potato wilt in Wisconsin, *J. Agric. Res.,* 63, 495, 1941.
51. **Gross, R. W.,** Relation of environment and other factors to potato wilt caused by *Fusarium oxysporum, Nebr. Agric. Exp. Stn. Res. Bull.,* 23, 84, 1923.
52. **Boyd, A. E. W.,** Dry-rot disease of potato, *Ann. Appl. Biol.,* 39, 322, 1952.
53. **Leach, S. S. and Nielsen, L. W.,** Elimination of fusarial contamination on seed potatoes, *Am. Potato J.,* 52, 211, 1975.
54. **Bhargava, S. N.,** Studies on charcoal rot of potato, *Phytopathol. Z.,* 53, 25, 1965.
55. **Bengall, R. H., Larson, R. H., and Walker, J. C.,** Potato virus M, S and X in relation to interveinal mosaic of the Irish cobbler variety, *Wis. Agric. Exp. Stn. Res. Bull.,* 198, 45, 1956.
56. **Barterls, R.,** Potato virus A, No. 54 in Descriptions of Plant viruses, *Commonw. Mycol. Inst. Assoc. Appl. Biol.,* Kew, Surrey, England, 1971.
57. **Clark, R. L.,** Leaf roll resistance in some tuberous solani under controlled aphid inoculations, *Am. Potato J.,* 40, 115, 1963.
58. **Davidson, T. R. and Sanford, G. B.,** Expression of leafroll phloem necrosis in potato tubers, *Can. J. Agric. Sci.,* 35, 42, 1955.
59. **Black, L. M.,** Potato yellow dwarf virus, No. 35, in Descriptions of plant viruses, *Commonw. Mycol. Inst., Assoc. Appl. Biol.* Kew, Surrey, England, 1970.
60. **Howard, H. W.,** Chimaeras, in *Genetics of the Potato, Solanum tuberosum,* Howard, H. W., Ed., Logos Press Ltd., London, 1970, 176.
61. **Harkett, P. J. and Burton, W. G.,** The influence of low oxygen tension on tuberization in the potato plant, *Potato Res.,* 18, 314, 1975.
62. **Cunningham, H. H., Zachringer, M. V., Bransen, G., and Sparks, W. C.,** Internal quality of Russet Burbank potatoes following chilling, *Am. Potato J.,* 53, 177, 1976.
63. **Hooker, W. J.,** Sublethal chilling injury of potato leaves, *Am. Potato J.,* 45, 250, 1968.
64. **Bennett, J. P. and Bartholomew, E. T.,** The respiration of potato tubers in relation to the occurrence of black heart, *Calif. Agric. Exp. Stn. Tech. Pap.,* 14, 41, 1924.
65. **Nelson, L. W.,** The susceptibility of seven potato varieties to bruising and bacterial soft rot, *Phytopathology,* 44, 30, 1954.
66. **Zimmerman-Gries, S.,** The occurrence of potato heat-necrosis symptoms in Israel and the use of affected tubers as seed, *Eur. Potato J.,* 7, 112, 1964.

67. **Lugt, C., Bodleander, C. B. A., and Goodijk, G.,** Observations on the induction of second-growth in potato tubers, *Eur. Potato J.,* 7, 218, 1964.
68. **Iritani, W. M. and Weller, L.,** The development of translucent end tubers, *Am. Potato J.,* 50, 223, 1973.
69. **Crumbly, I. J., Nelson, D. C., and Duysen, M. E.,** Relationships of hollow heart in Irish potatoes to carbohydrate reabsorption and growth rate of tubers, *Am. Potato J.,* 50, 266, 1973.
70. **Smittle, D. A., Thornton, R. E., Peterson, C. L., and Dean, B. B.,** Harvesting of potatoes with minimum damage, *Am. Potato J.,* 51, 152, 1974.
71. **Smith, O.,** Internal black spot of potatoes, in *Potatoes: Production, Storing Processing,* Smith, O., Ed., AVI Publishing, Westport, CT, 1968, 642.
72. **Akelcy, R. V., Houghland, G. V. C., and Shark, A. E.,** Genetic differences in potato-tuber greening, *Am. Potato J.,* 39, 409, 1962.
73. **Wien, H. C. and Smith, O.,** Influence of sprout tip necrosis and russett sprout formation on internal sprouting of potatoes, *Am. Potato J.,* 40, 29, 1969.
74. **Catchpole, A. H. and Hillman, J. R.,** Studies on the coiled sprout disorder of the potato, Part 2 and 3, *Potato Res.,* 18, 539, 1975.
75. **Burton, W. G.,** The response of the potato plant and tuber to temperature, in *Crop Processes in Controlled Environments,* Rees, A. R., Cockshull, K. E., Hand, D. W., and Hurd, R. G., Eds., Academic Press, New York, 1972, 391.
76. **Snyder, W. C., Thomas, H. E., and Fairchild, S. J.,** Spindling or hair sprout of potato, *Phytopathology,* 36, 897, 1946.
77. **LeClerg, E. L.,** Non-virus leafroll of Irish potatoes, *Am. Potato J.,* 21, 5, 1944.
78. **Beresford, B. C.,** Effect of stimulated hail damage on yield and quality of potatoes, *Am. Potato J.,* 44, 347, 1967.
79. **Grace, J.,** *Plant response to wind,* Academic Press, New York, 1977, 204.
80. **Hooker, W. J.,** Unusual aspects of lightening injury in potatoes, *Am. Potato J.,* 50, 258, 1973.
81. **Heggested, H. E.,** Photo-chemical air pollution injury to potatoes in the Atlantic coastal states, *Am. Potato J.,* 50, 315, 1973.
82. **Thomas, M. D. and Hendricks, R. H.,** Effect of air pollution on plants, in *Air Pollution Handbook,* Magill, P. L., Holden, F. R., and Ackley, C., Eds., McGraw-Hill, New York, 1956.
83. **Murphy, H. J.,** Potato vine killing, *Am. Potato J.,* 45, 472, 1968.
84. **Folson, D. and Rich, A. E.,** Potato tuber net-necrosis and stem end browning studies in Maine, *Phytopathology,* 30, 313, 1940.
85. **Meisinger, J. J., Boulding, D. R., and Jones, E. D.,** Potato yield reductions associated with certain fertilizer mixtures, *Am. Potato J.,* 55, 227, 1978.
86. **Houghland, G. V. C.,** The influence of phosphorus on the growth and physiology of the potato plant, *Am. Potato J.,* 37, 127, 1960.
87. **Baerug, R. and Enge, R.,** Influence of potassium supply and storage conditions on the discoloration of raw and cooked potato tubers of cv. Pimpernell, *Potato Res.,* 17, 271, 1974.
88. **Dekock, P. C., Dyson, P. W., Hall, A., and Gabowska, F. B.,** Metabolic changes associated with calcium deficiency in potato sprouts, *Potato Res.,* 18, 573, 1975.
89. **Sawyer, R. L. and Dallyn, S. L.,** Magnesium fertilization of potatoes on Long Island, *Am. Potato J.,* 43, 249, 1966.
90. **Hawkins, A., Brown, B. A., and Rubins, E. J.,** Extreme case of soil toxicity to potatoes on a formerly productive soil, *Am. Potato J.,* 28, 563, 1951.
91. **Boawn, L. C. and Leggett, G. E.,** Zinc deficiency of the Russett Burbank potato, *Soil Sci.,* 95, 137, 1963.
92. **Robinson, D. B., Easton, G. D., and Larson, R. H.,** Some common stem streaks of potato, *Am. Potato J.,* 37, 67, 1960.
93. **Howard, H. W.,** The production of new varieties, in *The Potato Crop: The Scientific Basis for Improvement,* Harris, P. M., Ed., Chapman and Hall, New York, 1978, 607.
94. **Howard, H. W.,** *Genetics of the Potato, Solanum tuberosum.* Logos Press Ltd., London, 1970.
95. **Kennedy, J. S., Day, M. F., and Eastop, V. F.,** A consensus of aphids as vectors of plant viruses, Commonwealth Institute on Entomology, London, 1962, 114.
96. **Bottrell, D. G.,** Integrated Pest Management, Council on Environmental Quality, Washington, D.C., 1979, 120.
97. **Eastop, V. F.,** Worldwide importance of aphids as virus vectors in *Aphids as Virus Vectors,* Harris, F. and Maramorosch, K., Eds., Academic Press, New York, 1974, 3.
98. **Cisneros, F. H.,** Integrated pest control: new approaches to the priority components, in *Proc. Inst. Congress "Research for the Potato in the Year 2000",* International Potato Center, Lima, Peru, 1983.
99. **Shelton, A. M., Wyman, J. A., and Mayor, A. J.,** Effects of commonly used insecticides on potato tuberworm and its associate parasites and predators in potato, *J. Econ. Entomol.,* 74, 303, 1981.
100. **Krishna Prasad, K. S.,** Nematode problems of potato, in *Plant Parasitic Nematodes of India—Problems and Progress,* Swarup, G. and Dasgupta, D. R., Eds., I.A.R.I., New Delhi, 1986.

101. **Jatala, P. and Rowe, P. R.,** Reaction of 62 tuber-bearing *Solanum* species to root-knot nematodes, *Meloidogyne incognita acrita, J. Nematol.,* 8, 290, 1977.
102. **Spears, J. F.,** The Golden Nematode Handbook: Survey, Laboratory, Control and Quarantine Procedures, U.S. Department of Agriculture, Agricultural Handbook, 353, 81, 1968.
103. **Faulkner, L. R. and Darling, H. M.,** Pathological histology, hosts and culture of the potato rot nematode, *Phytopathology,* 51, 778, 1967.
104. **Koen, H.,** Notes on the host range, ecology and population dynamics of *Pratylenchus brachyurus, Nematologica,* 13, 118, 1967.
105. **Rohde, R. A. and Jenkins, W. R.,** Host range of a species of *Trichodorus* and its host-parasite relationships on tomato, *Phytopathology,* 47, 295, 1957.
106. **Winslow, R. D. and Willis, R. J.,** Nematode diseases of potatoes, in *Economic Nematology,* Webster, J. H., Ed., Academic Press, New York, 1972, 17.
107. **Miller, P. M. and Hawkins, A.,** Long term effects of preplant fumigation on potato fields, *Am. Potato J.,* 46, 387, 1969.
108. **Gray, D. and Hughes, J. C.,** Tuber quality, in *The Potato Crop—The Scientific Basis for Improvement,* Harris, P. M., Ed., Chapman and Hall, New York, 1978, 504.
109. **Moorby, J.,** The physiology of growth and tuber yield, in *The Potato Crop—The Scientific Basis for Improvement,* Harris, P. M., Ed., Chapman and Hall, London, 1978, 153.
110. **Burton, W. G.,** The potato, in *A Survey of Its History and of the Factors Influencing its Yield, Nutritive Value, Quality and Storage,* Veenman, H. and Zonen, N. V., Eds., European Association of Potato Research, Wageningen, The Netherlands, 1966.
111. **Sanderson, J. B. and White, R. P.,** Effect of in-row spacing on potato tuber yield, sizes and numbers measured at 10 weekly sampling dates, *Am. Potato J.,* 59, 484, 1982.
112. **Rioux, R., Gosselin, J., and Genereux, H.,** Effect of planting date on potatoes grown in short seasons, *Can. J. Plant Sci.,* 61, 417, 1981.
113. **Howard, H. W.,** Factors influencing the quality of ware potatoes. I. The genotype, *Potato Res.,* 17, 490, 1974.
114. **Arthey, V. D.,** *Quality of Horticultural Products,* Butterworths, London, 1975, 22.
115. **Schoenemann, J. A.,** Grading, packing and marketing potatoes, in *Potatoes: Production, Storing, Processing,* Smith, O., Ed., AVI Publishing, Westport, CT, 1968, 359.
116. **Anon.,** Self-Help Stabilization Programs with Use of Marketing Agreements and Orders, U.S. Department of Agriculture, Agric. Stabil., No. 479, 1961.
117. **Anon.,** United States Standards for Potatoes, U.S. Department of Agriculture, Agric. Market Serv., U.S. Government Printing Office, Washington, D.C., 1958.
118. **Anon.,** United States Standards for grades of potatoes for processing, Fed. Regist., No. 8179—8181, 1962.
119. **Perry, A. L.,** Potential profits from packing Maine potatoes to the U.S. Fancy grade, *Maine Agric. Exp. Stn., Bull.,* No. 545, 1956.
120. **Miller, C. J.,** Price and quality of table stock potatoes, *Am. Potato J.,* 43, 22, 1966.
121. **Smith, O. and Nash, L. B.,** Potato quality. I. Relation of fertilizers and rotational systems to specific gravity and cooking quality, *Am. Potato J.,* 17, 163, 1940.
122. **Kunkel, R., Gregory, J., and Binkley, A. M.,** Mechanical separation of potatoes into specific gravity groups show promise for the potato chip industry, *Am. Potato J.,* 28, 690, 1951.
123. **Anon.,** Potato Preferences Among Household Consumers, U.S. Department of Agriculture, 667, 1948.
124. **Eberhard, M. F. and Eke, P. A.,** Consumer preference for sized Idaho; Russet Burbank potatoes, *Idaho Agric. Exp. Stn. Bull.,* 282, 1951.
125. **Smith, J. W.,** The effect of weather on the yield of potatoes, *U.S. Monthly Rev.,* 43, 222, 1915.
126. **Driver, C. M. and Hawkes, J. C.,** Photoperiodism in the potato, Imperial Bureau Plant Breeding and Plant Genetics, School of Agric., Cambridge, England, 1943.
127. **Krug, H.,** On the photoperiodic response of potato varieties, *Eur. Potato J.,* 3, 47, 107, 1960.
128. **Bodlaender, K. B. A.,** Influence of temperature, radiation, and photoperiod on development and yield. The growth of potato, *Proc. 10th Easter School Agric. Sci.,* University of Nottingham, 199, 1963.
129. **Bushnell, J.,** Sensitivity of the potato plant to soil reaction, *J. Agron.,* 27, 4, 1935.
130. **Struchtemeyer, R. A.,** Efficiency of the use of water by potatoes, *Am. Potato J.,* 38, 22, 1961.
131. **McDole, R. E., Stallknecht, G. F., Dwelle, R. B., and Pavek, J. J.,** Response of four potato varieties to potassium fertilization in a seed growing area of eastern Idaho, *Am. Potato J.,* 55, 495, 1978.
132. **Mosher, P. N.,** For better potato quality, *Maine Ext. Ser. Bull.,* 474, 1959.

Chapter 4

POSTHARVEST HANDLING AND STORAGE

S. J. Jadhav, G. Mazza, and U. T. Desai

TABLE OF CONTENTS

I. INTRODUCTION

Potatoes are generally stored before they are used for either table purpose or for processing.[1,2] After harvest, these are transported to storage by bullock carts, trucks (Figure 1), rails, or ships, depending upon the quantity of potatoes and the distance to the storage facilities. A large number of tubers are bruised, skinned, or injured during harvesting and transportation.[3] Immediately after harvest, potatoes are stored at 10 to 16°C and high humidity for curing purposes. This process stimulates the growth of periderm which helps in the healing of wounds. It also stimulates the process of thickening the periderm. This reduces weight loss and rotting of potatoes due to storage rot organisms. It has been shown that at a temperature of 12.8°C and 75 to 85% relative humidity (RH), suberization occurs within a week.[4] The harvest and postharvest operations for potato tubers are shown in the diagram in Figure 2.

The production of potatoes is virtually impossible throughout the year in most of the developing countries. Potatoes are frequently stored to prevent seasonal gluts and to increase the availability of potatoes to consumers throughout the year. The potato storage facilities should maintain tubers in their most edible and marketable conditions by preventing large moisture losses, spoilage by pathogens, attacks by insects and animals, and sprout growth. Several changes occur in nutrients during storage.[5-9] The changes vary according to storage conditions and potato variety. Potatoes are normally stored at temperatures of 4 to 20°C. Lower temperatures inhibit sprout growth. However, lower temperature storage results in an accumulation of sugars. Before storage, tubers are generally allowed to undergo wound-healing processes.[5] Potatoes used for domestic consumption should be stored at about 5°C to avoid sprouting and accumulation of sugars. Tubers for later use in the food-processing industry should be maintained at about 10°C, which avoids disease and excessive sprouting and prevents a high accumulation of reducing sugars.

II. CHANGES IN NUTRIENT COMPOSITION DURING STORAGE

A. SUGAR

It is established that the sugar content of potatoes increases when stored at comparatively low temperatures (Figure 3).[10-12] Both sucrose and reducing sugars accumulate in different proportions at different temperatures. This has been attributed to the conversion of starch into sugars by enzyme phosphorylase. The amount of sugar formed at low temperature depends upon cultivar, maturity of tubers, prestorage treatments, and storage temperature.[13-23] A large increase in sugars, primarily reducing sugars, occurs in potatoes stored at 1.1 to 2.2°C. The increase of CO_2 content in the storage atmosphere reduces the accumulation of sugars at low-temperature storage. Samatous and Schwimmer[24] reported that when potatoes are stored in nitrogen at 0°C, there is a complete suppression of sugar accumulation. Irradiation increases sugar content in tubers stored at low temperatures.

A large portion of stored potatoes goes to industries making chips and French fries in developed countries. Hence, potatoes stored at low temperatures and having high amounts of sugars are reconditioned at temperatures of 15.0 to 26.0°C. The RH is maintained between 75 to 90%. The conditioning of potatoes results in a decrease in phosphorylase activity and sugars (Figure 4), which are converted to starch by starch synthetase (Figure 5).[2]

B. STARCH

The starch content of potatoes decreases with a lowering of the storage temperature. The starch content may increase in potatoes by conversion of sugars to starch at high temperatures (above 10°C). The potatoes stored at 1.1 to 13.3°C for 2 to 3 months contained only about 70% of their original starch. The changes of starch to sugars at low temperatures

FIGURE 1. A truck container van to transport potatoes.

and subsequent partial resynthesis of starch from sugars at high temperatures affect the quality of starch and the texture of cooked potatoes. The degradation of starch into sugars at low temperatures is catalyzed by enzyme phosphorylase, whereas the reverse process at high temperature is catalyzed by enzyme starch synthetase.

C. PROTEINS

It is known that potatoes contain low amounts of proteins ranging from 1.5 to 2.5% on the fresh weight basis. Storage conditions, especially temperature, may have an effect on proteins. Potatoes stored at room temperature have a higher content of amide nitrogen than those stored at 0, 4.4, or 10°C. Although the total nitrogen of stored potatoes has generally been reported to change very little, there are many reports indicating changes in individual nitrogenous constituents, but these have been studied mainly in NPN fraction. Miča[25,26] found that protein nitrogen decreased with the length of storage at both 2 and 10°C, although changes in total nitrogen were small. The free amino acid content was higher at the end of storage. However, Habib and Brown[27] reported little or no changes in free amino acid composition of four cultivars stored at about 4°C, but reconditioning at 23°C caused a marked decrease in total free amino acids and complete loss of arginine, histidine, and lysine. Fitzpatrick et al.[28] found a general increase in all free amino acids after cold storage and reconditioning. They attributed this to metabolic degradation of protein occurring as tubers sprouted during later stages of reconditioning. Many reports present data on decreases in NPN or free amino acid content during cold storage.

D. VITAMINS

The loss in ascorbic acid during storage has been reported.[29] Linnemann et al.[11] reported losses of vitamins under farm storage conditions in developing countries. Thomas et al.[30] reported that ascorbic acid is stable during and after irradiation. The loss of ascorbic acid during cold storage has been reported.[8,19,23,31,32] Losses have been found to take place most

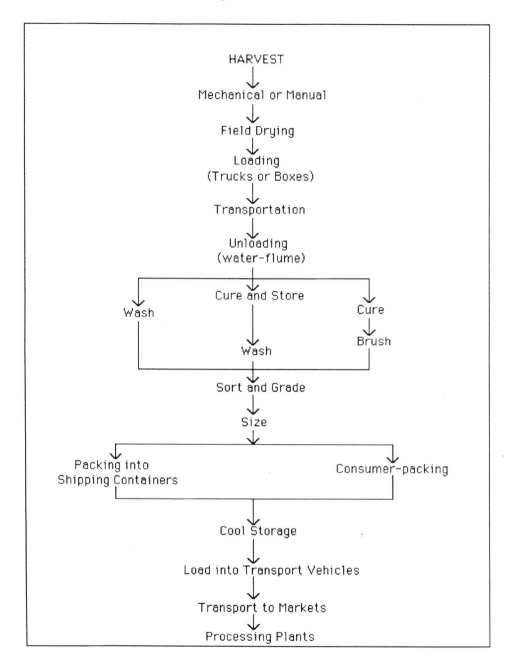

FIGURE 2.   Harvest and postharvest operations for potato tubers.

rapidly during the early part of storage.[10,33] Linnemann et al.[11] studied the effects of high storage temperatures on the reduced ascorbic acid content of potatoes. It was indicated that losses from potatoes in traditional stores in developing countries are likely to be lower than those during low-temperature storage. Losses of vitamins of B groups, such as folic acid, have been reported during storage.[31] However, an increase in pyridoxine in potatoes during storage has been demonstrated[34] (Figure 6). Augustin et al.[31] have pointed out that it is not known whether this increase is due to synthesis of the vitamin or its release from a bound form during the early stages of storage. Addo and Augustin[35] provided further evidence to support the synthesis of B_6 during storage. Barker and Mapson[36] stored potatoes in nitrogen

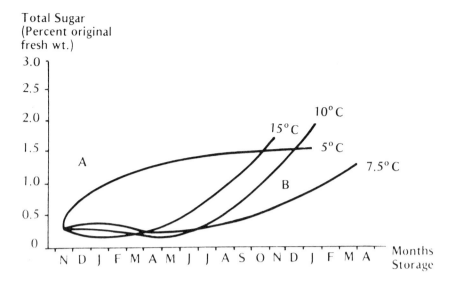

FIGURE 3.   Change in amount of total sugar during storage at different temperatures.

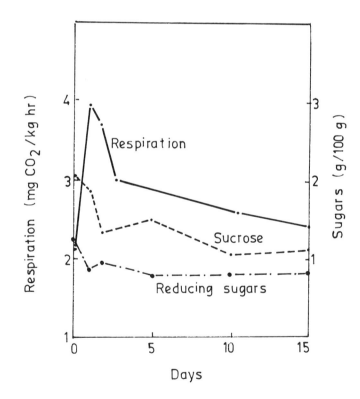

FIGURE 4.   Changes in amounts of reducing and nonreducing sugars and the rate of respiration when potatoes previously stored at 10°C are moved to 21°C.[2]

and found that the content of ascorbic acid was almost stabilized by the exclusion of oxygen. Storage of potatoes at low temperature has little effect on thiamine and riboflavin.

E. FATTY ACIDS AND LIPIDS

Storage of potatoes at room temperature increases fatty acid content.[37] This is followed by a marked decrease when storage is extended.

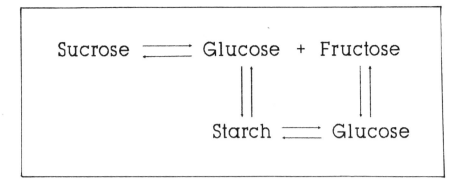

FIGURE 5.   Conversion of sucrose to glucose plus fructose during cold (5°C) storage of potato tuber and after conditioning of the tubers at high (20—25°C) storage temperature. Sugars are converted to starch.

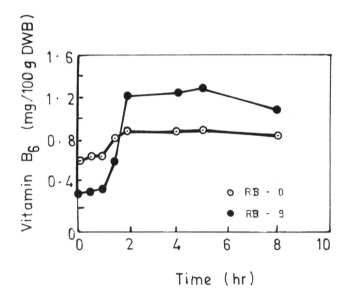

FIGURE 6.   Amounts of vitamin B_6 extracted from potato samples with increasing time from freshly harvested (RB-0) and 9 months of storage (RB-9).[34]

F. ORGANIC ACIDS

Low temperature of potatoes results in an increase in malic acid and citric acid in tubers.

G. MINERALS

No significant changes are reported in the mineral content of potatoes during storage. Yamaguchi et al.[23] observed no significant changes in the contents of calcium, iron, or phosphorus in White Rose potatoes held at 5 or 10°C for 30 weeks.

III. EFFECTS OF STORAGE ON PROCESSING QUALITY OF POTATOES

Mazza et al.[38] attempted to relate changes in sucrose, reducing sugars, ascorbic acid, protein, and nonprotein nitrogen contents with the processing quality (chip and French fry

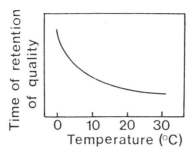

FIGURE 7.   Influence of temperature on time of retention of quality of tubers.

color) of Alberta (commercially grown and stored), Kennebec and Norchip (both potato chippers), and Russet Burbank (a French fry cultivar) potatoes during growth and long-term storage. Correlation analysis of chip color, dry matter, sucrose, reducing sugars, ascorbic acid, protein, and storage temperature data showed that while dry matter, reducing sugars, and sucrose were significant in determining chip color of freshly harvested potatoes, reducing sugars, tuber temperature, and sucrose were important in determining the chip color of stored tubers. The relative importance of each parameter varied with the cultivar and the age of the potato tubers. Mazza et al.[38] reported data collected on multiple correlation analysis over a period of 3 years. The multiple correlation coefficient among chip color, dry matter, sucrose, reducing sugars, and tuber weight was 0.901 for fresh Russet Burbank, 0.839 for fresh Norchip, 0.909 for fresh Kennebec, and 0.790 for the three cultivars combined. Similarly, for stored potatoes, the multiple correlation coefficients among chip color, reducing sugars, sucrose, and tuber temperature was 0.866 (Russet Burbank), 0.731 (Norchip), 0.914 (Kennebec), and 0.790 (for combined stored material). The regression equations varied from cultivar to cultivar and from one season to another. The data reported by these authors indicated that the quantitative relationship between the factors assayed was not sufficiently stable to serve as a general measure of predicting processing quality of potatoes.

The reducing sugar content of potatoes, in particular, and processing quality, in general, are markedly affected by cultivar and environmental conditions during the growing season and storage. Mazza[39] described the effect of several cultural and environmental factors on potato processing quality, including date of planting, soil type, soil reaction (pH), soil moisture, season, location, mineral nutrition of the plants, cultivation and weed control, control of disease and insect pests, temperatures during growing season, time and method of killing vines, time of harvest, degree of bruising and other mechanical injuries, temperature of tubers going into storage, storage temperature (Figure 7), RH (Figure 8), and ventilation, length of storage, and cultivar.

Optimum storage conditions for potatoes vary depending upon the cultivar, the length of storage, tuber maturity, and other prestorage and storage factors[40] as well as the intended end use of the tubers. Mazza[39] illustrated the following processes taking place in the stored potatoes:

1.   Respiration, which utilizes sugars by converting them to carbon dioxide, water, and energy
2.   Conversion of starch to sugar by amylolytic enzymes
3.   Conversion of sugar to starch, presumably by starch-synthesizing enzymes

The starch content of potatoes decreased with lowering of the storage temperature through the process of starch hydrolysis by an amylolytic enzyme. Potatoes stored at 1 to 3°C for 2

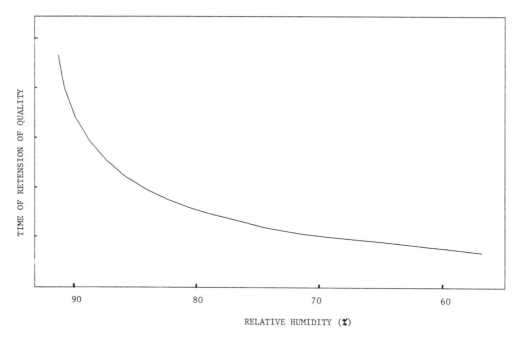

FIGURE 8.   Time of retention of quality in relation to relative humidity (%).

TABLE 1
Correlation Coefficients Between Chip Color and Other Quality Parameters of Russet Burbank, Norchip, and Kennebec Potatoes Taken from Commercial Fields in Southern Alberta (1979 to 1981)[39]

Cultivar	Parameter	Correlation coefficient (r)	Level of significance (%)
Russet Burbank	Dry matter	0.805	0.1
	Sucrose	−0.766	0.1
	Reducing sugars	−0.808	0.1
Norchip	Dry matter	0.786	0.1
	Sucrose	−0.755	0.1
	Reducing sugars	−0.625	0.1
Kennebec	Dry matter	0.808	1.0
	Sucrose	−0.869	0.1
	Reducing sugars	−0.657	3.0

to 3 months may lose as much as 30% of their starch. At higher temperatures, on the contrary, the starch content of potatoes may increase, resulting from synthesis of starch from sugars. These changes from starch to sugars at low temperature and the subsequent resynthesis of starch from sugars at high temperature may significantly alter the structure of starch granules.

Optimum storage conditions for controlling the color of chips or French fries also require stringent control on weight loss of potatoes during storage. Cargill et al.[40] reported the influence of RH on weight loss during 300 d of storage of Kennebec potatoes held at 4.4°C and 95% RH or 4.4°C and 80% RH. The weight loss at 80% RH was more than double the weight loss at 95% RH. A difference in weight loss during a 200-d storage, between 80 and 95% RH, represented a 10% loss, corresponding to a monetary loss of about $16,000 to the grower. Statistical data presented in Tables 1, 2, and 3 illustrate the relationship of

TABLE 2
Correlation Coefficients Between Chip Color and Other Quality Parameters of Russet Burbank, Norchip, and Kennebec Potatoes Taken from Commercial Storages in Southern Alberta (1979 to 1982)[39]

Cultivar	Parameter	Correlation coefficient (r)	Level of significance (%)
Russet Burbank	Tuber temperature	0.603	0.1
	Sucrose	−0.220	5.0
	Reducing sugars	−0.685	0.1
Norchip	Tuber temperature	0.368	0.1
	Sucrose	−0.526	0.1
	Reducing sugars	−0.647	0.1
Kennebec	Tuber temperature	−0.134	N.S.[a]
	Sucrose	−0.799	0.1
	Reducing sugars	−0.776	0.1

[a] N.S. = Not significant.

TABLE 3
Correlation Coefficient Between Potato Sucrose Content at Harvest and Poststorage Chip Color, Reducing Sugars, and Dry Matter Content of Russet Burbank, Norchip, and Kennebec Potatoes (Production/Storage Seasons, 1979 to 1980, 1980 to 1981, 1981 to 1982)[39]

Cultivar(s)	Time of sampling from storge	No. of storage bins sampled	Correlation coefficients (r)		
			Chip color	Reducing sugars	Dry matter
Russet Burbank	Nov.	10	−0.590[c]	0.408	0.814[d]
	Jan.	10	−0.049	−0.175	−0.904[b]
	March	7	−0.106	0.022	−0.798[b]
Norchip	Nov.	15	0.080	0.074	−0.437[d]
	Jan.	12	−0.200	0.102	0.437
	March	6	0.034	0.749[c]	0.976[a]
Russet Burbank, Norchip	Nov.	25	−0.587[b]	0.645[a]	−0.095
	Jan.	22	−0.509[b]	0.260	−0.201
	March	18	−0.578[b]	0.495[c]	0.258
Russet Burbank, Norchip, Kennebec	Nov.	26	−0.614[a]	−0.625[a]	−0.135
	Jan.	23	−0.514[b]	−0.239	−0.276
	March	14	−0.567[b]	0.440[c]	0.247

[a,b,c,d] Significant at 0.001, 0.01, 0.05, and 0.1 level of probability, respectively.

processing quality (chip color) with the chemical composition of fresh and stored potatoes in respect of dry matter, reducing sugars, and sucrose contents. A significant correlation with chip color was obtained for dry matter, reducing sugars, and sucrose levels of all potatoes tested.[39] The signs of the correlation coefficients indicated that increased levels of sugars corresponded to darker chip color, and increased dry matter content corresponded to lighter and more desirable chip color (Table 1). The correlation coefficients among chip color, tuber temperature, sucrose, and reducing sugar levels of Russet Burbank, Norchip, and Kennebec potatoes (Table 2) showed distinct cultivar differences. The correlation among sucrose content of potatoes at harvest and poststorage chip color, reducing sugars, and dry matter contents of Russet Burbank, Norchip, and Kennebec (Table 3) indicated that harvest sucrose ratings of a given cultivar is not a very good predictor of poststorage processing performance.[39]

IV. GREENING OF POTATOES DURING STORAGE

Synthesis of chlorophyll in the peridermal layers of tubers exposed to light leads to their "greening" which markedly reduces acceptability of the product. Sometimes the greened tubers taste bitter when cooked owing to the parallel synthesis of glycoalkaloids. Greening may occur in the retail outlets.[41] Gull and Isenberg[42] demonstrated that light intensities as low as 3 to 11 W m^{-2} for as short a period as 24 h could induce greening. The development of the green color is influenced by variety,[43] stage of maturity,[44] and temperature.[42] No greening was found at 5°C, and it was extensive at 20°C, the greater effect being observed in immature tubers.[45] Lewis and Rowberry[46] noted that the major cause of greening in the field was insufficient cover over the tubers at planting. The effect of greening was aggravated in potato varieties whose tubers were formed near the surface. Clumping of stems resulted from the use of large seed planted at wide spacings.[47] Less severe competition at wide spacing probably allowed more tubers to set, some of which were forced to the surface, exposing them to light and thus turning them green.

Greening in potatoes is very often associated with formation of steroidal alkaloids, which can cause off flavors on cooking at concentrations of 15 to 20 mg/100 g.[48] This concentration is five to ten times higher than that occurring in normal potatoes. Most of the glycoalkaloids are concentrated in the skin, and in prepared potatoes it is usually too low to cause any nutritional hazards or poisoning. The synthesis of the glycoalkaloids is markedly influenced by variety. Zitnak and Johnston[48] recorded concentrations of glycoalkaloids as high as 35 mg/100 g in cultivar "Lenape", which was withdrawn from commercial cultivation for this reason. Pallman and Schindler[49] noted that immature potato tubers contained higher amounts of the glycoalkaloids than the mature tubers. The glycoalkaloid concentration was not influenced by the level of fertilization.

Patil et al.[50] reported that various cultivars of potato differed significantly in chlorophyll and glycoalkaloid (Table 4). Salunkhe et al.[51] studied the effects of temperature and light on glycoalkaloid formation in potato slices. The results are summarized in Figure 9. At low temperatures (0 and 8°C), there was a slow but significant increase in the glycoalkaloid content during a 48-h period in the dark, while storage temperatures of 15 and 24°C vigorously stimulated solanine formation. In the 48-h exposure to 2152 lx light at 24°C, the solanine concentration increased up to 7.4 mg/100 g of slices. In general, the light increased the rate of solanine synthesis nearly three to four times more than the dark. Hilton[52] investigated the effect of storage temperature on tuber bitterness and concluded that low-temperature storage maintained or caused more bitterness of tubers than when the tubers were held above 10°C (Table 5). Zitnak[56] noted that high glycoalkaloid concentrations could develop at low temperatures (4 to 8°C) during storage of Netted Gem tubers either illuminated or in the dark (Table 6). Yamaguchi et al.[53] found that the freshly harvested potatoes greened slowly at 5°C, at an intermediate rate at 10°C, and at a much higher rate at 15, 20, and 25°C (Table 7). The potatoes stored for 18 d in the dark at various temperatures were similarly exposed to light at the corresponding temperatures. These tubers developed chlorophyll slowly at 5°C. However, the rate of greening was comparatively higher than the freshly harvested tubers. The potatoes stored at 10°C had less chlorophyll after 80 h of exposure to light than did the tubers held at 15 or 20°C. Buck and Akeley[54] stored potatoes at 4.4, 12.8, and 21.2°C for 2 and 4 months and observed more greening in tubers stored at 4.4°C than those stored at 12.8 or 21.1°C. The tubers greened less after 4 months of storage than did those stored for 2 months (Table 8). Yamaguchi et al.[53] reported that tubers stored at higher

TABLE 4
Chlorophyll and Glycoalkaloid Content
after Exposure to White Fluorescent Light

| Cultivar[a] | mg/100 g fresh peel | |
	Chlorophyll[b]	Glycoalkaloid[c]
La Chipper	0.691	44.81
Platte	0.720	55.86
Cascade	0.966	65.69
LaRouge	1.175	73.06
Sioux	1.247	58.93
Norchip	1.395	79.20
Red LaSoda	1.566	69.99
Shurchip	1.808	44.50
Russet Burbank[d]	2.083	69.07
Kennebec	2.228	96.40
Bounty	2.431	70.30

Note: Potatoes were exposed to white fluorescent light (100 fc) for a period of 5 d.

[a] Initial glycoalkaloid contents were not reported; initial chlorophyll content was negligible.
[b] Determined by AOAC method (Assoc. Official Agric. Chemists).
[c] Determined by the sulfuric acid-formaldehyde reagent method.
[d] 24 mg of initial glycoalkaloid content per 100 g of fresh peel.

From Patil, B. C., Salunkhe, D. K., and Singh, B., *J. Food Sci.*, 36, 474, 1971. With permission.

temperatures had a reduced rate of chlorophyll accumulation (Table 9). These investigators[127] further noted that reconditioning of tubers at 20°C for a longer time decreased the rate of chlorophyll formation of stored potato tubers exposed to light at room temperature (Table 10). Liljemark and Widoff[55] studied the effect of light of different intensities from 25 to 36 lx on the development of chlorophyll. The chlorophyll content increased with an increase in light intensity (Table 11). While the chlorophyll content rose in proportion to the logarithm of light intensity values in lux, the content of glycoalkaloid almost paralleled chlorophyll development, despite the higher alkaloid content of tubers at the onset of the illumination. Patil et al.[50] reported that chlorophyll synthesis up to a light intensity of 1076 lx slowed and degraded gradually up to 1614 lx and degraded rapidly at 2152 lx during 5 d of light exposure (Figure 10). These authors postulated that the insignificant differences in the high glycoalkaloid content after exposure to four light intensities (538, 1076, 1614, and 2152 lx) could be a result of storage at low temperatures for a long period of time. Patil et al.[57] further observed that the glycoalkaloid and chlorophyll content of tubers enhanced linearly up to the 6th and 10th days, respectively (Figure 11). Jeppsen et al.[58] reported that orange and yellow cellophane reduced greening by 20 and 30% of clear (control) cellophane, respectively, while glycoalkaloid formation under green cellophane was 27% less than that under clear (control) cellophane (Figure 12).

Jadhav and Salunkhe[59] reviewed the research work done on the control of chlorophyll and glycoalkaloid formation, describing several physicochemical methods employed. Various measures that can be used include cultivar selection; packaging; treatment with wax, oil, soap and surfactants, and various chemicals; controlled and hypobaric storage; and

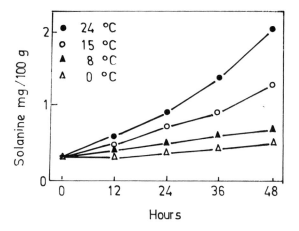

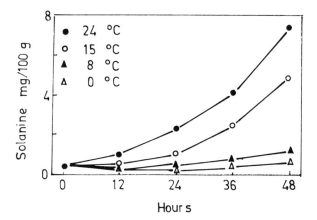

FIGURE 9.   Effects of temperature on solanine (glycoalkaloids) formation in potato slices stored in the dark (top) and exposed to a fluorescent light (2152 lx) (bottom). (From Salunkhe, D. K., Wu, M. T., and Jadhav, S. J., *J. Food Sci.*, 37, 969, 1972. With permission.)

ionizing radiation. Wu and Salunkhe[60] reported that hot paraffin wax effectively controlled the formation of chlorophyll and glycoalkaloid in the potato (Figure 13). The potatoes were treated with paraffin wax at 60, 80, 100, 120, 140, and 160°C for ½ s and exposed to fluorescent light (200 fc) for 10 d at 160°C and 60% RH. The synthesis of chlorophyll and glycoalkaloid was not inhibited at 60 and 80°C; it was significantly inhibited at 100 and 120°C and was almost completely inhibited at 140 and 160°C. Heating of tubers at 160°C in air for 3 to 5 min and subsequent exposure to light did not prevent formation of chlorophyll and glycoalkaloid. The combined treatment of waxing and heating effectively retarded the development of chlorophyll and alkaloid. Wu and Salunkhe[61] dipped potato tubers in corn oil at 20, 60, 100, and 160°C for ½ s and removed excess oil with tissue paper. Oiling at 22°C reduced chlorophyll by 93 to 100% and reduced glycoalkaloid formation by 92 to 97%. At the elevated temperatures (60, 100, and 160°C), the synthesis of chlorophyll and alkaloid was completely inhibited (Figure 14). These authors further noted that treatment of potatoes with corn oil, peanut oil, olive oil, vegetable oil, or mineral oil at 22°C was equally

TABLE 5
Effects of Storage Temperature on Tuber Bitterness

Storage temp[a] (°C)	Taste score		Mean
	Tubers originally not bitter	Tubers originally bitter	
0—5	2.3 (9)[b]	2.9 (9)	2.60
5—10	3.0 (18)	2.5 (18)	2.75
10—15	2.1 (18)	2.2 (18)	2.15
15—20	1.9 (9)	1.7 (9)	1.80

Note: Effects reported for cooked potatoes (cultivar Netted Gem). Gly-coalkaloid content was evaluated by organoleptic tests based on taste score such as 0, not discernible; 1, trace; 2, slight; 3, moderate; 4, marked; 5, very marked.

[a] Stored for 5 months at the indicated temperature.
[b] Values in parentheses indicate the number of readings.

From Hilton, R. J., *Sci. Agric.*, 31, 61, 1951. With permission.

TABLE 6
Effects of Various Storage Conditions on Glycoalkaloid Content

Storage time (weeks)	Cold, humid storage at 4 to 8°C		Dry, warm storage at 12 to 15°C	
	Dark	Light[a]	Dark	Light[a]
1	7.93	19.83	7.50	7.20
2	5.63	19.01	7.96	5.04
3	11.30	18.51	3.52	3.20
4	13.75	17.38	5.78	7.26
5	11.09	16.34	4.01	3.65
6	15.42	23.50	8.72	6.98
Average	10.86	19.09	6.26	5.86

Note: Effects reported for Netted Gem potatoes. Determined colorimetrically using sulfuric acid-formaldehyde reagent and expressed in mg % of untreated tissue; 5.9 mg % average value of normal Netted Gem tuber.

[a] From weak Mazda light (15 W) at a distance of 30 in from tubers.

From Zitnak, A., The Influence of Certain Treatments upon Solanine Synthesis in Potatoes, M.S. thesis, University of Alberta, Edmonton, 1953. With permission.

effective,[61] but the tubers appeared oily and the possibility of the development of oxidative rancidity of oil was indicated. To decrease the amount of oil used, the corn oil was diluted with acetone. Treatment with one half, one fourth, and one eighth oil significantly and effectively inhibited the formation of chlorophyll and glycoalkaloids.

The treatments with $^1/_{16}$, $^1/_{32}$, and $^1/_{64}$ oil had 95, 72, and 22% inhibition on chlorophyll and 82, 49, and 28% inhibition on glycoalkaloid formation, respectively. The tubers treated with acetone alone or with $^1/_{128}$ oil behaved in the same way as the control or untreated tubers. Jadhav and Salunkhe[62] investigated the effects of mineral oil at different concentrations. The efficiency of treatment increased with increasing concentration up to 10% (w/v in petroleum ether) and then remained almost constant and maximum up to 100%.

TABLE 7
Effects of Temperature During Light Exposure on Greening

	Chlorophyll[a] (μg/100 cm^2 tuber surface)					
	40 h		80 h		120 h	
Exposure temp (°C)	Freshly harvested	18-d storage[b]	Freshly harvested	18-d storage[b]	Freshly harvested	18-d storage[b]
5	5	12	5	18	13	26
10	44	37	87	66	168	142
15	79	68	210	158	—[c]	—[c]
20	95	81	—[c]	142	—[c]	—[c]
25	103	66	—[c]	134	—[c]	—[c]

Note: Effects of temperature during light exposure (1076 lx) on greening of White Rose potatoes.

[a]   Determined by AOAC method.
[b]   Storage temperature is the same as exposure temperature.
[c]   Values not reported.

From Yamaguchi, M., Hughes, D. L., and Howard, F. D., *Proc. Am. Soc. Hortic. Sci.*, 75, 529, 1960. With permission.

TABLE 8
Storage Time and Temperature and Greening of Tubers
Following Exposure to Light

	Time in storage (months)		Storage temperature (°C)		
Cultivar	2	4	4.4	12.8	21.1
Blanca	68[a]	69	60	72	74
B 4523-8	64	68	54	70	74
Russet Burbank	52	54	43	58	58
Katahdin	51	59	42	60	64
Kennebec	36	46	32	42	50

Note: Light exposure was 35 fc for 100 h at 21.1°C.

[a]   Optical density values (inversely proportional to greening) as measured by Ratiospect, mean of four replications.

From Buck, R. W., Jr. and Akeley, R. V., *Am. Potato J.*, 44, 56, 1967. With permission.

Numerous chemicals have been known to prevent greening and/or glycoalkaloid synthesis in potato tubers. Jadhav and Salunkhe[62] reported that postharvest applications of Phosfon®, Phosfon-S®, Amchem 72-A42®, Amchem 70-334®, and Telone® (250, 500, and 100 ppm) significantly inhibited glycoalkaloid and chlorophyll formation. Amchem-72-A42® was the most effective chemical in preventing the synthesis of both glycoalkaloids and chlorophyll.

Forsyth and Eaves[63] evaluated the effect of CO_2 on the greening of Sebago potatoes in controlled atmosphere storage (Table 12). They did not notice immediate greening in an atmosphere of 100% N_2 or 75% CO_2, and the storage time could be extended up to 1 week without obvious development of greening. Potatoes stored under 100% N_2, but not those stored under 75% CO_2, developed off flavor due to formation of glycoalkaloids. It was concluded that 15% or higher concentrations of CO_2 prevented greening without affecting

TABLE 9
Storage Temperature and Greening During Exposure to Light at Room Temperature

Storage temp (°C)	Chlorophyll[a] (µg/cm² tuber surface)		
	25 h	48 h	72 h
5	68	216	279
10	36	121	224
15	28	85	158
20	20	63	121
25	12	44	111
30	12	32	55

Note: White Rose tubers were stored in the dark for 18 d at the indicated temperatures. Light exposure was 968 lx.

[a] Determined by AOAC method.

From Yamaguchi, M., Hughes, D. L., and Howard, F. D., *Proc. Am. Soc. Hortic. Sci.,* 75, 529, 1960. With permission.

TABLE 10
Effects of Reconditioning on Greening During Exposure to Light at Room Temperature

Previous storage temp (°C)	Duration of reconditioning[a]					
	0 days		10 days		17 days	
	36 h	72 h	36 h	72 h	36 h	72 h
5	125	237	97	219	23	119
10	100	239	75	225	18	102
15	58	175	58	163	18	86
20	36	137	36	102	12	50
25	35	106	25	66	18	47
30	25	75	25	58	13	50

Note: White Rose tubers had been stored in dark for 14 d at the indicated temperatures. Light exposure was 968 lx.

[a] Reconditioning was done at 20°C and duration of light exposure is in hours. Values in columns are in micrograms of chlorophyll per 100 cm² of tuber surface.

From Yamaguchi, M., Hughes, D. L., and Howard, F. D., *Proc. Am. Soc. Hortic. Sci.,* 75, 529, 1960. With permission.

palatability. Patil[64] observed no significant effect of CO_2 on the formation of potato glycoalkaloids, while chlorophyll content was decreased by 33% of values for control tubers. Jadhav and Salunkhe[59] reported that potato tubers stored at 126 mmHg did not turn green. The treatments at 253, 380, 507, and 633 mmHg pressure were ineffective in controlling formation of chlorophyll. Patil[64] found no differences in glycoalkaloid levels among tubers subjected to any of the storage pressure treatments and the control (Table 13). Ziegler et

TABLE 11
Effects of Light Intensities on Greening and Glycoalkaloid Content

Light intensity	Chlorophyll (and glycoalkaloid) content[a]				
(fc)	0 d	5 d	7 d	11 d	17 d
2.3	1.2 (387)	1.7 (437)	3.2 (537)	4.0 (475)	4.5 (550)
14	1.2 (387)	2.2 (450)	5.2 (625)	8.2 (562)	6.9 (662)
56	1.2 (387)	3.0 (532)	5.6 (675)	9.5 (600)	11.9 (750)
334	1.2 (387)	—[b]	8.0 (720)	21.6 (1057)	17.5 (1200)

Note: Potato tubers (cultivar Majestic) were exposed to light (daylight lamp) at room temperature.

[a]   Expressed in mg/100 g of dry weight of surface disks (3 mm thick) as determined by AOAC method. Values in parentheses represent glycoalkaloid content as determined by the method of Baker et al.[151]

[b]   Values not reported.

From Liljemark, A. and Widoff, E., *Am. Potato J.*, 37, 379, 1960. With permission.

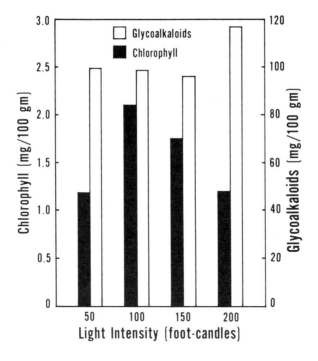

FIGURE 10.   Chlorophyll and glycoalkaloid contents of Kennebec potato tubers exposed to 4 light intensities for 5 days at 21.1°C and 90% RH. (Data expressed as fresh weight peels.) (From Patil, B. C., Salunkhe, D. K., and Singh, B., *J. Food Sci.*, 36, 474, 1971. With permission.)

al.[65] exposed tubers to 3012 lx of continuous illumination after irradiation for 12 d and found increasing inhibition of greening with increasing doses of irradiation from 0 to 400 krd (Table 14). The irradiation decreased greening irrespective of CO_2 treatment after 4 d of illumination, and the greening decreased with increasing CO_2 in the atmosphere irrespective of irradiation after 12 d of illumination. Madsen et al.[66] reported that potatoes

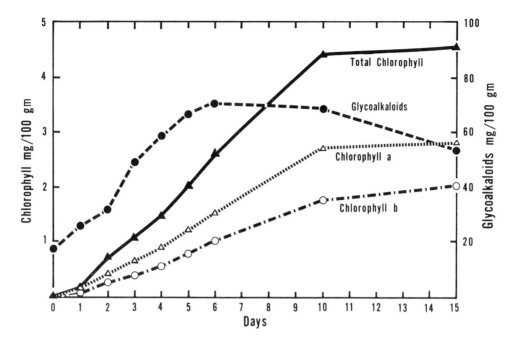

FIGURE 11.   Chlorophyll and glycoalkaloid contents of White Rose potato tubers exposed to light (1076 lx) up to 15 d at 21.1°C and 90% RH. (Data expressed on fresh weight of peels.) (From Patil, B. C., Singh, B., and Salunkhe, D. K., *Lebensm. Wiss. Technol.*, 4, 123, 1971. With permission.)

irradiated with more than 1×10^4 rd gamma radiation failed to sprout during 6 months of storage at 10°C. Wu and Salunkhe[67] reported inhibitory effects of gamma radiation on wound-induced glycoalkaloid formation in 13 potato cultivars. Whereas a dose of 25 to 100 krd resulted in 11 to 79% inhibition in Russet Burbank, at 200 krd inhibition was 81 to 92%. Gamma radiation had no effect on existing alkaloids nor on light-induced glycoalkaloid formation in tubers.

V. SPROUTING OF POTATOES DURING STORAGE

Potatoes exhibit sprouting if stored at temperatures of 10 to 20°C.[68] Sprout growth is slow at temperatures of 5°C and below. Above 5°C, increasing temperature causes increased sprout growth up to about 20°C; at even higher temperatures, sprout growth rate decreases. However, storage of potatoes below 10°C causes an increase in sugar content. This increases browning of heat-processed products. It has been shown that ascorbic acid content changes in tubers during sprouting.[69,90] Sprouting results in a decrease in ascorbic acid during the early stages of sprout growth, followed by a temperature increase and by another decrease.[69] Bantan et al.[70] found a higher ascorbic acid content in sprouts than in the rest of the tuber after 8 months of storage. However, conflicting reports are available on other vitamin contents of tubers during sprouting.[71] Yamaguchi et al.[53] found no changes in thiamin in "White Rose" potatoes even after 30 weeks of storage. Leichsenring[10] noted a slight increase in thiamin over 24 weeks of storage even on a dry weight basis, although the overall trend for one of the four varieties (Chippewa) studied showed a slight decrease.

Several chemical inhibitors have been found to inhibit sprouting in potatoes.[72-74] The use of maleic hydrazide in the control of potato sprouting is well documented.[75-84] Other chemicals used for control of sprouting include chloroisopropyl carbamate (CIPC), isopropyl-*N*-chlorophenyl carbamate (Chloroprophan, IPC),[85-87] tetra-chloronitrobenzene (tecnazene),[88] alcohols,[89] methyl ester of naphthalene acetic acid (MENA),[90,91] and other inhibitors such

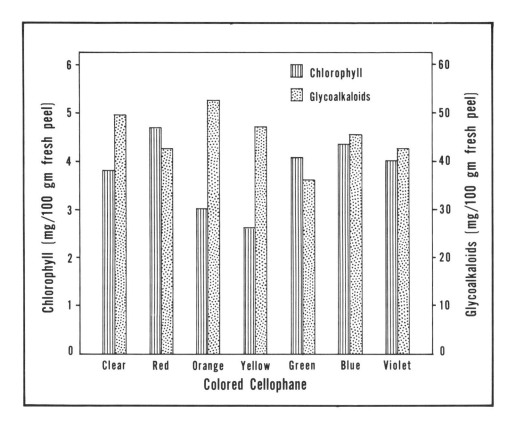

FIGURE 12.   Effects of fluorescent light passing through colored cellophane filters on the formation of chlorophyll and glycoalkaloid in Russet Burbank tubers after exposure to 538 lx light intensity through each filter for 10 d at 16°C and 60% RH. (Data expressed on fresh weight of peels.) (From Jeppsen, R. B., Salunkhe, D. K., and Jadhav, S. J., *33rd Annu. Int. Food Technol. Meet.*, 153, 1973. With permission.)

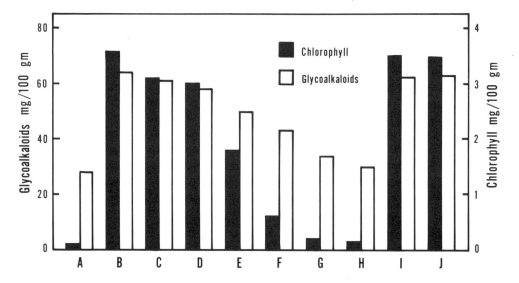

FIGURE 13.   Effects of waxing and heating at different temperatures on chlorophyll and glycoalkaloid formation in peels (fresh) of Russet Burbank tubers after exposure to 2152 lx light intensity for 10 d at 16°C and 60% RH. (A) Original (zero time) sample; (B) control (nonwaxed); (C) to (H) waxing at 60, 100, 120, 140, and 160°C; (I) heating with air at 160°C for 3 min; (J) heating with air at 160°C for 5 min. (From Wu, M. T. and Salunkhe, D. K., *J. Food Sci.*, 37, 629, 1972. With permission.)

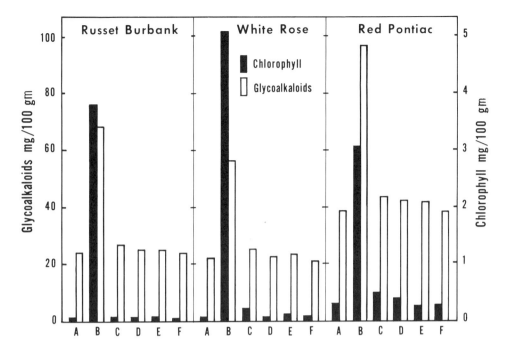

FIGURE 14. Effects of oil dipping on chlorophyll and glycoalkaloid formation in fresh peels of potato tubers after exposure to 2152 lx light intensity for 10 d at 16°C and 60% RH. (A) Original (zero time) sample; (B) control (nontreated); (C) to (F) oiling at 22, 60, 100, and 160°C. (From Wu, M. T. and Salunkhe, D. K., *J. Am. Soc. Hortic. Sci.,* 97, 614, 1972. With permission.)

TABLE 12
Effects of CO$_2$ on Greening in Controlled Atmosphere Storage

Present CO$_2$			
Mean	Range	Greening	Market grade
0.03	—	Severe	Below grade
7.5	4—10	Medium	Below grade
12.0	7—15	Slight	Below grade
14.9	9—22	Very slight	33.3% below grade
20.3	12—26	None	Canada grade 1[a]
20.3	18—36	None	Canada grade 1[a]

Note: Potatoes were of cultivar Sebago. Greening scores were based on eye observations after 48 h of illumination (2260 lx).

[a] Graded by Canada Department of Agriculture Inspector 168 h after commencement of experiment. Rejections are done to an excess of green coloration.

From Forsyth, F. R. and Eaves, C. A., *Food Technol.,* 22, 48, 1968. With permission.

as abscisic acid.[92] The most widely used sprout inhibitors in the U.S. are maleic hydrazide (MH-30) and CIPC. Canada has approved the use of both of these chemicals except that CIPC cannot be used on table stock. Field applications are made of MH, while CIPC is applied to potatoes placed in airtight bulk storages. The timing is critical for applying MH

TABLE 13
Effects of Subatmospheric Pressure Storage and Light
Intensity on Chlorophyll and Glycoalkaloids

Treatment pressure (mmHg)	Chlorophyll[a] (mg/100 g fresh peel)	Glycoalkaloid[b] (mg/100 g fresh peel)
Control	9.43	46.34
633	7.52	42.66
507	10.89	40.82
380	9.79	51.87
253	9.83	35.82
126	0.26	52.94

Note: Potato tubers (cultivar Russet Burbank) were exposed to 2260 lx light intensity for 15 d.

[a] Determined by AOAC method.
[b] Determined by the sulfuric acid-formaldehyde reagent method.

From Patil, B. C., Formation and Control of Chlorophyll and Solanine in Tubers of *Solanum tuberosum* L. and Evaluation of Solanine Toxicity, Ph.D. thesis, Utah State University, Logan, 1972.

TABLE 14
Effects of Gamma Irradiation on Chlorophyll
Formation

Irradiation dose (krd)	Chlorophyll content (% of control)		
	4 d	8 d	12 d
0	100	100	100
50	50	59	80
100	29	70	65
200	24	24	35
400	14	24	26

Note: Chlorophyll formation of potatoes (cultivar Kennebec) during 12 d of illumination (3012 lx).

From Ziegler, R., Schanderl, S. H., and Markakis, P., *J. Food Sci.*, 33, 533, 1968. With permission.

(2 to 3 weeks after full bloom). The CIPC treatment should not be applied until after the potatoes have cured, since its interference with periderm formation can lead to increased rot of the tubers.

Irradiation is a very potent sprout inhibitor.[93-102] Sparrow and Christensen[103] observed that certain irradiation doses gave excellent sprout control for as long as 15 months at 4.4°C. Dallyn and Sawyer[104] found that a 10-krd dose inhibited sprouting at storage temperatures up to 21.1°C (Table 15). At the storage temperature of 4.4°C, sprouting was inhibited by a 5-krd dose. The differences found in shrinkage and sprouting for four potato varieties are presented in Table 16. It has been suggested that irradiation might be economically attractive in warm, tropical countries where it could be used in combination with cold storage at 10 to 15°C rather than conventional cold storage at 2 to 4°C.

TABLE 15
Sprouting of Irradiated Potatoes after 6 Months of Storage

Storage temp (°C)	Average sprout length (in)					
		Irradiation dose (krd)				
	Control	1.25	5	10	20	30
4.4	$1/_8$	$1/_8$	None	None	None	None
10.0	$1/_2$	6	1	None	None	None
21.1	30	30	30	None	None	None

From Dallyn, S. and Sawyer, R. L., Cornell University, Contract No. -DA 19-129-QM-755. Final report, January, 1959.

TABLE 16
Losses of Potatoes After 8 Months of Storage Following Irradiation

Dose (krd)	Sprouting[a]	Shrinkage[b]	Sprouting[a]	Shrinkage[b]
	Katahdin		Green Mountain	
0.0	31.4	10.0	120.4	8.8
5.0	10.6	7.5	32.5	13.4
7.5	2.5	9.9	13.6	10.5
10.0	0.8	8.2	3.2	10.8
12.5	0.5	4.4	—	9.1
	Kennebec		Russet Burbank	
0.0	38.3	12.2	62.0	12.8
5.0	30.5	12.2	37.2	12.8
7.5	26.2	12.2	35.1	10.0
10.0	10.5	11.8	22.8	8.1
12.5	3.5	12.0	6.0	6.1
15.0	0.6	9.9	0.1	5.4

[a] Grams of sprouts per kilogram of potato.
[b] Percentage of original weight.

From Dallyn, S. and Sawyer, R. L., Cornell University, Contract No. -DA 19-129-QM-755. Final report, January, 1959.

VI. STORAGE LOSSES

A. EXTENT OF LOSSES

Potatoes, like other tuber crops, are subjected to postharvest losses owing to their continuing metabolism and damage during harvesting and handling, and rotting, shriveling, and sprouting.[105] In the Dominican Republic, Mansfield[105] estimated that 7.5% of weight loss was due to dehydration and pathogenic infection over a 1- to 2-d period with the total losses reaching about 30% in less than 15 d. In contrast, in the U.S., with improved storage systems involving control of temperature and humidity, the total loss (including weight loss) was below 13% after 11 months of storage. The weight loss in storage due to dehydration of immature tubers is comparatively higher than in mature tubers (Figure 15). Postharvest losses of potatoes have been estimated to vary from 5 to 40%.[106] According to an FAO

Percent Weight Loss (dehydration)

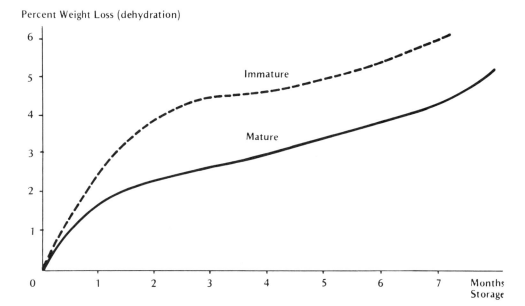

FIGURE 15.   Water loss. Average weight loss in well-stored mature and immature tubers at 4°C to 10°C.

report,[107] losses of potatoes in cold storage were about 8%, while on the farm the losses were to the extent of 20 to 40%.

The tuber characteristics of 21 potato cultivars have been studied with reference to duration of dormancy; intensity of sprouting; disease resistance; weight losses resulting from respiration, evaporation, sprouting, and rotting; and quality characteristics.[108] The tubers were stored for 2 weeks at 15°C and 90 to 95% RH and then at 6°C over the next 2 weeks, followed by storage for 6 months at 2, 4, 6, or 8°C or in a stack. Weight losses due to respiration and transpiration ranged from 7.2 to 18.3% of the initial weight. The losses were smaller at 2 to 4°C than at 6 to 8°C. The losses due to diseases were also lower at 2 to 4°C, ranging from 0.3 to 3.5% in different cultivars. The dry matter losses increased from 1.3 to 2.3% at 2°C, 2.0 to 3.0% at 8°C, and 2.0 to 3.2% in the stack. Reducing-sugar content increased from 0.42 to 0.63% fresh weight at harvest to 1.10 to 1.87% after 7 months at 2°C, the increases being smaller at the higher temperatures. Blackbeard[108] described a procedure for assessing superficial damage to potato tubers by staining in catechol solution and discussed the possible causes of damage during the harvesting. A "hot box" or accelerated damage test to assess internal bruising was also described.[109]

B. CAUSES OF LOSSES

1. Internal Black Spot

Substantial grade-outs and market losses of potatoes occur due to internal black spot, affecting both freshly harvested and stored potatoes.[110] A small area of cortex tissue blackens, deeper tissue being affected occasionally. The black spot may be caused by a deficiency of potassium[111] and mechanical harvesting may increase the incidence of black spot at each step. The susceptibility of tubers to black spot during prolonged bin storage is related to the loss of moisture during the storage. Howard et al.[110] reported that high CO_2 concentrations in potato tubers increased susceptibility to black spot.

2. Blackheart

Inadequate aeration or deficiency of O_2 during storage is the primary cause of the blackheart of potatoes, which is characterized by dark gray to purplish or black internal discoloration.[110] Optical nondestructive quality evaluation (NDQE) measures, such as high-transmittance measurements of external and internal colors at near-infrared and difference in optical density at 600 to 740 or 690 to 815 nm, are used to assess internal flesh color, dryness, freeze damage, and contents of total solids, and to predict certain disorders like development of black spot, internal browning and darkening, and hollow heart of potatoes.

3. Chilling Injury

Most cultivars of potatoes are susceptible to chilling injury when they are exposed to low temperatures (0 to 1.1°C) for a prolonged time. A disorder called "Mahogany browning" develops in the susceptible cultivars such as Katahdin.[110] The reddish-brown areas or the blotches in the flesh of affected tubers are the main symptoms of chilling injury. The freezing point of potatoes varies from -2.1 to $-0.06°C$, depending primarily upon the content of total solids. Potato tubers exposed to a prolonged freezing collapse quickly upon thawing, becoming soft and watery.

4. Heat Injury (Scald)

Heat injury is caused by exposing potatoes to excessive temperatures in the field during hauling or at the packing shed. It is usually associated with radiated heat from direct sunlight, but any source of heat raising the surface tissue to 48.9°C or higher may cause heat injury.

5. Mechanical Injury

Potatoes are susceptible to mechanical injury during harvesting and transportation due to brushing, cutting, dropping, puncturing, and hammering. Wu and Salunkhe[112] reported that mechanical injuries of potato tubers greatly stimulated glycoalkaloid synthesis in both the peel and the flesh of tubers. The extent of glycoalkaloid formation depend upon the cultivar, the type of mechanical injury, the storage temperature, and the duration of storage. According to Proctor et al.,[102] the physical damage during harvest is the main cause of loss during subsequent storage, since it facilitates fungal infection and stimulates physiological deterioration and moisture loss.

6. Greening

Potatoes exposed to light either before or after harvest develop chlorophyll pigment.[113] Greening is a serious problem in potatoes not only because it adversely affects the market quality of potatoes, but because the development of chlorophyll is often associated with formation of glycoalkaloids, which are potentially toxic alkaloids. However, the appearance of chlorophyll on tubers is not always a positive indication of an increase in the glycoalkaloid content.[114] Greening develops even at low light intensities (about 20 fc) that are readily exceeded on the surface of the produce counters in food stores.[115]

7. Postharvest Diseases

Ryall and Lipton[109] described several diseases of potatoes with the possible preventive measures: bacterial soft rot *(Erwinia carotovora* and *Pseudomonas marginalis),* Fusarium rot (various *Fusarium* sp.), late blight *(Phytophthora infestans),* leak *(Pythium debaryanum, P. ultimum,* and others), and ring rot *(Corynebacterium sepedonicum).* According to Eckert et al.,[115] three groups of organisms attack potatoes, causing severe postharvest losses:

1. *Phytophthora* sp. (These originate from the field and continue to develop in storage, causing blight and pink rot.)
2. *Pythium ultimum* (water wound-rot), *Fusarium* sp. (dry rot), and *Phoma* sp. (gangrene). (These form the second important group of fungi and attack tubers through wounds.)
3. A third group of organisms which render the tuber unattractive are, e.g., skin spots *(Cercospora pustulans),* silver scurf *(Helminthosporium solani),* etc.

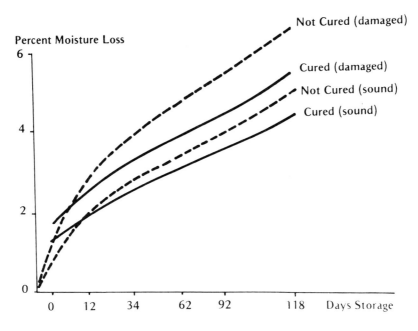

FIGURE 16.   Percent of moisture loss in storage from cured and uncured tubers.

8. Sprouting
Potatoes stored at higher temperatures exhibit sprouting, causing significant losses. The sprouted potatoes are not suitable for processing and domestic consumption. Significant losses due to sprouting have been reported from developing countries.

VII. CONTROL OF LOSSES DURING HANDLING AND STORAGE

A. HARVESTING, HANDLING, AND MATURITY
Commercial plantings of potatoes are harvested mechanically using potato diggers or combines of various types. A spading fork is the best equipment for hand digging. In potatoes harvested with machines, gathering of potatoes is facilitated by conveyor belts. Modern potato harvesters convey potatoes to low-level trailers to minimize damage.[116] Others are supplemented with grading and bagging attachments in conjunction with mechanical harvesters.[117] Potatoes should be picked up as soon as possible to prevent sunburn and similar injuries. Pantastico et al.[121] stated that skin slipping from the tuber, starch and total solids contents, and leaf senescence are the important harvest indices of potatoes. Careless digging and handling increase storage loss and reduce the grade and market value of potatoes. Potatoes thus should not be thrown, dropped, or walked upon to prevent physical injuries.[118] Smith[119] studied the effect of handling on the percentage weight loss of potatoes. The carefully handled tubers showed a weight loss of 5.5% after a 7-month storage period in comparison with potatoes handled under normal commercial conditions, which showed an 8.5% weight loss. Proctor et al.[102] recommended harvesting of potatoes when mature with a thick outer skin using careful digging and handling, protecting the newly harvested potatoes from exposure to wind and sun, and adopting rapid drying and curing processes, whether the tubers are to be stored or marketed immediately.

B. CURING
One of the most simple and effective ways to reduce water loss (Figure 16) and decay during postharvest storage of potatoes is curing. Injured or bruised surfaces are allowed to

A — Open uncured wound

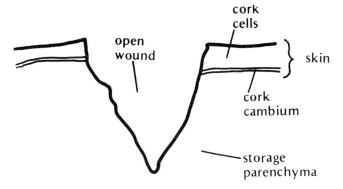

B — Cured wound

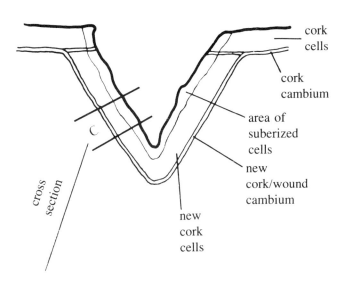

FIGURE 17.   Healing of wound during curing. (A) open uncured wound; (B) cured wound; (C) cross section of wound surface.

heal and form periderm thickness on the potato. During the process of curing, wounds are healed by producing a new cork cambium, thus preventing infections by pathogenic organisms and reducing loss (Figure 17). Some water loss takes place during curing. The removal of decaying products prior to curing and storage ensures a greater percentage of usable product after storage. Curing of potatoes effectively reduces postharvest losses in storage by providing protection against infections by pathogenic organisms. Successful curing of potatoes can be achieved by drying potatoes at 8 to 20°C with at least 85% RH.[105] Care should be taken to avoid the condensation of water on the tubers during curing. The optimum conditions for curing root, tuber, and bulb vegetables are given in Table 17.

C. LOW-TEMPERATURE STORAGE

To prevent excessive shrinkage and rotting of potatoes, they are generally stored for the

cork
cells

C.   Detailed cross section of wound surface from Figure 17

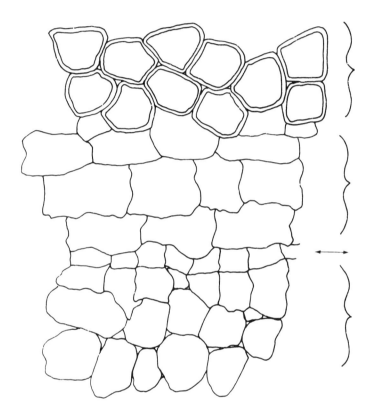

FIGURE 17C.

TABLE 17
Optimum Conditions for Curing Root, Tuber, and Bulb
Vegetables

| | Temperature | | | |
| | Celsius | Fahrenheit | Relative humidity | Duration |
Commodity	(degrees)		(%)	(days)
Potato	15—20	59—68	85—90	5—10
Sweet potato	30—32	85—90	85—90	4—7
Yam	32—40	90—104	90—100	1—4
Cassava	30—40	86—104	90—95	2—5
Onion and garlic	35—45	95—113	60—75	0.5—1

first week at a temperature of 10.0 to 15.6°C and a RH of 85 to 95% to permit suberization and formation of wound periderm.[120] Most of the shrinkage takes place during the first month of storage. It is therefore important to allow the cuts and bruises to heal rapidly and then to lower the temperature of the storage. Following the curing period, the temperature is lowered to 3.3 to 4.4°C to prevent sprouting after the rest period is over. Most cultivars of potatoes can be stored for 6 months or longer without sprouting if held at 4.4°C. Although potatoes freeze at about -1.7°C, low-temperature injuries may be encountered from prolonged storage below -1.7°C. Storage at 4.4°C is not desirable for tubers that are to be used for chips or frozen French fries, which develop an undesirable dark brown color due to sugars accumulated at 4.4°C. Temperatures of about 10.0 to 12.8°C and 90% RH are suitable conditions for storing potatoes to be used for manufacturing chips of French fries. The internal discoloration ("Mahogany browning") develops in all potato cultivars when stored for 20 weeks or longer at 0 to 1.1°C. Pantastico et al.[121] recommended temperatures of 3.3 to 4.4°C with 85% RH to store potatoes for about 34 weeks.

D. CONTROLLED ATMOSPHERE (CA) STORAGE

According to Ryall and Lipton,[109] CA is not advantageous for storage of potatoes destined for table use. The concentrations of 5% or less oxygen inhibit periderm formation and wound healing, and oxygen levels of 1% or less cause off flavors[122] and greatly increase decay and surface mold and blackheart during 1 week at moderate temperatures of 15 and 20°C. At 5°C, the deleterious effects of a low temperature were less pronounced or absent.[112] High CO_2 (10% or more) enhanced decay even at 4.4°C and aggravated effects of low oxygen at higher temperatures.[123,124] Oxygen at 10% or below increased sprouting of potatoes in storage, but had no discernible effect on the subsequent field performance of the seed potatoes.[125] Use of 12% CO_2 (8% or more) and low O_2 for 6 months resulted in the complete failure of seed potatoes. The combination of high CO_2 (8% or more), low O_2 (5% or less), and low storage temperature (0°C) had the most serious adverse effects.[126] Although high CO_2 (15 to 20%) prevented greening of prepackaged potatoes at 10.0 to 15.6°C, the incidence of bacterial soft rot increased even under a lower CO_2 (10%) level and at a lower temperature, 4.4°C.[123] Yamaguchi et al.[127] reported that susceptibility of potatoes to black spot increased with even lower CO_2 concentrations (5%) and a very short exposure period of just once a day. Thus, CA storage of potatoes hardly appears to be beneficial.

E. CHEMICAL CONTROL OF LOSSES

Several chemicals have been found to suppress sprouting effectively in the potato and onion during storage at higher temperatures. Van Niekerk[128] demonstrated that effective suppression of sprouting and consequent reduction in weight loss could be obtained over an 8-month storage period in a thatched-roof structure with a temperature range of 10 to 20°C and a 34 to 70% RH using a mixture of propham (isopropyl-N-phenylcarbamate [IPPC]) and chloropropham (3-chloroisopropyl-N-phenylcarbamate [CIPC]). The most important suppressant for ware marketing is CIPC, which is widely used. Others in commercial use are maleic hydrazide, nonylalcohol, MENA, and tecnazene (2,3,4,6-tetrachloronitrobenzene [TCNB]). TCNB, although the weakest inhibitor, has the advantage that it does not inhibit suberization in curing and may be used for seed potatoes.[105] CIPC is a strong sprout inhibitor and is probably the most widely used chemical sprout inhibitor for the storage of potatoes. It may be applied as a dust, water dip, vapor, or aerosol. Since CIPC interferes with periderm formation, it should be applied only after curing. In the U.S. the tolerance of CIPC in raw and processed potatoes is 50 µg/g. Kennedy and Smith[76] found that 3 lb of maleic hydrazide (MH) per acre used as a foliage spray prevented sprouting of potatoes stored at 10.0°C. It is applied when the lowest leaves begin to turn yellow and die.[118] The application of CIPC to the potato tubers in storage prevented their sprouting for several months.[85] Salunkhe and Wu[129] reviewed the literature on the chemical modifications in several fruits and vegetables, including the effects of MH and other growth inhibitors on the potato. Arteca[130] found that vacuum infiltration of 1 and 2% $CaCl_2$ reduced the chlorophyll accumulation in Katahdin potatoes by 50 to 60% and by 70 to 80%, respectively, related to controls, and $CaCl_2$ at 3 or 4% induced internal breakdown. Maximum uptake of $CaCl_2$ was obtained by placing the tubers under a vacuum (-0.9 atm) for 30 min with a soak time of 15 min. The infiltration of calcium ions helped to maintain tuber quality in light, allowing the consumers maximum visibility of the product.

The effects of gibberillic acid (GA), indoleacetic acid (IAA), and kinetin on ascorbic acid and the total sugar content of potatoes, associated with sprouting during storage, were investigated by Kumar et al.[131] The ascorbic acid content decreased and the total sugar content increased progressively during storage at 5 to 10°C, but the increase in ascorbic acid and total sugar content did not change later on, when the tubers were transferred to room temperature (30 $\pm$ 2°C). Both the ascorbic acid and the sugar contents were higher in potatoes treated with GA, corresponding to earlier visible growth of sprouts, than the IAA- and kinetin-treated tubers which corresponded to delayed visible growth of sprouts.

F. IRRADIATION

The irradiation of potatoes produced no detrimental effects on flavor when irradiated at the low doses.[132] Panalaskas and Pelletier[133] found that specified levels of gamma radiation did not cause consistent variations in the ascorbic acid content of potatoes. Mikaelsen and Roer[134] reported that the vitamin C content decreased in both the irradiated and the non-irradiated potatoes during the first 7 months of storage at 5°C, but was restored after this period, and that the ascorbic acid levels of the irradiated potato samples were higher than those under control. Ogata et al.[135] and Tatsumi et al.[136] reported the results of their experiments on the mechanism of browning. These investigators[136] showed that the O-diphenol content increased in irradiated potato tubers (Table 18) and the rate of increase was greater in the cortex and vascular bundles than in the pith. The ascorbic acid content decreased with increasing levels of irradiation dose, the rate of decrease being greater in the cortex and vascular bundles than in the pith. Sparrow and Christensen[93] also demonstrated that the storage life of potatoes could be extended by the use of ionizing radiation. Their observation on the inhibition of sprouting of potatoes by gamma irradiation has been confirmed by other investigators.[137,138]

Salunkhe[139] summarized the results of the effects of gamma radiation on several bulb, tuber, and root crops and assessed the possibility of using irradiation to minimize sprouting losses of these commodities. The potato tubers were irradiated at 0 (control), 1, 3.7, 7.4, 9.3, 14.9, 29.8, and 59.5 $\times$ 10^3 rd and stored for 5 months at 10°C and 85% RH. Higher doses of radiation (over 14.9 $\times$ 10^3 rd) were found to either retard or inhibit sprouting of potatoes, although a slight acceleration of sprouting was noticed at the 3.7 $\times$ 10^3 rd dose. A dose of 9.3 $\times$ 10^3 rd effectively controlled the sprouting losses of potatoes and doses higher than 9.3 $\times$ 10^3 rd did not result in an additional significant decrease of sprouting. The radiation did not affect the sensory quality of potatoes during the first month of storage. However, subsequent to storage for 4 months at 10°C and 85% RH, there was an increase in the quality scores with an increase in radiation dose which was attributed to the reduction in sprouting losses of potatoes. In addition, Salunkhe[139] noticed that the greening of irradiated potatoes (over 9.3 $\times$ 10^3 rd) was inhibited.

Chachin and Iwata[140] reported the effects of irradiation on the respiratory metabolism and potassium release of potatoes. Gamma irradiation at low doses (10 to 100 krd) for sprout inhibition increased the respiration rate of potatoes immediately after the irradiation, as determined with whole potatoes, tissue slices, and mitochondria. The increased respiration was lowered to the level of unirradiated potatoes during the subsequent storage period. The

TABLE 18
Ascorbic Acid and *O*-Diphenol Contents after
Irradiation

Dose (krd)	Irradiated 1 d after harvest[a]			Irradiated 3 months after harvest[a]		
	A	B	C	A	B	C
Ascorbic acid						
0	18.2	18.5	19.3	9.3	9.8	7.6
10	11.3	12.5	15.5	8.9	8.3	6.7
20	15.4	11.0	11.7	7.2	7.1	7.3
40	9.9	10.4	10.1	7.6	7.6	7.3
O-Diphenol						
0	3.2	3.4	2.0	5.7	3.6	1.2
10	8.4	7.2	4.4	6.0	5.7	0.6
20	10.8	7.6	4.8	5.4	6.6	0.9
40	11.7	8.4	4.6	3.9	6.0	0.0

Note: A, cortex; B, vascular bundle; C, pith.

[a] mg/100 g fresh weight of potato tubers.

From Tatsumi, Y., Chachin, K., and Ogata, K., *J. Food Sci. Technol.*, 19, 508, 1972. With permission.

potassium leakage (or release) from disks of irradiated potatoes increased as compared to that of unirradiated ones, showing the changes in the membrane function. This effect also declined in the stored sample. The tentative results of the experiment indicated that the cellular radiation damage was induced in the tissues of irradiated potatoes, even at the dose level of sprout inhibition, and the influence may be eliminated during storage through some repair processes operating in the irradiated potatoes. The effects of the level of gamma irradiation on the CO_2 production of potatoes and the influence of gamma irradiation on the CO_2 production of potatoes at different stages of maturity are shown in Figures 18 and 19 respectively. The highest respiration (CO_2 production) was found 24 h after irradiation and the stimulative effect increased with the level of the dose. Even at the 10-krd level, the CO_2 production in irradiated potatoes at the time was three times that in control (unirradiated) potatoes. During subsequent storage, the increased respiration of irradiated potatoes lowered, reaching the magnitude of CO_2 production normally found in unirradiated samples. This decline in respiration was much faster in potatoes irradiated with lower dosages than at higher dosages (Figure 18). The effect of gamma irradiation on less mature potatoes (as classified by the weight of potatoes when harvested) was larger than in more mature potatoes (Figure 19). Irradiation of potatoes, even at the low dose for sprout inhibition, increased potassium release from disks of potatoes. Figure 20 depicts changes of the total amount of potassium released into water in 3 h from disks taken from potatoes at 20°C. Potassium release increased in irradiated samples immediately after irradiation, and its extent tended to be greater with higher dosages of gamma irradiation. During storage, the increased potassium release declined to the initial value of control, although potassium release markedly increased in control samples which were stored for 10 months and were extremely shrunken. Chachin and Iwata[140] also noted that the increase of respiration rate with irradiation was considered suppressed in low temperatures and stated that due care must be taken for irradiation dosage and stage of maturity of potatoes to be used for radurization.

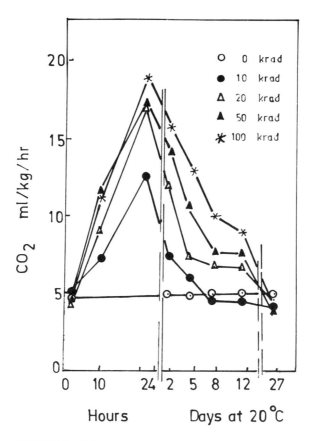

FIGURE 18.   Effects of gamma-irradiation on CO_2 production of potatoes of different maturity. (From Chachin, K. and Iwata, T., Paper read at seminar on Food Irradiation for Developing Countries in Asia and the Pacific, IAEA/FAO, Tokyo, November 9 to 13, 1981.)

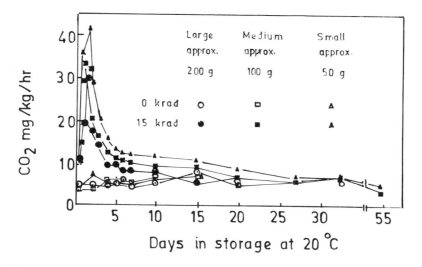

FIGURE 19.   Effects of gamma-irradiation on CO_2 production of potatoes of different maturity. (From Chachin, K. and Iwata, T., Paper read at seminar on Food Irradiation for Developing Countries in Asia and the Pacific, IAEA/FAO, Tokyo, November 9 to 13, 1981.)

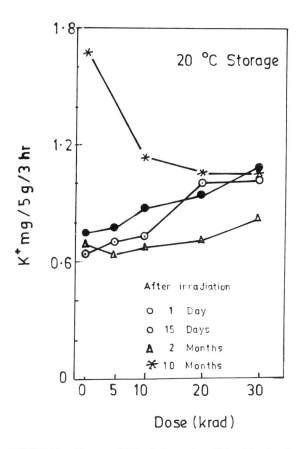

FIGURE 20. Changes of K ion leakages from disks of irradiated and unirradiated potatoes during storage at 20°C. (From Chachin, K. and Iwata, T., Paper read at seminar on Food Irradiation for Developing Countries in Asia and the Pacific, IAEA/FAO, Tokyo, November 9 to 13, 1981.)

Palmer et al.[141,142] studied the wholesomeness of irradiated potatoes. They investigated the effect of feeding of an irradiated potato diet on reproduction and longevity in rats. No significant effects attributable to the feeding of irradiated potatoes were observed on growth, food consumption, fertility, and reproductive performance. The values for litter data, fecal losses, and the incidence of malformations were comparable in all groups. No evidence of adverse effect on spermatogenesis (dominant lethal assay) or on the incidence of chromosomal aberrations was noted. A list of irradiated potato tubers cleared for human consumption in different countries is given in Table 19.[150]

By 1977, 10 countries had issued 30 clearances covering the irradiation of 15 different food items; 14 of these clearances were for sprout inhibition of the potato and onion.[143] At present, Japan allows irradiation of the potato to suppress sprouting. Sprouting in the potato and the onion during long-term storage is inhibited by 0.02 to 0.15 krd. This dose has little effect on other aspects of potato and onion quality such as sugar level, rate of wastage and water, texture, and flavor. According to Wills et al.,[143] irradiation of the potato and onion is more expensive than treatment with chemical sprout inhibitors like CIPC and MH.

VIII. SYSTEMS OF POTATO STORAGE

The storage of potatoes is intended to conserve them with minimum losses in quantity and to preserve their quality for use, processing, or propagation. Potatoes are subjected to

TABLE 19
Irradiated Potato Tubers Cleared for Human
Consumption in Different Countries[150]

Country	Product	Year
Belgium	Potatoes**	1980
Bulgaria	Potatoes*	1971
Canada	<u>Potatoes</u>	1965
Chile	Potatoes* ***	1974
Czechoslovakia	Potatoes*	1976
Denmark	<u>Potatoes</u>	1970
France	Potatoes**	1972
Federal Republic of Germany	Potatoes*	1974
Hungary	Potatoes***	1969
Israel	<u>Potatoes</u>	1967
Italy	<u>Potatoes</u>	1973
Japan	<u>Potatoes</u>	1972
Netherlands	<u>Potatoes</u>	1970
	Peeled potatoes***	1976
Philippines	Potatoes**	1972
South Africa	<u>Potatoes</u>	1977
Spain	<u>Potatoes</u>	1969
U.S.S.R.	<u>Potatoes</u>	1958
U.S.	<u>Potatoes</u>	1964
World Health Organization	Potatoes**	1969

Note: *, Experimental batches; **, temporary acceptance; ***, test
marketing; <u>underlined,</u> unlimited clearance.

storage losses due to respiration, transpiration, sprouting, rotting, and damage from pests and diseases. The storage requirement of ware potatoes is to preserve their culinary properties and to prevent them from sprouting. Seed potatoes should be able to sprout normally by the end of the storage period. Proper storage aims at reducing wastage of both ware and seed potatoes to the minimum. This in turn depends upon the extent to which the storage conditions meet the optimum environmental requirement of the potato tuber, such as low (but not freezing) temperatures (2 to 3°C) and high humidity (85 to 90% RH). Seed storage methods should provide desired development of sprouts prior to planting in terms of both number and size. Factors influencing the number of stems per plant are variety, tuber size, and apical dominance, which is influenced by storage temperature and light. In practice, storage involves considerations of climate and weather, design of the storage equipment, control of the environment in the store, economics, and related factors.[139] Research work carried out at the International Potato Center (CIP) has shown that diffused light storage (Figure 21) as compared to conventional seed tuber storage in the dark, improves seed quality so that potato yields are increased by about 18%. The average results of all on-station trials carried out at CIP/Huancayo for 24 cultivars are summarized in Table 20.

The factors which affect the inherent storage and market life of potatoes at the time of harvest are variations in potential length of dormancy, cultural practices, the selection of harvest maturity in relation to dry matter content, the presence or absence of disease, and the method of harvesting adopted.[105] Booth and Proctor[144] reviewed storage problems and handling practices of ware potatoes in the tropics. The storage methods in the warm plains fall into two categories: (1) storage in cool, dry rooms with proper ventilation on the floor or on bamboo racks and (2) storage in pits. It has been found that tubers keep well in sand or in sand plus naphthalene. The surface of the sand is lightly sprinkled with water in the hot months without dampening the tubers. Recent studies on the storage of potatoes in the

FIGURE 21.   Diffused light storage of potatoes. (Courtesy of CIP, Lima, Peru).

TABLE 20
Effects of Natural Diffused Light on Seed Potato Tubers

Parameter	Diffused light stored	Dark stored
A. Condition after 6 months storage		
Sprout length	1.8	21.7
Sprout number per tuber	3.4	1.4
Total storage loss (%)	9.9	20.3
B. Field performance		
Days to full emergence	30.6	38.1
Total yield (t/ha)	28.8	24.6

From CIP Annual Report, 1981, International Potato Centre, Lima, Peru, 1981. With permission.

plains indicate that an insulated, ventilated, cool store (constructed at the Central Potato Research Station, Daurala, Meerat by Central Potato Research Institute, in collaboration with the Warehousing Corporation of India, Ltd.) is an ideal storage house for potatoes. The store is located in a grove of trees. It is a masonry structure, measuring 9.15×4.57 m with 13 air inlets and 22 ventilators, each 0.91×0.30 m, on the walls. The air inlets and the ventilators are located 0.30 and 2.13 m above the floor level and are provided with wooden shutters to exclude light from the store. A door 1.83×1.22 m is provided to serve as an antechamber. The ceiling, walls, doors, ventilators, and air inlets have been insulated with fiberglass. A coating of bitumen is used on the wall to act as a vapor barrier. Thus, the store incorporates all the principles of durability, insulation, and ventilation, as well as the other ancillary devices to reduce the heat load. Records of temperature and humidity in the store have revealed the beneficial effect of the structure on the storage environment and the behavior of stored potatoes. Rotting and drying percents were generally lower, and fewer eyes sprouted per tuber with CIPC treatment.[145]

Modern cold storage consists of an enclosed chamber, which has its floor, walls, roof, and door insulated with a suitable insulant of appropriate thickness and is mechanically cooled with liquid freon or NH_3. The former is preferred as a refrigerant, because the latter, in event of leakage, could damage tubers irreparably, apart from other hazards. Small- and medium-sized potatoes for consumption can be stored in well-insulated cheap stores with provision for ventilation and cooling. Insulation with straw, paddy husk, or shreds of pith in maize cobs, held in position with polythene-lined board or matting, can be provided on the inner faces of walls. The store should be situated on high-lying ground, preferably in a grove of trees. As an alternative or addition, a quick-growing live hedge or climbers should be grown about 30 cm away from the walls, on all sides of the store except the north, to protect them from direct sun. The ceiling and roof must be thickly thatched and the latter may be covered with tiles or any other waterproof material. Vents, which can be opened at night and closed during the daytime, should be provided on the walls 30 cm above ground level and also in the gabled ends above ceiling level for natural ventilation. During the dry period of summer, it may be necessary to keep down the temperature and increase humidity in the store with an electrically operated water cooler mounted on the wall. In the rainy season when the humidity is high, the fans are operated to provide the required aeration. The use of such a cooling system brings down the storage temperature to 20 to 30°C and increases humidity, creating environmental conditions comparable to those in the precooling room attached to a cold store. This would reduce rotting, but is likely to induce sprouting, which can be effectively controlled by the use of CIPC or other sprout inhibitors.

Rustowski[146] reported that the main factors affecting the rate of moisture loss from potato tubers in storage are temperature, humidity, heat production by tubers in store, tuber quality, and storage management. Rate of ventilation is a minor factor. In storages cooled by fresh air, increasing the rate of ventilation results in a lowering of moisture loss. Rustowski[146] further stated that ventilation was needed only to maintain the desired environmental condition, since unventilated tubers had the lowest moisture loss, and that the infiltration of fresh air through leakages in the storage structure needed to be reduced as much as possible. Metlitskii[147] examined the causes of high losses of potatoes during storage and discussed methods of reducing them. According to this author, greater attention should be given to the biochemical mechanisms of resistance in plant tissues to diseases, dormancy, and transition to active growth, maturation and aging, and methods of controlling these mechanisms for reducing losses due to diseases and sprouting of potatoes. The combination of the following different methods was suggested: forced ventilation, cold storage, regulation of gas exchange, and the use of physiologically active substances and ionizing radiation.

Von Gierke[148] outlined different types of storage and summarized conditions for satisfactory storage of potato tubers, including adequate maturity of crop; healthy, dry, and undamaged tubers; and a slow lowering of temperature and careful control of the storage environment. The improved postharvest handling and storage of potatoes increased final tuber yields by about 20 to 30%.[149] The researchers at the International Potato Center, Lima, Peru, are investigating simple stores with ambient air ventilation rather than complex storage with refrigerated air. A successful use of outside air requires that it be cooler than the temperature required for 6 h daily. An adequate and timely curing is important to heal wounds caused during harvest and handling and can be achieved with controlled ventilation at about 13 to 15°C and high RH (85% RH). The ideal storage structure invented at the Center has been illustrated in Figure 22.

A joint research project on storage of potatoes between the Kenya Ministry of Agriculture, the CIP, and the Department of Agricultural Economics of the University of Nairobi is currently in progress in Kenya, with the objective of evaluating alternatives for farm-level storage. Preliminary results of this project indicate that the use of pits and the use of loose piles in houses are the least effective storage methods, pits being totally uneconomical, with wastage levels as high as 50% within 2 months of storage. With purpose-built wooden stores

FIGURE 22. Improved Potato Storage in Developed Countries by CIP, Lima, Peru. From CIP Annual Report, IPC, Lima, Peru, 1981.

offering convective ventilation (Figure 23), total loss of potatoes after 4 months of storage ranged from 5 to 16%, depending upon the location.[149] Underground pits are used at the subsistence level in several East African countries to hold up to 15 t of potatoes with a lower level of loss of about 20% after 3 months.

Air-cooled structures referred to above are simply insulated structures, aboveground or partly underground, which are cooled by circulation of colder, outside air when the temperature of the produce is above the desired level; and if the temperature of the outside air is lower, air is circulated through the stack in the store by convectional or mechanical means through bottom inlet vents and top outlets with dampers. Fans, if fitted, are controlled manually or automatically with differential thermostats. The air may be humidified, which can also be automated. According to Wills et al.,[143] air-cooled stores are cheap to construct and operate and are still widely used for storage of the potato and sweet potato, both of which need relatively higher storage temperatures to avoid accumulation of sugars and chilling injury, respectively. Potatoes are commonly stored in bulk piles in stores with air-delivery ducts under the floor or at floor level and with suitably spaced air outlets. An on-farm potato storage system used by the farmers is shown in Figure 24.

FIGURE 23.   Simple naturally ventilated stores. Courtesy of CIP, Lima, Peru.

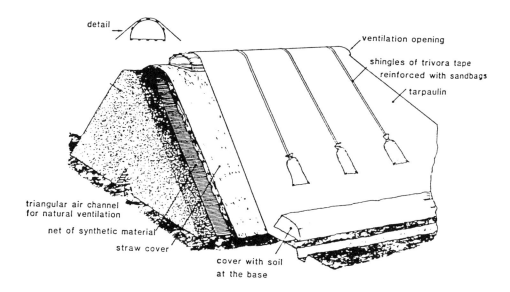

FIGURE 24.   Growers potato storage on farm. Courtesy of CIP, Lima, Peru.

REFERENCES

1. **Salunkhe, D. K. and Desai, B. B.,** *Postharvest Biotechnology of Vegetables,* Vol. I, CRC Press, Boca Raton, FL, 1984, 108.
2. **Talburt, W. F. and Smith, O.,** *Potato Processing,* 4th ed., Van Nostrand Reinhold, New York, 1987, 203.
3. **Woolfe, J. A.,** *The Potato in the Human Diet,* Cambridge University Press, London, 1987, 83.
4. **Smith, O.,** Studies of potato storage, Cornell University Agricultural Exp. Station Bulletin, 553, 1933.

5. **Thomas, P.,** Wound induced suberization and periderm development in potato tubers as affected by temperature and gamma irradiation, *Potato Res.,* 25, 155, 1982.

6. **Burton, W. G.,** Postharvest behavior and storage of potatoes, in *Applied Biology,* Vol. 2, Coaker, T. H., Ed., Academic Press, New York, 1966.

7. **Booth, R. H. and Shaw, R. L.,** *Principles of Potato Storage,* International Potato Centre, Lima, 1981.

8. **Faulks, R. M., Griffith, N. M., White, M. A., and Tomlins, K. I.,** Influence of site, variety and storage on nutritional composition and cooked texture of potatoes, *J. Sci. Food Agric.,* 33, 589, 1982.

9. **Fitzpatrick, T. J. and Porter, W. L.,** Changes in the sugars and amino acids in chips made from fresh, stored and reconditioned potatoes, *Am. Potato J.,* 43, 238, 1966.

10. **Leichsenring, J. M.,** Factors influencing the nutritive value of potatoes, Minnesota Technical Bulletin No. 96, University of Minnesota, Agricultural Experiment Station, Minneapolis, 1951.

11. **Linnemann, A. R., Van Es, A., and Hartmans, K. J.,** Changes in the content of L-ascorbic acid, glucose, fructose, sucrose and total glycoalkaloids in potatoes stored at 7, 16 and 28°C, *Potato Res.,* 28, 271, 1985.

12. **Picha, D. H.,** Influence of storage duration and temperature on sweet potato sugar content and chip color, *J. Food Sci.,* 51, 239, 1986.

13. **Plaza, S. G., Sueldo, R. J., Crupkin, M., and Barassi, C. A.,** Changes in composition of potatoes stored in clamps, *J. Food Sci.,* 50, 1254, 1985.

14. **Willis, R. B. H., Lim, J. S. K., and Greenfield, H.,** Variation in nutrient composition of Australian retail potatoes over a 12 month period, *J. Sci. Food Agric.,* 35, 1012, 1984.

15. **Augustin, J., Johnson, S. R., Teitzel, C., Toma, R. B., Shaw, R. L., True, R. H., Hogan, J. M., and Deutsch, R. M.,** Vitamin composition of freshly harvested and stored potatoes, *J. Food Sci.,* 43, 1566, 1978.

16. **Porter, W. L. and Heinze, P. H.,** Changes in composition of potatoes in storage, *Potato Handbook,* 10, 5, 1965.

17. **Shekhar, V. C., Iritani, W. M., and Arteca, R.,** Changes in ascorbic acid content during growth and short-term storage of potato tubers (*Solanum tuberosum* L.), *Am. Potato J.,* 56, 663, 1978.

18. **Singh, M. and Verma, S. C.,** Postharvest technology and utilization of potato, in *Postharvest Technology and Utilization of Potato,* Kishore, H., Ed., International Potato Center, Region VI, New Delhi, 1979.

19. **Sweeney, J. P., Hepner, P. A., and Libeck, S. Y.,** Organic acid, amino acid and ascorbic acid content of potatoes as affected by storage conditions, *Am. Potato J.,* 46, 463, 1969.

20. **Talley, E. A., Toma, R. B., and Orr, P. H.,** Amino acid composition of freshly harvested and stored potatoes, *Am. Potato J.,* 61, 267, 1984.

21. **Toma, R. B., Augustin, J., Shaw, R. L., True, R. H., and Hogan, J. M.,** Proximate composition of freshly harvested and stored potatoes, *J. Food Sci.,* 43, 1702, 1978.

22. **Werner, H. O. and Leverton, R. M.,** The ascorbic acid content of Nebraska grown potatoes as influenced by variety, environment, maturity and storage, *Am. Potato J.,* 23, 265, 1946.

23. **Yamaguchi, M., Perdue, J. W., and MacGillivray, J. H.,** Nutrient composition of "White Rose" potatoes during growth and after storage, *Am. Potato J.,* 37, 73, 1960.

24. **Samatous, B. and Schwimmer, S.,** Changes in carbohydrate and phosphorus content of potato tuber during storage in nitrogen, *J. Food Sci.,* 28, 163, 1963.

25. **Miča, B.,** Changes in content of glucose, fructose, sucrose and lysine in potatoes during storage and boiling, *Rostl. Vyroba,* 24, 35, 1978.

26. **Miča, B.,** Effect of storage and boiling on the content and free amino acids in potatoes, *Rostl. Vyroba,* 24, 731, 1978.

27. **Habib, A. T. and Brown, H. D.,** Factors influencing the color of potato chips, *Food Technol.,* 10, 332, 1956.

28. **Fitzpatrick, T. J., Talley, E. A., Porter, W. L., and Murphy, H. J.,** Chemical composition of potatoes. III. Relationships between specific gravity and the nitrogenous constituents, *Am. Potato J.,* 41, 75, 1964.

29. **Burton, W. G.,** *The Potato,* Veenman, H. and Zonen, N. V., Eds., Drukkerij Veenman BV, Wageningen, Holland, 1966.

30. **Thomas, P., Srirangarajan, A. N., Joshi, M. R., and Janave, M. T.,** Storage deterioration in gamma-irradiated and unirradiated Indian potato cultivars under refrigeration and tropical temperatures, *Potato Res.,* 22, 261, 1979.

31. **Augustin, J., Johnson, S. R., Teitzel, C., True, R. H., Hogan, J. M., Toma, R. B., Shaw, R. L., and Deutsch, R. M.,** Changes in nutrient composition of potatoes during home preparation. II. Vitamins, *Am. Potato J.,* 55, 653, 1978.

32. **Roine, P., Wichmann, K., and Vihavainen, Z.,** The content and stability of ascorbic acid in different potato varieties in Finland, *Suom. Maataloustiet. Seuran Julk.,* 83, 71, 1955.

33. **Mareschi, J. P., Belliot, J. P., Fourlon, C., and Gey, K. F.,** Decrease in vitamin C content in "Bintje" potatoes during storage and conventional cooking procedures, *Int. J. Vitamin Nutr. Res.,* 55, 402, 1983.

34. **Page, E. and Hanning, F. M.,** Vitamin B_6 and niacin in potatoes, *J. Am. Diet. Assoc.,* 42, 42, 1963.

35. **Addo, C. and Augustin, J.,** Changes in the vitamin B_6 content in potatoes during storage, *J. Food Sci.,* 53, 749, 1988.

36. **Barker, J. and Mapson, L. W.,** The ascorbic acid content of potato tubers. III. The influence of storage in nitrogen, air and pure oxygen, *New Phytol.,* 51, 90, 1952.

37. **Cheng, F. C. and Muneta, D.,** Lipid composition of potatoes as affected by storage and potassium fertilization, *Am. Potato J.,* 55, 441, 1978.

38. **Mazza, G., Hung, J., and Dench, M. J.,** Processing/nutritional quality changes in potato tubers during growth and long term storage, *Can. Inst. Food Sci. Technol.,* 16(1), 39, 1983.

39. **Mazza, G.,** Correlations between quality parameters of potato during growth and long term storage, *Am. Potato J.,* 60, 145, 1983.

40. **Cargill, B. F., Heldman, D. R., and Bedford, C. L.,** Influence of environmental storage conditions on potato chip quality, *Potato Chipper,* 30, 10, 1971.

41. **Smith, O.,** Culinary quality and nutritive value of potatoes, in *Potatoes: Production, Storing, Processing,* Smith, O. Ed., AVI Publishing, Westport, CT, 1968, 498.

42. **Gull, D. D. and Isenberg, F. M.,** Light and off-flavor development in potato tubers exposed to fluorescent light, *Proc. Am. Soc. Hortic. Sci.,* 71, 446, 1958.

43. **Akeley, R. V., Houghland, G. V. C., and Schark, A. E.,** Genetic differences in potato tuber greening, *Am. Potato J.,* 39, 409, 1962.

44. **Yamaguchi, M., Hughes, D. L., and Howard, F. D.,** Effect of season, storage temperature during light exposure on chlorophyll accumulation of "White Rose" potatoes, *Proc. Am. Soc. Hortic. Sci.,* 75, 529, 1960.

45. **Larsen, E. C.,** Investigations on cause and prevention of greening in potato tubers, *Idaho Agric. Exp. Stn. Res. Bull.,* No. 16, 1949.

46. **Lewis, W. C. and Rowberry, R. G.,** Some effects of depth of planting and time and height of hilling on 'Kennebec' and 'Sebago' potatoes, *Am. Potato J.,* 50, 301, 1973.

47. **Bleasdale, J. K. A. and Thompson, R.,** Potatoes, *Annu. Rep. Natl. Veg. Res. Stn., Wellesbourne, England,* 37, 1964.

48. **Zitnak, A. and Johnston, G. R.,** Glycoalkaloid content of "B5141-6" potatoes, *Am. Potato J.,* 47, 256, 1970.

49. **Pallman, H. and Schindler, K.,** Beeinflusst die Duengung den Solaningehalt der Kartoffeln, *Schweiz Landwirtsch. Monatsh.,* 20, 21, 1942.

50. **Patil, B. C., Salunkhe, D. K., and Singh, B.,** Metabolism of solanine and chlorophyll in potato tubers as affected by light and specific chemicals, *J. Food Sci.,* 36, 474, 1971.

51. **Salunkhe, D. K., Wu, M. T., and Jadhav, S. J.,** Effects of light and temperature on the formation of solanine in potato slices, *J. Food Sci.,* 37, 969, 1972.

52. **Hilton, R. J.,** Factors in relation to tuber quality in potatoes: preliminary trials on bitterness in Netted Gem potatoes, *Sci. Agric.,* 31, 61, 1951.

53. **Yamaguchi, M., Hughes, D. L., and Howard, F. D.,** Effect of season, storage temperature and temperature during light exposure on chlorophyll accumulation of "White Rose" potatoes, *Proc. Am. Soc. Hortic. Sci.,* 75, 529, 1960.

54. **Buck, R. W., Jr. and Akeley, V.,** Effect of maturity, storage, temperature and storage time on greening of potato tubers, *Am. Potato J.,* 44, 56, 1967.

55. **Liljemark, A. and Widoff, E.,** Greening and solanine development of white potato in fluorescent light, *Am. Potato J.,* 37, 379, 1960.

56. **Zitnak, A.,** The Influence of Certain Treatments upon Solanine Synthesis in Potatoes, M.S. thesis, University of Alberta, Edmonton, 1953.

57. **Patil, B. C., Singh, B., and Salunkhe, D. K.,** Formation of chlorophyll and solanine in Irish potato (*Solanum tuberosum* L.), tubers and their control by gamma radiation and CO_2-enriched packaging, *Lebensm. Wiss. Technol.,* 4, 123, 1971.

58. **Jeppsen, R. B., Salunkhe, D. K., and Jadhav, S. J.,** Formation and anatomical distribution of chlorophyll and solanine in potato tubers and their control by chemical and physical treatments, *The 33rd Annu. Int. Food Tech. Meet.,* 153, 1973.

59. **Jadhav, S. J. and Salunkhe, D. K.,** Formation and control of chlorophyll and glycoalkaloids in tubers of *Solanum tuberosum* L. and evaluation of glycoalkaloids toxicity, *Adv. Food Res.,* 21, 307, 1975.

60. **Wu, M. T. and Salunkhe, D. K.,** Control of chlorophyll and solanine synthesis and sprouting of potato tubers by hot paraffin wax, *J. Food Sci.,* 37, 629, 1972.

61. **Wu, M. T. and Salunkhe, D. K.,** Control of chlorophyll and solanine formation in potato tubers by oil and diluted oil treatments, *HortScience,* 7, 466, 1972.

62. **Jadhav, S. J. and Salunkhe, D. K.,** Effects of certain chemicals on photo-induction of chlorophyll and glycoalkaloid synthesis and on sprouting of potato tubers, *Can. Inst. Food Sci. Technol. J.,* 7, 178, 1974.

63. **Forsyth, F. R. and Eaves, C. A.,** Greening of potatoes, CA cure, *Food Technol.,* 22, 48, 1968.

64. **Patil, B. C.,** Formation and Control of Chlorophyll and Solanine in Tubers of *Solanum tuberosum* L. and Evaluation of Solanine Toxicity, Ph.D. thesis, Utah State University, Logan, 84322, 1972.

65. **Ziegler, R., Schanderl, S. H., and Markakis, P.,** Gamma irradiation and enriched CO_2 atmosphere storage effects on the light induced greening of potatoes, *J. Food Sci.,* 33, 533, 1968.

66. **Madsen, K. A., Salunkhe, D. K., and Simon, M.,** Morphological and biochemical changes in gamma irradiated carrots and potatoes, *Radiat. Res.,* 10, 48, 1959.
67. **Wu, M. T. and Salunkhe, D. K.,** Effects of gamma-irradiation on wound induced glycoalkaloid formation in potato tubers, *Lebensm. Wiss. Technol.,* 10, 141, 1977.
68. **Bogucki, S. and Nelson, D. C.,** Length of dormancy and sprouting characteristics of ten potato cultivars, *Am. Potato J.,* 57, 151, 1980.
69. **Burton, W. G. and Wilson, A. R.,** The sugar content of sprout growth of tubers of potato cultivar Record, grown in different localities when stored at 10, 15 and 20°C, *Potato Res.,* 21, 145, 1978.
70. **Bantan, S., Krapez, M., and Vardjan, M.,** Variation in ascorbic acid during development and storage of tubers of potato Cv. Vesna and Bintje, *Biol. Vestn.,* 25, 1, 1977.
71. **Meiklejohn, J.,** The vitamin B_1 content of potatoes, *Biochem. J.,* 37, 349, 1943.
72. **Boyd, I. M. G., Dalziel, J., and Duncan, H. J.,** Studies on potato sprout suppressants. 5. The effect of chlorpropham contamination on the performance of seed potatoes, *Potato Res.,* 25, 51, 1982.
73. **Struckmeyer, B. E., Weis, G. G., and Schoenemann, J. A.,** Effect of two forms of maleic hydrazide on the cell structure at the midsection, stem and bud ends of the cortical and perimedullary regions of Russet Burbank tubers, *Am. Potato J.,* 58, 611, 1981.
74. **Sawyer, R. L. and Malagamba, J. P.,** Sprouting inhibition, in *Potato Processing,* Talburt, W. F. and Smith, O., Eds., Van Nostrand Reinhold, New York, 1987, 183.
75. **Kennedy, E. J. and Smith, O.,** Response of the potato to field application of maleic hydrazide, *Am. Potato J.,* 28, 701, 1951.
76. **Kennedy, E. J. and Smith, O.,** Response of seven varieties of potatoes to foliar application of maleic hydrazide, *Proc. Am. Soc. Hortic. Sci.,* 61, 395, 1953.
77. **Franklin, E. W. and Thompson, N. R.,** Some effects of maleic hydrazide on stored potatoes, *Am. Potato J.,* 30, 289, 1953.
78. **Salunkhe, D. K. and Wittwer, S. H.,** The influence of a preharvest foliar spray of maleic hydrazide on the specific gravity of Irish cobbler potatoes and quality of their chips, *Summary U.S. Rubber Co. MHIS,* 6B, 21, 1952.
79. **Weis, G. G., Schoenemann, J. A., and Groskopp, M. D.,** Influence of time of application of maleic hydrazide on the yield and quality of Russet Burbank potatoes, *Am. Potato J.,* 57, 197, 1980.
80. **Kunkel, R., Holsted, N. M., Mitchell, D. C., Russell, T. S., and Thornton, R. E.,** Maleic hydrazide studies on potatoes in Washington's Columbia basin, *Am. Potato J.,* 56, 470, 1979.
81. **Davis, J. R. and Groskopp, M. D.,** Yield and quality of Russet Burbank potato as influenced by interactions of rhizoctonia, maleic hydrazide and PCNB, *Am. Potato J.,* 58, 227, 1981.
82. **Mondy, N. I., Tymiak, A., and Chandra, S.,** Inhibition of glycoalkaloid formation in potato tubers by the sprout inhibitor, maleic hydrazide, *J. Food Sci.,* 43, 1033, 1978.
83. **Covsini, D., Stallknecht, G. F., and Sparks, W. C.,** A simplified method for determining sprout-inhibiting levels of chloropropham (CIPC) in potato, *J. Agric. Food Chem.,* 26, 990, 1978.
84. **Van Vliet, W. F. and Sparenberg, H.,** The treatment of potato tubers with sprout inhibitors, *Potato Res.,* 13, 223, 1970.
85. **Marth, P. C. and Schultz, E. S.,** A new sprout inhibitor for potato tubers, *Am. Potato J.,* 29, 268, 1952.
86. **Kim, M. S., Swing, E. E., and Siezcka, J. B.,** Effects of cloropropham (CIPC) on sprouting of individual potato eyes on plant emergence, *Am. Potato J.,* 49, 420, 1972.
87. **Van Vliet, W. F. and Sparenberg, H.,** The treatment of potato tubers with sprout inhibitors, *Potato Res.,* 13, 223, 1970.
88. **Ellison, J. H.,** Inhibition of potato sprouting by 2, 3, 5, 6 tetrachloronitrobenzene and methyl ester of naphthaleneacetic acid, *Am. Potato J.,* 29, 176, 1952.
89. **Sawyer, R. L. and Dallyn, S.,** Vaporized chemicals control sprouting in stored potatoes, *Farm Res.,* 23(3), 6, 1957.
90. **Ellison, J. H. and Cunningham, H. S.,** Effect of sprout inhibitors on the incidence of fusarium dry rot and sprouting of potato tubers, *Am. Potato J.,* 30, 10, 1953.
91. **Thompson, N. R. and Isleib, D. R.,** Sprout inhibition of bulk stored potatoes, *Am. Potato J.,* 36, 32, 1959.
92. **Sukumaran, N. P., Grewal, S. S., and Virk, M. S.,** Inhibition of mustard seed germination as a bioassay for growth inhibitory substances in potato tubers, *J. Indian Potato Assoc.,* 5, 13, 1978.
93. **Sparrow, A. H. and Christensen, E.,** Improved storage quality of potato tubers following exposure to gamma irradiation of Cobalt 60, *Nucleonics,* 12(3), 16, 1954.
94. **Brownell, L. E., Burns, C. H., Gustafson, F. G., Isleib, D. R., and Hooker, W. J.,** Storage properties of gamma irradiated potatoes, *Proc. 17th Annu. Potato Utilization Conf.,* 1956, 2.
95. **Sparks, W. C. and Iritani, W. M.,** The effect of gamma rays from fission product wastes on storage losses of Russet Burbank potatoes, *Idaho Agric. Exp. Stn. Res. Bull.,* 60, 1964.

96. **Bellomonte, G., Gaudiano, A., Sanzini, E., Boniforti, L., Civalleri, S., Giammarioli, S., Gilardi, G., Leeli, L., Massa, A., and Mosca, M.,** Treatment of food with gamma radiation. I. Biochemical studies on potato, *Riv. Soc. Ital. Sci. Aliment.,* 7, 157, 1978.

97. **Bergers, W. W. A.,** Investigation of the contents of phenolic and alkaloidal compounds of gamma irradiated potatoes during storage, *Food Chem.,* 7, 47, 1981.

98. **Guo, A. X., Wang, G. Z., and Wang, Y.,** Biochemical effect of irradiation on potato, onion and garlic in storage. I. Changes of major nutrients during storage, *Yuang Tzu Neng Nung Yeh Ying Yang,* 1, 16, 1981.

99. **Janave, M. T. and Thomas, P.,** Influence of postharvest storage temperature and gamma irradiation on potato carotenoids, *Potato Res.,* 22, 365, 1979.

100. **Mazon-Matanzo, M. P. and Fernandez, G. J.,** Effects of gamma radiation on potato (*Solanum tuberosum* L.) tuber preservation during storage periods, *Junta Energ. Nucl.,* 354, 1976.

101. **Thomas, P., Srirangarajan, A. N., Joshi, M. R., and Janave, M. T.,** Storage deterioration in gamma-irradiated and unirradiated Indian potato cultivars under refrigeration and tropical temperatures, *Potato Res.,* 22, 261, 1979.

102. **Proctor, F. J., Goodliffe, J. P., and Coursey, D. G.,** Postharvest losses of vegetables and their control in the tropics, in *Vegetable Productivity,* Spalding, C. R. W., Ed., Macmillan, London, 1981, 139.

103. **Sparrow, A. H. and Christensen, E.,** Improved storage quality of potato tubers following exposure to gamma radiation of Cobalt 60, *Nucleonics,* 12, 16, 1954.

104. **Dallyn, S. and Sawyer, R. L.,** Physiological effects of ionizing radiation on onions and potatoes, Cornell University, Contract No. - DA 19-129-QM-755, Final Report, January, 1959.

105. **Mansfield, J.,** Postharvest losses of potatoes quantified during marketing process, in Report of the Seminar on the Reduction of Postharvest Food Losses in the Caribbean and Central America, August 8 to 11, Doc. VI-I, Instituto Interamericano de Ciencias Agricolas, Santo Domingo, Dominican Republic, 1977.

106. NRC, Report on the Steering Committee for Study on Postharvest Food Losses in Developing Countries, National Research Council, National Academy of Sciences, Washington, D.C., 1978.

107. FAO, Analysis of an FAO Survey on Postharvest Crops Losses in Developing Countries, Food and Agriculture Organization, Rome, 1977.

108. **Blackbeard, J.,** Do it yourself damage testing, *Arable Farming,* 8, 16, 1981; *Field Crop Abstr.,* 35, 3595, 1982.

109. **Ryall, A. L. and Lipton, W. J.,** *Handling, Transportation and Storage of Fruits and Vegetables, Vol. 1., Vegetables and Melons,* AVI Publishing, Westport, CT, 1972.

110. **Howard, F. D., Yamaguchi, M., and Knott, J. E.,** Carbon dioxide as a factor in the susceptibility of potatoes to black spot from bruising, *Proc. 16th Intl. Hortic. Congr.,* 582, 1962.

111. **Lorenz, O. A., Takatori, F. H., Timm, H., Oswald, J. W., Bowman, T., Fullmer, F. S., Snyder, M., and Hall, H.,** *Potato Fertilization and Black Spot Studies,* Vegetable Series No. 88, Santa Maria Valley, Department of Vegetable Crops, University of California, Davis, 1957.

112. **Wu, M. T. and Salunkhe, D. K.,** Control of chlorophyll and solanine formation in potato tubers by oil and diluted oil treatments, *HortScience,* 7, 466, 1972.

113. **Isenberg, F. M. and Gull, D. D.,** Potato greening under artificial light, *Cornell Univ. Exp. Bull.,* 1033, 1959.

114. **Hardenburg, R. E.,** Greening of potatoes during marketing—a review, *Am. Potato J.,* 41, 215, 1964.

115. **Eckert, J. W., Rubio, P. P., Mattoo, A. K., and Thompson, A. K.,** Diseases of tropical crops and their control, in *Postharvest Physiology, Handling and Utilization of Tropical and Subtropical Fruits and Vegetables,* Pantastico, Er. B., Ed., AVI Publishing, Westport, CT, 1975, 415.

116. **Thompson, A. K., Bhatti, M. B., and Rubio, P. P.,** Harvesting and handling: harvesting, in *Postharvest Physiology, Handling and Utilization of Tropical and Subtropical Fruits and Vegetables,* Pantastico, Er. B., Ed., AVI Publishing, Westport, CT, 1975, 236.

117. **Ware, G. W. and McCollum, J. P.,** *Raising Vegetables,* Interstate Publishers, Danville, IL, 1959.

118. **Thompson, H. C. and Kelly, W. C.,** *Vegetable Crops,* 15th ed., McGraw-Hill, New York, 1957.

119. **Smith, O.,** Studies of potato storage, *Cornell Agric. Exp. Stn. Bull.,* 553, 1933.

120. **Smith, O.,** Transport and storage of potatoes, in *Potato Processing,* Talburt, W. F. and Smith, O., Eds., Van Nostrand Reinhold, New York, 1987, 7.

121. **Pantastico, Er. B., Chattopadhyay, T. K., and Subramanyam, H.,** Storage and commercial storage operation, in *Postharvest Physiology, Handling and Utilization of Tropical and Subtropical Fruits and Vegetables,* Pantastico, Er. B., Ed., AVI Publishing, Westport, CT, 1975, 314.

122. **Lipton, W. J.,** Some effects of low oxygen atmospheres on potato tubers, *Am. Potato J.,* 44, 292, 1967.

123. **Nielsen, L. W.,** Accumulation of respiratory CO_2 around potato tubers in relation to bacterial soft rot, *Am. Potato J.,* 45, 174, 1968.

124. **Butchbaker, A. F., Nelson, D. C., and Shaw, R.,** Controlled atmosphere storage of potatoes, *Trans. ASAE,* 10, 534, 1967.

125. **Workman, M. N. and Twomey, J.,** The influence of oxygen concentration during storage on seed potato respiratory metabolism and on field performance, *Proc. Am. Soc. Hortic. Sci.,* 90, 268, 1967.

126. **Workman, M. N. and Twomey, J.,** The influence of storage atmosphere and temperature on the physiology and performance of Russet Burbank seed potatoes, *J. Am. Soc. Hortic. Sci.,* 94, 260, 1969.

127. **Yamaguchi, M., Flocker, W. J., Howard, F. D., and Timm, M.,** Changes in the CO_2 levels with moisture in fallow and cropped soil and susceptibility of potatoes to blackspot, *Proc. Am. Soc. Hortic. Sci.,* 85, 446, 1968.

128. **Van Niekerk, B. P.,** Potatoes can be stored for eight months, *Farming S. Afr.,* 36, 20, 1960.

129. **Salunkhe, D. K. and Wu, M. T.,** Developments in technology of storage and handling of fresh fruits and vegetables, in *Storage, Processing and Nutritional Quality of Fruits and Vegetables,* Salunkhe, D. K., Ed., CRC Press, Boca Raton, FL, 1974, 121.

130. **Arteca, R. N.,** Calcium infiltration inhibits greening in ''Katahdin'' potatoes, *HortScience,* 17(1), 79, 1982.

131. **Kumar, P., Baijal, B., and Alka, D.,** Effects of some growth regulators on ascorbic acid and total sugar content of potatoes associated with sprouting during storage, *Acta Bot. Indica (India),* 8, 235, 1980.

132. **Pederson, S.,** The effects of ionizing radiations on sprouts prevention and chemical composition of potatoes, *Food Technol.,* 10, 532, 1956.

133. **Panalaskas, T. and Pelletier, O.,** The effect of storage on ascorbic acid content of gamma radiated potatoes, *Food Res.,* 25, 33, 1960.

134. **Mikaelsen, K. and Roer, L.,** Improved storage ability of potatoes exposed to gamma radiation, *Acta Agric. Scand.,* 6, 145, 1956.

135. **Ogata, K., Tatsumi, Y., and Chachin, K.,** Studies on the browning of potato tubers by gamma radiation. I. Histological observation and the effects of the time of irradiation after harvest, low temperature and polyethylene packaging, *J. Food Sci. Technol.,* 17, 298, 1970.

136. **Tatsumi, Y., Chachin, K., and Ogata, K.,** Studies on the browning of potato tubers by gamma radiation. II. The relationship between the browning and the changes of o-diphenol ascorbic acid and activities of polyphenol oxidase and peroxidase in irradiated potato tubers, *J. Food Sci. Technol.,* 19, 508, 1972.

137. **Smith, O.,** *Potatoes: Production, Storing, and Processing,* AVI Publishing, Westport, CT, 1968.

138. **Workman, M., Patterson, M. E., Ellis, N. K., and Heilligman, F.,** The utilization of ionizing radiation to increase the storage life of white potatoes, *Food Technol.,* 14, 395, 1960.

139. **Salunkhe, D. K.,** Gamma radiation effects on fruits and vegetables, *Econ. Bot.,* 15, 28, 1961.

140. **Chachin, K. and Iwata, T.,** Respiratory Metabolism and Potassium Release of Irradiated Potatoes, paper read at Seminar on Food Irradiation for Developing Countries in Asia and Pacific, Tokyo, November 9 to 13, 1981, IAEA/Food and Agriculture Organization, Rome, 1981.

141. **Palmer, A. K., Cozens, D. D., Prentice, D. E., Richardson, J. C., and Christopher, D. H.,** Reproduction and longevity of rats fed an irradiated potato diet, *Intl. Project in the Field of Food Irradiation Final Tech. Rep.,* IFIP-R, 25, 1975.

142. **Palmer, A. K., Cozens, D. D., Prentice, D. E., Richardson, J. C., Cristopher, D. H., Gottschalk, H. M., and Elias, P. S.,** Reproduction and longevity of rats fed on irradiated potato diet, *Toxicol. Lett.,* 3, 163, 1979.

143. **Wills, R. H. H., Lee, T. H., Graham, D., McGlasson, W. B., and Hall, E. G.,** Postharvest, *An Introduction to Physiology and Handling of Fruits and Vegetables,* South Wales University Press, Kensington, Australia, 1981.

144. **Booth, R. H. and Proctor, F. J.,** Considerations relevant to the storage of ware potatoes in the tropics, *PANS (Pest Artic. News Summ.),* 18, 409, 1977.

145. **Anon.,** Research and development. Potato mid-week review, *Econ Times (India),* April 15, 1981.

146. **Rustowski, A.,** Potato storage and storage environment, in Survey Papers, European Assoc. for Potato Research, 7th Triennial Conf. Warsaw, June 26 to July 1, 1978, Bonin (Poland): *Field Crop Abstr.,* 35, 5433, 1979.

147. **Metlitskii, L. V.,** Biological aspects of protection of yield of potatoes, vegetables and fruits from losses during storage, *Izv. Akad. Nauk SSSR Ser. Biol.,* 1, 73, 1980; *Field Crop Abstr.,* 34, 1251, 1981.

148. **Von Gierke, K.,** Potato storage, in *Potato Research in Pakistan,* Shah, M. A., Ed., Pakistan Agriculture Research Council, Islamabad, 1978, 69.

149. CIP, Annual Report, 1981. International Potato Centre, Lima, Peru, 1981.

150. **Goresline, H. E.,** Historical aspects of the radiation preservation of food, in *Preservation of Food by Ionizing Radiation,* Vol. 1., Josephson, E. S., and Peterson, M. S., Eds., CRC Press, Boca Raton, FL, 1982, 8.

151. **Baker L. C., Lampitt, L. H., and Meredith, O. B.,** Solanine glycoside of the potato. III. An improved method of extraction and determination, *J. Sci. Food. Agric.,* 6, 197, 1955.

Chapter 5

PROCESSING

S. S. Kadam, B. N. Wankier, and R. N. Adsule

TABLE OF CONTENTS

I. INTRODUCTION

Potatoes used for various types of processing need to have certain quality characteristics. This has resulted in development of potato varieties grown solely for the purpose of processing, e.g., Dutch variety Record, having yellow flesh, which is used to manufacture crisps (chips) in the U.K. The biochemical composition of the potato tuber, especially the dry matter (DM) content, carbohydrate (reducing and nonreducing sugars), discoloration of raw flesh, textural characteristics of the tuber, browning, and after-cooking blackening, constitute the important quality parameters for processing of potatoes.[1]

Harris[2] stressed the importance of the quality aspects required by four main parts of the potato-processing industry, viz., crisps or chips, French fries, dehydrated potatoes, and canned potatoes. The effects of individual aspects of quality on the finished product have been described. Varieties of potatoes have a considerable influence on the cooked product. Important factors in determining acceptability of the cooked products are absence of defects such as excessive discolorations, disintegration, wetness, and off flavors.[3] According to French,[3] the DM content of potatoes is one of the most important components of cooking and processing qualities over a range of uses. Increased stability of DM content would probably substantially improve textural characteristics in practice and processing suitability. French[3] stated the following specific utilization requirements of potato processing:

1. Domestic use: Variable as between individuals, time of year, region, age group, etc. Medium, firm slightly mealy potatoes which do not disintegrate are most popular for multipurpose use.
2. Chips and French fries: Moderately high DM content with low reducing sugars and large, long, oval tubers are preferred.
3. Crisps (England): High DM and low reducing sugars are essential; moderate-sized oval tubers are preferred.
4. Dehydrated potatoes: High DM content preferred.
5. Canned potatoes: Moderately low DM and small tubers are the essential requirements.

During the past 3 decades, the proportion of potato crop which is processed for domestic consumption has increased considerably (Figures 1A and 1B). The potatoes processed were divided into chips (crisps), 42%; frozen French fries, 36%; dehydrated products, 4%; miscellaneous uses, 4%; and only 2% for canned new potatoes.

II. FACTORS INFLUENCING PROCESSING QUALITY OF POTATOES

A. TEXTURE

Texture of food depends on the interaction of a number of physical properties of the food material such as hardness, cohesiveness, viscosity, elasticity, and adhesiveness.[5] Various terms used to describe texture of cooked potatoes include mealy, floury, smooth, grainy, coarse, waxy, soapy, soggy, etc. Mealiness or flouriness has three main components: friability (disintegration, crumbliness), mouth feel, and dryness (from appearance). It is, however, difficult to measure the relative weightage of disintegration and palatability in the subjective assessment of mealiness. The European Association of Potato Research has therefore introduced the following categories in the newly defined system: consistency, mealiness, dryness, and structure.[5] Howard[1] also split textured cooked potato tubers into five characteristics: disintegration, consistency, mealiness, dryness, and structure. Texture of a potato is associated with its starch content and because starch is the major component of total solids, with high DM, which is usually measured as specific gravity. The methods based on weight loss after cooking, puncture, and tensile measurements and extrusive characteristics, have been used to measure the texture of cooked potatoes objectively.

Mealy potato cultivars are those that cook to give dry-appearing tissue that crumbles readily, are preferred by consumers for baking and mashing and, by industry, for chipping and frying. Waxy potato cultivars when cooked have a moist appearance and disintegrate less. They are appropriate for preparations such as potato salad and creamed potatoes.[6] Many studies have attempted to elucidate the causes of the different textural characteristics of various potato cultivars.[7-10] Cell wall distension or "rounding off" of cells caused by swelling of starch granules has been described as a possible cause of extensive cell separation in mealy potatoes. Other researchers have disagreed with the cell "rounding off" theory.[9-11] Cell wall and middle lamellar constituents have been studied as a cause of differences.[9]

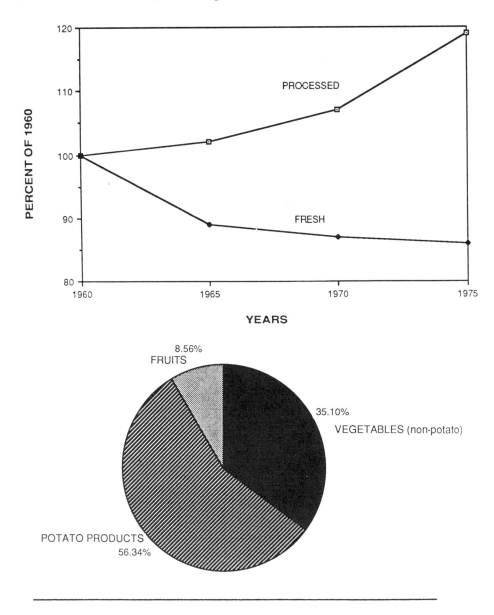

FIGURE 1.   (Top) Percent of potato crop consumed as fresh and processed products. (Bottom) Percent of potato products consumed as compared to fruits and vegetables.

Reeve[8] has pointed out that both pectic materials and starches are significant in the texture of potatoes.

Briant et al.[12] noted that a high percentage of small starch granules (diameter 20 μ) was associated with lower mealiness scores. Barrios et al.[13] found that the mealy Russet had higher specific gravity, larger tuber cells, higher starch and amylose, and a higher percentage of large starch granules (diameter 50 μm) than the waxy LaSoda cultivar. However, Unrau and Nytund[14] found no correlation between mealiness and granule size.

Cell separation is believed to increase the tendency of potatoes to break down on cooking (slough), which plays an important role in determining the suitability of potatoes for processing or for domestic use as salad, boiled, or baked potatoes.[15] The relative role of starch and pectic substances in cell separation is still controversial,[16,17] and it is not known why certain samples of potatoes disintegrate or slough more than others. Although specific gravity appears to be correlated with mealiness and breakdown,[18,19] particularly within a variety, there are reports quoting instances where such a relationship was not noticed.[20] It is postulated that cells separate and rupture on cooking due to the exertion of a swelling pressure, produced by gelling of starch granules,[5] but there are instances of reverse effects, i.e., increased firmness or retained weight being associated with higher amounts of starch or amylose.[18] Warren and Woodman,[21] however, reported that the association of increased firmness with higher levels of starch was a reflection of enhanced viscosity and not of cell adhesion.

The solubilization during cooking of pectic substances in the middle lamellae plays a role in determining cell adhesion of cooked potatoes.[18,19] According to Gray and Hughes,[5] however, this is difficult to demonstrate because (1) the cooked potato is very high in gelled starch, which may mask the effect of pectic substances and give difficulty in analysis; and (2) large variations occurring in the cell size of potatoes[19] may conceal the effect of pectic substances. Hughes et al.,[22] however, demonstrated that when starch and cell size were kept constant in the starting material (by using discs), compressive strength (altered by varying cooking time and adding ions) was closely related to the degradation of pectic substances as inflected by their release into the cooking medium.

Gray and Hughes[5] described a number of mechanisms of hydrolysis and breakdown of pectic substances during cooking, e.g., removal of calcium and magnesium bridges between galacturonic acid chains by chelation or ion exchange; cleavage, within a galacturonic acid polymer, of glycosidic bonds adjacent to methoxyl groups by transelimination; and breakage of the hydrogen bond. The relative role of these mechanisms in the breakdown of pectic substances in potatoes during cooking has not been fully established. Conditions such as methoxylation of pectic substances, pH, and ions have a profound effect on the extent and type of degradation the pectic substances undergo. Presoaking of potatoes at room temperature or the addition of calcium salts can reduce sloughing on cooking.[5]

Hughes and Faulks[23] demonstrated that the composition of polymers (the relative amounts of galacturonic acid and neutral sugars) released from the potato tissue on cooking is altered during the course of cooking, possibly by the presence of monovalent ions. Changes in polymer structure occurring on cooking in relation to the molecular structure of the pectic substances in the raw potato cell wall need to be understood before the complex relationship between pectic substances and intercellular adhesion can be evaluated.

It is not known whether or not starch can directly increase cell separation. The relationship found between specific gravity and increased disintegration as measured by sloughing or mealiness is probably indirect, and reflects differences in cell wall composition or cell size.[5] Starch may, however, directly influence the palatability.[23] The release of starch from cells plays a major role in the texture of reconstituted potato products. This depends upon the amount and quality of starch and also on the presence of ions which affect starch swelling, composition of cell walls, and the changes taking place in the cell walls on cooking.[24]

Increased cell size, which is influenced by variety, maturity, and environmental conditions, has been shown to be related to the reduction of cell adhesion on cooking.[17] Whether or not this is a direct effect of smaller cells giving a greater surface area per unit volume of tissue and greater strength, or whether or not it reflects differences in cell wall composition between small and large cells remains to be understood. Gray and Hughes[5] concluded that possibly all these factors play some part in influencing the texture of the cooked potato, many of which are determined genetically and by type and development of growth patterns of the tuber, most consistent differences in texture being produced by variety, maturity, and plant spacing.[19,24] Although maturity is closely associated with specific gravity and cell size,[25] differences in cell wall composition may play a role.[5] A greater understanding of intrinsic factors involved (cell size, cell wall composition) and of the effect of environment on the growth pattern of the tuber with which composition is related is necessary before precise control of texture by environmental manipulation is possible.[5]

B. SPECIFIC GRAVITY AND TUBER COMPOSITION

The specific gravity of the tuber most significantly influences the processing quality of potatoes, and is extensively used by processors to assess the suitability for the manufacture of French fries, chips, and dehydrated products. The yield of the processed product is greater per unit fresh weight from tubers with high total solids content.[5] It is also used as an indicator of mealiness,[22] breakdown of tissues on cooking,[4] and susceptibility to internal bruising.[26] Gray and Hughes,[5] however, stated that the relationship between DM and these attributes of processing quality do not often hold true, and some of the effects may be indirect.

Salunkhe et al.[27] investigated the effects of various environmental factors on the suitability of potatoes for chip making. In trials conducted in Michigan at two locations during 1950 to 1952, potatoes were grown under different fertilizer and irrigation treatments and four different planting dates. The specific gravity of potatoes was found to be closely related to chip color. Significant differences were noted among varieties, planting dates, location, and years and between fertilized and unfertilized crops. Early planted crops produced tubers with low content of reducing sugars, which made lighter-colored chips than those from late-planted crops. Also, early planted tubers had high specific gravity and best chip color, although certain exceptions were noted to this correlation. Best chip color was produced by no irrigation treatments, combined with 100 lb 3-12-12 (NPK) fertilizer. The high specific gravity was found where irrigation, but no fertilizer, was applied. Irrigation appeared to reduce specific gravity and darken chip color on comparable plots.[28] The preharvest foliar application of maleic hydrazide had a significant influence on the specific gravity and quality of potato chips.[29] Long storage tended to increase chip quality differences among varieties, but largely eliminated variations caused by different planting dates.

Starch being the largest (60 to 80% of the solids) component of potatoes, and starch and total solids being little affected by the environmental conditions, the DM content of potatoes is very often equated with its starch content. Variety influences relationship between starch and DM content as a result of differences in air spaces between cells in the tuber and by internal disorders such as hollow heart. Owing to its ease of measurement, DM is used widely by the processing industry to assess quality for a particular use.[5] Potatoes for the manufacture of chips (crisps in U.K.), frozen French fries, and dehydrated products should have a high DM content, which in turn is determined by variety and environment. Howard[1] stated that by including suitable control varieties, it was possible to select varieties for high DM content by measuring it as specific gravity. Potatoes for canning, however, need low DM content, and should be waxy rather than mealy.[1] The tubers of early maturing potatoes usually, but not invariably, have lower DM content than those of late main crop varieties.[30] Cultural and environmental factors modify the genetic control of tuber DM content,[31] but the interactions between these factors and variety are usually small.[32] The application of nitrogen and potassium can reduce the tuber DM content,[33] but the effects are small over the range of levels of fertilizer giving optimum yields.

Mazza and associates[34-36] reported changes in sucrose, reducing sugars, ascorbic acid, protein, and nonprotein nitrogen contents with chip and French fry color of several commercial potato cultivars during growth and long-term storage. Sucrose concentration of potato tubers was high during the early stages of growth, decreasing rapidly with later dates of sampling, and reaching a minimum at maturity. Growth, location and season, and storage site and temperature had little influence on the sucrose content of potatoes. However, the early harvested Norchip potatoes always had the least amount of sucrose. The ascorbic acid content increased with growth and maturity of the tuber, but steadily decreased by storage. True and crude proteins increased only slightly with maturity and storage. Russet Burbank potatoes had the lowest crude protein, but the highest DM content.

TABLE 1

Average Ratings on Distinctness of Lamellae of Starch Grains from Various Regions of Potato Tubers

Variety	Central		Pith			Cortex	
	Pith	Cortex	Apical	Basal	"Eye"	Apical	Basal
Russett Burbank	2.63	2.29	2.32	2.72	2.60	1.67	2.64

Note: 1 = least distinct; 3 = very distinct.

From Salunkhe, D. K. and Pollard, L. H., *Proc. Am. Soc. Hortic. Sci.,* 64, 331, 1954. With permission.

TABLE 2

Average Ratings on Distinctness of Lamellae of Starch Grains from the Central Cortex Region of Kennebec and Russet Rural Tubers of Specific Gravity Under 1.060 and of Over 1.090, from May 10 and June 16 Plantings

Variety	Planting date	Specific gravity groups	
		Under 1.060	Over 1.090
Kennebec	May 10	1.86	2.39
	June 16	1.53	2.01
Russet Rural	May 10	2.17	2.51
	June 16	1.81	2.24

Note: 1 = least distinct; 3 = very distinct.

From Salunkhe, D. K. and Pollard, L. H., *Proc. Am. Soc. Hort. Sci.,* 64, 331, 1954. With permission.

C. STARCH PROPERTIES AND CELLULAR STRUCTURE
1. Starch Properties

Reeve[24] published a review of the work done on various aspects of the combined characteristics of cellular structure and starch properties related to culinary qualities of potato varieties. According to Reeve, larger tissue cells and larger average starch granules are associated with mealiness. Smaller cells and starch granules characterize the less mealy and "waxy" varieties. Within a variety, proportionately larger numbers of large starch granules are associated with tubers of high specific gravity, and more smaller granules with low specific gravity. The percent of small starch granules is reduced markedly during storage of tubers. The microscopic examination of starch grains from various anatomical parts of a potato tuber by Salunkhe and Pollard[37] indicated variations in distinctness of the lamellae. The lamellae of starch grains of Russet Rural tubers were more distinct than those of Kennebec. The lamellae and hylum of starch grains of tubers having high specific gravity were more distinct than those having low specific gravity. The lamellae and hylum of starch grains of tubers from early planting were more distinct than those from late planting (Tables 1 and 2).

Differences in the size of starch granules are associated with differences in amylose and amylopectin content. Small granules contain less amylose and gel at higher temperatures than the larger starch granules. The amylose content of potatoes is strongly influenced by the variety. The varietal differences in the amylose content reflect fundamental differences

TABLE 3
Size Distribution of Starch
Granules in 'Russet Burbank'
Potato

Length (μm)	Percent of total
Less than 10	11.2
10—19	27.2
20—29[a]	22.4
30—39	14.4
40—49	8.4
50—59	6.9
60—69	4.2
70—79	3.0
More than 79	2.3

[a]   Average arithmetic length = 28 μm.

From Reeve, R. M., *Econ. Bot.*, 21, 294,
1967. With permission.

TABLE 4
Starch Granule Sizes in Different Tuber Zones of 'White Rose' and 'Russet
Burbank' Potatoes

Tissue zones	'White Rose'		'Russet Burbank'	
	Average length (μm)	Percentage of average (μm)	Average length (μm)	Percentage of average (μm)
Cortex	15	34	24	29
Vascular "ring"	7.4	33	7.3	33
Inner parenchyma	22	30	32	40
"Water core"	17	33	27	45

Note: Tubers about $1^3/_4 \times 2^1/_4 \times 4^1/_2$ in.

From Reeve, R. M., *Econ. Bot.*, 21, 294, 1967. With permission.

in the properties of starch gels formed when different varieties of potatoes are cooked. The starches within the different tissue zones of individual tubers also vary significantly. Similarly, cell size varies characteristically within different tuber regions.

The variations in starch in respect to size distribution of granules in individual variety (Russet Burbank) and in different tuber zones of two commercial potato varieties of Russet Burbank are shown in Tables 3 and 4. Over 50% of the starch granules of Russet Burbank potatoes were less than the average size of 28 μm. Larger starch granules occur most abundantly in the large storage parenchyma cells associated with the internal phloem, and in the "water core" or pith parenchyma, but in much less abundance in the storage parenchyma.[24] They are, however, very numerous in the somewhat smaller cortical parenchyma cells between the vascular bundle ring and in the skin. This area contains the most starch per unit volume. Size distribution of starch in these four areas, reported by Reeve,[38] is presented in Table 4 for White Rose and Russet Burbank tubers of similar size.

Storage temperature markedly influences the size distribution of starch granules of potato (Table 5). When potatoes were stored at 21.1°C or above, the average length of starch granules increased as the numbers of smaller starch granules reduced concomitantly. This shift in distribution was attributed to more rapid digestion of smaller granules and compositional differences between large and small granules affecting ease of digestion.

TABLE 5

Influence of Temperature upon Starch Granule Size Distribution in Potatoes Stored $1^1/_2$ and $2^1/_2$ Weeks

Variety	Starch granules	Storage temperature (°C)			
		4.4 ($2^1/_2$ week)	10.0 ($2^1/_2$ week)	21.1 ($2^1/_2$ week)	26.6 ($1^1/_2$ week)
'Kennebec'	Average length (μm)	31	33	32	36
	Percent <15 μm	20.5	18.4	15.5	12.5
'Red Pontiac'	Average length (μm)	26	30	33	—
	Percent <15 μm	17.0	20.2	13.9	—
'Russet'	Average length (μm)	28	32	32	34
	Percent <15 μm	25.5	25.5	17.1	15.0

From Reeve, R. M., *Econ. Bot.*, 21, 294, 1967. With permission.

TABLE 6

Cell Size Difference Between Tissues within Individual Tubers of 'Russet Burbank' and 'White Rose'

Tuber zone	'Russet Burbank'		'White Rose'	
	Range (μ^3)[b]	Av. (μ^3)[c]	Range (μ^3)[b]	Av. (μ^3)[c]
Cortex	2.3 to 2.6 × 10⁶	2.4 × 10⁶	1.6 to 1.8 × 10⁶	1.7 × 10⁶
Inner parenchyma	3.8 to 4.6 × 10⁶	4.1 × 10⁶	2.3 to 2.7 × 10⁶	2.4 × 10⁶
"Water core"	3.9 to 5.3 × 10⁶	4.4 × 10⁶	2.1 to 2.4 × 10⁶	2.3 × 10⁶

[a] Tubers about $1^3/_4 \times 2^1/_4 \times 4^1/_2$ in.
[b] Range of individual reticule count.
[c] 300 to 500 total cells counted per sample.

From Reeve, R. M., *Econ. Bot.*, 21, 294, 1967. With permission.

A direct relationship between specific gravity and starch size has been noted in a number of potato varieties.[39-41] A greater percentage of large starch granules was found to be associated with high specific gravity and mealiness of potato tubers. Higher amylose content of starch was found to have a highly significant correlation with mealiness.[42,43]

2. Cellular Structure

In addition to variations in the size and composition of starch granules, the cells within which they are contained exhibit marked differences in size among different potato varieties as well as within different zones.[24] Most of the tissue volume of any individual tuber is composed of starch-containing cells of rather large size in the internal phloem area. Reeve[38] reported the data obtained on cell size differences between tissues within individual tubers of Russet Burbank and White Rose (Table 6), showing proportional differences in cell size between the two varieties. Barrios et al.[41] similarly noted a close association of large cells and starch granules with mealy tubers.

Dehydration of cooked, mashed potatoes in the manufacture of either granule or flake forms results in an appreciable cell shrinkage, the cells assuming a resin-like appearance.[24] The full size of the cell is restored upon reconstitution. A typical cell size distribution for the reconstituted product of a granule manufactured from Russet Burbank potatoes is shown in Table 7. About one third of the cells ranged from 180 to 220 μm in diameter, or with a mean diameter of 200 μm.

TABLE 7
Cell Size Distribution of a
Dehydrated 'Russet
Burbank' Potato Granules
after Reconstitution

Diameter (μm)	Percent
Less than 101	3.9
101—140	14.0
141—180	30.0
181—220	30.0
221—260	15.5
261—300	4.0
More than 300	2.6

From Reeve, R. M., *Econ. Bot.*, 21,
294, 1967. With permission.

3. Structure and Composition of Cell Wall

Swelling of cooked potatoes without rupturing their walls depends to some extent on cell wall properties. In raw potatoes, the cell walls of parenchyma consist principally of interwoven cellulosic microfibrils, between which occur amorphous pectic substances and the hemicelluloses.[44] Reeve[24] regarded this as a chemically laminated structure. Dissolution of the amorphous, noncellulosic components by cooking weakens the wall, reducing its tensile strength. There are natural thin areas in the walls of parenchyma cells, the primary pits in which microfibrils are fewer in number[44] and in which the cell rupturing of cooked potatoes has been observed to begin.[45]

Textural qualities of cooked potatoes also appear to be related to cell wall thickness. Thinner walls of Canso and Netted Gem were probably responsible for a more mealy character of these potatoes as compared to that of Irish Cobbler and Warba.[46] Reeve[24] opined that the observed difference in wall thickness represented some difference in hydration due to compositional differences in noncellulosic constituents.

Changes in the pectic substances of the cell wall have been related to textural properties of potatoes. Dastur and Agnihotri[47] reported that pectic fractions of potatoes increased from the beginning of tuber growth to harvest. On storage, free soluble pectins continued to increase and other fractions decreased. Pronounced changes in pectic substances were noted when hard-cooking potatoes became soft cookers after storage.[48] Water-soluble pectins increased while both oxalate-soluble and acid-soluble pectins decreased during storage, accompanied by a marked decrease in hemicelluloses. Sharma et al.[48] also noticed that the soft-cooking tubers contained more water-soluble pectin and less insoluble pectin than in hard-cooking tubers. These differences in pectins and hemicellulose contents of potatoes relate to sloughing upon cooking. In canned potatoes and other potato products that require piece integrity, sloughing may be controlled to some extent by calcium treatment,[49] which reduces cell separation by rendering pectic substances less soluble and by decreasing swelling of gelled starch to increase its firmness.[24]

D. PIGMENTATION

The processing quality factors associated with internal structure, composition, and culinary properties also include pigmentation in the form of enzymic and nonenzymic browning and after-cooking blackening.

1. Enzymatic Browning

Enzymatic browning occurs when fresh potatoes are cut or peeled, due to the oxidation of phenolic compounds by enzyme phenolase.[50] Out of two major phenolic substrates found in potatoes, namely, chlorogenic acid and the amino acid tyrosine, the former is readily oxidized in the presence of the enzyme, but its products are not deeply colored. The products of oxidation of tyrosine, however, are dark brown in color, with subsequent formation of melanin pigments (see Figure 12 in Chapter 2). The formation of melanin from tyrosine takes place in the following sequence:

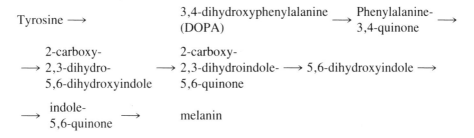

The first two reactions are catalyzed by potato phenolase; the others are or can be nonenzymatic, while chlorogenic acid can also activate the enzyme, but its concentration is not limiting.[51] Tyrosine is therefore generally considered to be the most important factor in controlling the extent of pigmentation caused by enzymatic browning.[5]

Potato varieties vary distinctly in their properties related to pigmentation during processing, and as in the case of after-cooking blackening, high- and low-browning potato varieties are available,[51] the extent of pigmentation being related to the amount of tyrosine they contain. The amount of browning produced is also influenced by season, cultural practices, growing and storage conditions employed, and varying the extent of pigmentation produced by a given amount of tyrosine. Clark et al.[52] reported that polyphenol chlorogenic acid, a caffeic tannin, constituted 0.095 to 0.150% of the dry weight of the potato tuber and was concentrated in a thin layer in the periderm tissue next to the skin. Mapson et al.[51] investigated the relative contribution of soil and climatic factors in controlling browning in potatoes. Soil was found to have a smaller effect than climate on enzymatic browning and tyrosine content. In controlled water regime experiments, the amount of browning and tyrosine were found to increase with an increase in the amount of water given to the plants. Mapson et al.[51] further suggested that rainfall itself, and not reduction in sunlight and temperature, can affect enzymatic browning.

Enzymatic browning of potatoes is significantly influenced by the preharvest mineral nutrition of the crop. The levels of enzymatic browning, tyrosine, and phenolase activity were found to decrease when higher levels of potassium were applied to the plants.[52-55] Although nitrogen fertilizers were shown to increase the total soluble amino acid pool considerably (sometimes twofold),[56] the level of tyrosine was only slightly reduced.[56,57] Nitrogen has therefore rarely been found to increase the level of enzymatic browning except under the conditions of very low potassium.[5] The direct relationship between tyrosine and browning has been found to vary according to variety and environmental conditions.[51] Since the enzymatic browning does not occur unless the cells are cut or bruised, the ease with which the cells under the cut layer may be damaged may contribute significantly to browning, but some other factors such as ascorbic acid may be involved.

Grewal and Sukumaran[58] reported that maleic hydrazide (MH)-treated tubers of two potato cultivars grown at Jullunder, Punjab, contained higher tyrosine contents and showed more enzymatic browning than untreated tubers. Storage of these tubers at Simla for 90 d increased enzymatic browning and total phenol and tyrosine contents; the increase was smaller in treated tubers than in untreated ones. There were no differences in true protein, reducing sugars, or starch contents of the treated and untreated tubers. The latter had higher DM contents. The lipid content decreased in both types with storage.

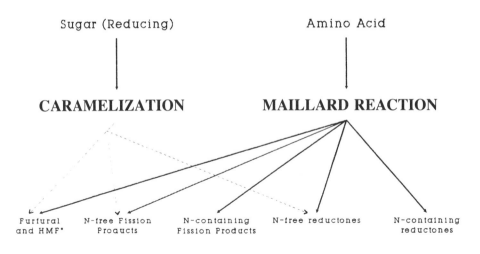

FIGURE 2.   Possible interactions between caramelization and Maillard action in potato chips.

2. Nonenzymatic Browning

There are two types of nonenzymatic browning reactions causing pigmentation in processed potato products:[50,59] (1) caramelization and (2) Maillard reaction (based on carbonyl-amino reactions) (Figure 2). According to Gray and Hughes[5] browning of potatoes due to caramelization is not likely to be of great importance because of the greater energy requirements of this reaction. At temperatures usually used for frying (165 to 170°C), browning of sugars takes place rapidly only in the presence of amino acids.[60]

The first-formed products of the Maillard reaction are the unsaturated carbonyl compounds which are converted to complex melanoid through a series of reactions involving conjugation followed by polymerization. The magnitude of pigmentation (browning) produced depends not only upon the amount of sugars and amino acids present, but also upon the type of chemical reactions leading to the development of chromophore, which are influenced by both pH and temperature.[5] According to Gray and Hughes,[5] in model systems the rate of browning depends upon alpha-amino N, and not on sugars, but in practice alpha-amino N is rarely limiting and so it does not usually determine the rate of reaction or the color developed. The extent of pigmentation is thus largely controlled by the amount of reducing sugars present in potatoes,[61,62] except at high temperatures when sucrose hydrolyzes to sugars[51] when the disaccharide content may control the browning. However, the correlations between browning and reducing sugar content have been found to vary considerably (from 0.32 to 0.99).[49,50]

Ascorbic acid content of potatoes is also thought to play a role in browning. The discs of potatoes soaked in ascorbic acid turned brown when fried. The concentration of ascorbic acid in the potato tuber is, however, never high enough to cause an unacceptable level of browning.[49]

The acceptable upper limits of reducing sugar, so as to obtain processed potato products that are fairly free from browning, is usually about 0.10% on a fresh weight basis, but levels up to 0.25% or higher may sometimes be accepted depending upon the method of cooking.[63] Burton and Wilson[63] reported that the contents of reducing sugars and sucrose were at high levels at tuber initiation, which then fell, more rapidly for sucrose, as the potato tubers grew. These rose again towards harvest, the reducing sugars increasing more than sucrose, possibly due to promotion of invertase activity.[64] The reducing sugar content of potatoes at harvest is influenced by environment, especially the temperature, but significant varietal differences in the content of reducing sugars and their levels at harvest have been reported.[65]

It has been demonstrated that potato varieties with consistently good processing qualities depleted their reducing sugar content more rapidly during growth of the crop than those of poor processing performance.[32] Good processing performance of potato varieties after a period of storage was found to be related to low sucrose content at harvest. The relation of levels of sucrose and reducing sugars at harvest and after a period of storage has not been investigated completely.[5] Although invertase activity is important in causing increased reducing sugar levels at harvest,[32] it does not appear to be so important during storage.[66] Schwimmer et al.[67] noticed no invertase activity in freshly harvested White Rose potato tubers, but after long cold storage, a significant invertase activity occurred.

The levels of reducing sugars of potatoes, in practice, can be controlled by manipulating storage temperatures. Tubers with high sugar levels can be reconditioned by increasing the temperature, but all varieties do not respond in a similar manner, and there is often considerable variability from tuber to tuber.[5]

3. After-Cooking Blackening

After-cooking blackening appears, as its name implies, only after the cooking of the tuber, especially when the tubers of susceptible varieties are boiled or steamed. The defect also shows up in frozen French fries and in canned and dehydrated potato products.[5]

The pigment responsible for discoloration has been shown to be a complex of chlorogenic acid and iron, formed on cooking and oxidized on cooling to produce a colored complex of ferri-dichlorogenic acid.[68,69] Gray and Hughes[5] stated that formation of this pigment could be affected by pH and by the presence of naturally occurring chelating agents such as citric and malic acids and phosphates in the tubers. The distribution of blackening within the tubers depends upon the ratio of chlorogenic acid to citric acid.

The after-cooking blackening has been reported to be influenced by a number of factors such as variety, soils, fertilizers, and season. Owing to the interaction of these factors, often conflicting results have been obtained by different investigators. A high ratio of N to K and cool, wet seasons were found to make the tubers more susceptible to after-cooking blackening.

The differences between varieties are largely due to varying amounts of chlorogenic acid and to a lesser extent to citric acid.[5] The levels of both these acids are in turn influenced by soil and season. Unlike the varietal effect, the edaphic factors mainly affected the level of citric acid.[70] Hughes and Evans[70] demonstrated that potatoes grown on the fen soil (high organic matter with low K) blackened more than tubers grown in the clay soil (low organic matter with high K), and were low in citric acid.

The investigations of the relative role of chlorogenic and citric acids, and the extent to which they are influenced by variety and environmental conditions, have thrown some light on the effects of K and N on blackening. Low-chlorogenic acid varieties blacken little, even with very low levels of K, since the acid is the limiting factor. High ratios of N to K may be avoided in fertilizer application so as to control the after-cooking blackening in potatoes. Low-chlorogenic varieties may be chosen at sites known to produce blackening, where ratios of N/K may be high owing to soil structure and/or a cool and wet climate. The possible role of iron in causing after-cooking blackening in potatoes has not been investigated.

E. ASSESSMENT OF QUALITY PARAMETERS

Various methods have been employed to control the quality of processed products.[71] Glucose and reducing sugar contents of potatoes serve as the indicators of the resultant color of chips. Smith[72,73] adapted the "Tes-Tape" method to determine the amount of glucose in raw potatoes. The prepared tape now distributed to potato processors under the name "Chip color tester" is impregnated with enzyme, glucose oxidase, and orthotolidine as an indicator. When glucose of potato juice comes into contact with the tape, it is oxidized to gluconic

FIGURE 3A.   Measurement of chip color.

acid and the orthotolidine changes from yellow to various shades of green, depending upon the concentration of glucose in the juice. A high degree of correlations between glucose as determined by this method and the resultant chip color[72] indicates the suitability of this method. The Clinitest® method is a self-heating, copper-reduction test to quantify sugars in potatoes.[74] The specific gravity of potatoes can be determined rapidly using a potato hydrometer which also gives the DM content of the sample.[75,76] Similarly, potato chip color can be measured objectively by using the Agtron Model F reflectance colorimeter[75] or the Hunter Color and Color Difference Meter (Figure 3A), which measure the color rapidly and highly accurately.[72-77] Isleib[78] described an objective method of measuring chip color with a Model 610 Photovolt photoelectric reflectance meter, and compared the reading with subjective ratings on the National Potato Chip Institute color chart. Brady[79] reported that of the various methods tested, a vacuum oven at 93.3°C for 2 h or a conventional oven at 104.4°C for 6 h was best for determining the moisture content of potato chips. A refractometer is used to determine soluble solid content (organic acids and sugars) of potatoes and potato products, and acidity (pH meter) is determined to study the changes in total acids during storage and after processing (Figure 3B). There are various instruments which are being used to determine the quality of processed potatoes.

III. COMMERCIAL PROCESSING

A. PREPEELED POTATOES

Prepeeled potatoes refer to those potatoes that are preserved from discoloration and are cold stored.[80] In developing countries, the production of prepeeled potatoes has increased during the last decade. These are extremely perishable and have a relatively short shelf life.[70]

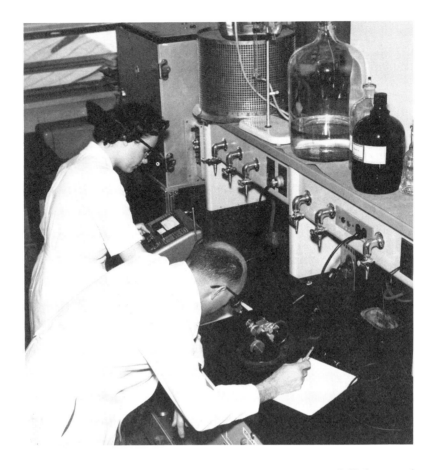

FIGURE 3B.   Determination of soluble solid content and measurement of pH of potato and potato products.

These are supplied to restaurants, canteens, and retail establishment. In this case, potatoes may be whole or cut into strips for fresh frying. The preprocessing operations in potato processing are outlined in Figure 4. Potatoes are washed to remove dirt and washed potatoes are peeled by various methods. These include abrasion peeling, lye peeling, steam peeling, or a combination of lye and steam peeling and infrared peeling.[81-84] The peeled potatoes are washed to remove any peel left on them and are trimmed to remove eyes, residual peel, and damaged, diseased, or green areas. The prepeeled potatoes are immersed in sulfite solution for a few minutes to prevent enzymatic browning reactions. These are then drained, packaged, and refrigerated. The losses in nutrients depend upon the method of peeling.[85-92] The losses also vary according to the size and shape of the tuber and the depth of the eyes and the length of storage. The weight loss varies from 5 to 24%, with an average weight loss of 8 to 12%. The damage in the form of penetrating cracks followed by superficial crushing and bruising results in significant loss of weight. Augustin et al.[180] found that peel fractions of the whole tuber when the peel was removed with a domestic peeling knife either before or after cooking were 6, 2, and 10% in cases of peeled raw potatoes, those peeled after boiling, and those peeled after baking, respectively. Peeling results in loss of ascorbic acid.[90] However, the degree of loss of ascorbic acid varies according to the method of peeling. Mondy and Barry[92] studied the effect of peeling on total phenols, total glycoalkaloids, discoloration, and flavor of cooked potatoes. During cooking, phenols migrated from the peel into both the cortex and the internal tissues of the potato (Tables 8A and 8B).

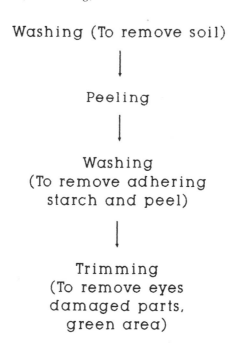

Washing (To remove soil)

↓

Peeling

↓

Washing
(To remove adhering
starch and peel)

↓

Trimming
(To remove eyes
damaged parts,
green area)

FIGURE 4.   Preprocessing operations.

TABLE 8A
Phenol Content of Potatoes Cooked With and Without Peel

Tissue/Treatment (S.D.)	Phenols mg/100 g wet wt			
	Katahdin		**Lemhi**	
	+	**−**	**+**	**−**
Cortex				
Uncooked	80.51 ± 0.01	64.19 ± 0.92	102.97 ± 0.40	67.91 ± 0.40
Water	92.89 ± 0.26	63.66 ± 1.93	103.76 ± 1.75	62.61 ± 1.92
Salt	91.78 ± 0.02	64.26 ± 0.93	108.71 ± 1.07	63.23 ± 0.40
Steam	96.92 ± 1.00	67.03 ± 2.64	112.97 ± 1.24	64.10 ± 1.65
Internal tissues				
Uncooked	54.84 ± 0.30	54.84 ± 0.30	55.75 ± 0.40	55.75 ± 0.40
Water	65.75 ± 1.04	63.36 ± 0.06	63.59 ± 1.65	61.24 ± 1.20
Salt	67.93 ± 1.99	63.01 ± 1.23	68.21 ± 0.78	62.91 ± 0.35
Steam	67.66 ± 1.24	65.36 ± 2.58	65.54 ± 1.62	58.33 ± 1.35

Note: +, Cooked with peel; −, cooked without peel. The least significant difference (LSD) in
phenols = 3.30 at $p < 0.01$; 2.28 at $p < 0.05$. S.D. = Standard deviation (based on
duplicate determinations).

From Mondy, N. I. and Barry, G., *J. Food Sci.,* 53, 756, 1988. With permission.

Glycoalkaloids were less mobile than phenols and migrated only into the cortex. The movement of phenols and glycoalkaloids into the cortex increased both discoloration and bitterness
in potatoes cooked with the peel. Information on the loss of other nutrients due to peeling
is very scanty. To reduce waste, potatoes need to be peeled after and not before boiling
when cooked domestically unless damage is extensive.

TABLE 8B
Total Glycoalkaloid (TGA) Content of Potatoes Cooked With and
Without Peel

Tissue/Treatment (S.D.)	TGA mg/100 g wet wt			
	Katahdin		Lemhi	
	+	−	+	−
Cortex				
Uncooked	10.76 ± 0.28	2.64 ± 0.04	7.84 ± 0.32	2.06 ± 0.05
Water	12.35 ± 0.25	3.44 ± 0.03	8.80 ± 0.12	2.08 ± 0.04
Salt	13.06 ± 0.06	3.43 ± 0.08	8.87 ± 0.39	2.10 ± 0.03
Steam	13.08 ± 0.04	3.46 ± 0.04	9.27 ± 0.15	2.11 ± 0.08
Internal tissues				
Uncooked	0.694 ± 0.011	0.694 ± 0.011	0.619 ± 0.021	0.619 ± 0.021
Water	0.759 ± 0.018	0.720 ± 0.008	0.643 ± 0.007	0.636 ± 0.008
Salt	0.765 ± 0.007	0.716 ± 0.011	0.635 ± 0.008	0.625 ± 0.006
Steam	0.800 ± 0.010	0.726 ± 0.008	0.676 ± 0.007	0.661 ± 0.021

Note: +, Cooked with peel; −, cooked without peel. LSD in TGA = 0.11 at $p < 0.01$; 0.08 at $p < 0.05$. S.D. = Standard deviation (based on duplicate determinations).

From Mondy, N. I. and Barry, G., *J. Food Sci.*, 53, 756, 1988. With permission.

B. POTATO CHIPS

Several factors influence the yield and quality of chips prepared from potatoes.[92-97] The specific gravity of potatoes is positively linked to the yield of chips. The factors influencing the specific gravity of potatoes have been already discussed in Section II of this chapter. The major problem associated with the chip industry is the maintenance of the desired color of the products.[98,99] Chip color is the result of the browning reaction between sugars and other constituents such as amino acid or ascorbic acid. Potatoes stored at 70°F for 3 weeks yielded darker chips than those held at room temperature. This is due to the conversion of starch to sugars by phosphorylase enzymes. The process for making potato chips is outlined in Figure 5. Several treatments to slices such as a hot solution of sodium bisulfite or a combination of sodium bisulfite, citric acid, and phosphoric acid, dilute HCl, hot water, HCl and phosphoric acid solution,[100-107] and glucose oxidase[103-105] have been recommended to improve the quality of chips. Potatoes with high reducing sugars produce darker chips when fried in oil. Such potato slices are partially fried and then subjected to infrared heat or microwave heat to remove excess moisture. Temperatures above 250°F induce browning and the degree of darkening increases with time. Hence the time and temperature of frying are critical factors influencing the color of the finished product.[106,107] The color of potato chips can be improved by using microwave processing and vacuum frying developed in the Netherlands.

Several oils can be used for frying potato slices. These include soybean oil, palm oil, and safflower or groundnut oil. The amount of oil absorbed by the chips is influenced by the amylose/amylopectin ratio in the starch. The amount and quality of oil present in the chip influence the shelf-life of chips. Other factors such as exposure to air, light, and high-temperature, contamination of oils with metals, and poor packages also influence the keeping quality of chips.[108-112] Antioxidants can be added to oil or to the salt which is applied to chips to delay rancidity and extend shelf life. A number of antioxidants such as BHA and BHT have been found suitable in the chip industry. The texture of chips can be improved by adding sodium acid pyrophosphate to the frying oil, which increases the crispness of

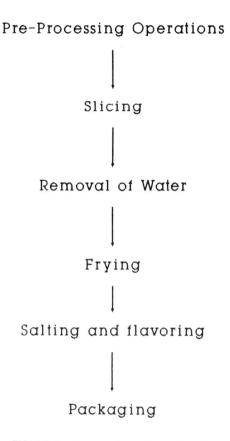

FIGURE 5.   Processes for making potato chips.

chips. However, it produces an acidic flavor. Details of the various steps involved in chip making from potatoes are outlined by Smith.[74] Several modifications are suggested to improve the quality of chips under commercial scale. Several reports[113-120] are available on changes in nutrient composition when chips are prepared from raw potatoes. Chip-making results in the loss of amino acid[114-116] and ascorbic acid.[117,118] The nutritional value of chips has been reviewed.[119-120]

C. FRENCH FRIES

French fries is one of the important products of potato-processing industries. A large percentage of frozen French fries are served in restaurants and institutions.[121,122] These are prepared for serving by finish frying in deep fat. Like potato chips, the reducing sugar content should be low to avoid dark fried pieces. After cooking, darkening is one of the problems in French fries. Sodium acid pyrophosphate has been reported to prevent darkening in French fries. Davis and Smith[75] reported that a low concentration of ammonia can cause frozen potato products to turn from a light color to gray or black. Products that are discolored from ammonia may be decolorized by a citric acid dip while they are frozen. The addition of sucrose prevents the acid taste in treated potato products.

French fries are prepared from good quality potatoes. The process of preparation of French fries includes washing and peeling, trimming, sorting and cutting, blanching, frying, defatting, cooling, freezing, and packaging (Figure 6). Peeling can be done with lye or steam or infrared treatment.[123-128] Blanching prior to frying results in a more uniform color of the fried products, reduced fat absorption by gelatinizing the surface layer of starch,

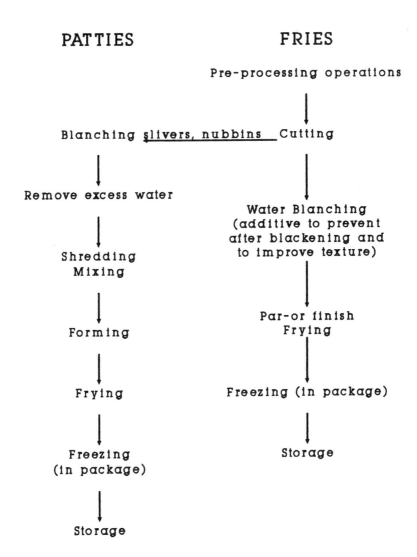

FIGURE 6.   Processes for making frozen patties and French fries.

reduced frying time, and improved texture of the fried product.[121] Blanching results in leaching of sugars and lighter and more likely uniform-color French fry. Sodium acid pyrophosphate or calcium lactate can be used during blanching to improve texture. The blanched product is fried in oil at 350 to 370°F. Excess fat is removed from the fried product and the product is air cooled. Freezing of the product can be performed before or after packaging.

D. OTHER FROZEN PRODUCTS

Products other than French fries prepared from potatoes include patties, puffs, hashed-brown potatoes, scalloped, mashed, and rounds. These are made from the slivers and nubbins from the French fry line, and from chopped or sliced small potatoes.[80]

1. Potato Patties

This product was developed in the early 1950s. Slivers and nubbins from the French

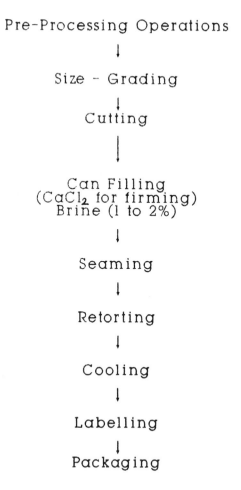

Pre-Processing Operations

↓

Size – Grading

↓

Cutting

↓

Can Filling
(CaCl$_2$ for firming)
Brine (1 to 2%)

↓

Seaming

↓

Retorting

↓

Cooling

↓

Labelling

↓

Packaging

FIGURE 7.   Processes involved in the canning of potatoes.

fry line and small potatoes are steam-blanched, cooled, and shredded or chopped. Small, whole, partially cooked potatoes may also be sliced or shredded. These are mixed with potato or rice flour, salt, and monosodium glutamate or other seasoning agents. Round or rectangular patties of 3 oz each are formed with the help of a patty-forming machine. They may be frozen before or after packaging (Figure 7).

2. Mashed or Whipped Potatoes
Pieces of potato from the French fry operation are diced and steam cooked. Rolls mash the material and it is then mixed with skimmed-milk solids and salt. After passing through a finisher, it is beaten vigorously and placed in film-lined packages and frozen in an airblast tunnel.[80]

3. Potato Puffs
Slivers and small pieces of potatoes are cooked with steam, mashed, and mixed with flour, eggs, shortenings, and seasoning.[80]

4. Potato Rounds
A relatively new product, termed "potato rounds", is made from the slivers and short

pieces from the French fry line. The pieces are blanched, shredded, and mixed with potato flour, salt, and spices. The mixture is extruded, fried in deep fat, frozen, and packaged.[128]

5. Diced Potatoes

This product is used for frying as hashed-brown potatoes or in potato salads. Slivers, nubbins, and small potatoes are diced to 3/8 × 3/8 × 3/16 in and blanched in steam at 100°C for 3 min. The product is cooled and onion flavoring is added if desired. It may be loose-frozen (IQF) on a belt before packaging or packaged and then frozen in a contact or blast freezer.[128]

6. Hashed-Brown Potatoes

Small, whole potatoes are cooked, cooled, shredded, and packed loose in cartons before freezing. Slivers or nubbins may be used by shredding or dicing followed by blanching or cooking.[116]

7. Rissole Potatoes

Small, whole potatoes are blanched, fried in deep fat, and frozen.[128]

8. Au-Gratin Potatoes

Cooked, diced potatoes are mixed with a sauce of milk, cheddar cheese, salt, monosodium glutamate, and pepper in the proportion of 2:1 (potato:sauce). Shortening, rice flour, and sugar may also be added. A topping consisting of cheddar cheese, toasted bread crumbs, and margarine may be used on the product before freezing.[128]

9. Potato Cakes

Beaten eggs and salt are mixed with cold mashed potatoes or grated raw potatoes. Small, blanched potato pieces from the French fry line with chopped parsley, celery, or celery seed and grated onion may be cooked in a continuous blancher and riced in a pulper. The mix is formed into cakes which are dipped either in fine bread crumbs, cracker crumbs, or flour and then fried to a brown color on both sides.[80] The cakes can be packed before freezing or may be served as part of a frozen, precooked dinner.

10. Dehydrofrozen Products

Potatoes can be processed into dehydrofrozen or frozen, "concentrated" mashed potato. Potatoes are boiled, cut into $1/2$-inch slabs, and cooked, and then mashed with nonfat skimmed milk solids added. The mash is extruded into a thin layer on a continuous belt drier and dehydrated to about 15% moisture. It is milled slightly, packaged, and frozen.[128]

11. Precooked Frozen Dinners and Dishes

A great variety of frozen potato dinners and dishes are available, including French fried, whipped, mashed, hashed-brown, au-gratin, roasted, baked, boiled, scalloped, cottage-fried, and stuffed-baked potatoes, potato cakes, and rissole potatoes.[129]

E. CANNED PRODUCTS

Potatoes are canned in different forms including whole, diced, sliced, strips, and julienne. A good canning potato should not slough or disintegrate during processing. Immature potatoes of less than $1^1/_2$-inch diameter and of low specific gravity (less than 1.075) are preferred for canning. Sloughing of the canned product can be prevented by adding calcium chloride, which tends to firm the tissue.[69,80] Smith and Davis[80] described the following canning procedure: potatoes are washed, then peeled with lye, steam, or abrasion, using both batch and continuous high-pressure steam-peeling equipment. They are then washed, employing reel-type washers with water sprays under pressure. The peeled potatoes are trimmed and sized if necessary. Most potatoes are canned whole, but they may also be sliced, diced, julienne stripped, or cut into shoestring size. Cans are filled with whole or cut potatoes by automatic or rotary hand-pack fillers, using either boiling water or brine of 1.3 to 3.0% salt. A salt tablet is added if boiling water is used. Calcium may be added in the form of calcium chloride, calcium sulfate, calcium citrate, monocalcium phosphate, or a mixture of these salts.[130] The cans are processed at 115.5 to 121.1°C or 20 to 55 min, and cooled in water to 37.7°C immediately. The canning process is outlined in Figure 7. Smith and Davis[80] described the preparation of the following canned potato products: potato pancakes, potato soup, canned French fried potatoes, or shoestring potatoes. For canned corned beef hash and beef stew, raw chopped or diced potatoes or reconstituted dehydrated diced potatoes are used.

F. FABRICATED FRENCH FRIES AND CHIPS

"Preformed French fries" (Figure 8A) is a relatively new product, sold in the U.S. and The Netherlands, which consists of a mixture of dehydrated potatoes and other ingredients which can be reconstituted rapidly in cold water to form a dough-like material. The dough is extruded in square cross sections and cut at the desired length while being extruded. The stimulated French fry strips of extruded material are deep-fat fried, resulting in a product very uniform in color, shape, form, and texture.[80]

Fabricated chips (Figure 8B) are also available in markets. Markakis et al.[131] described the process of making potato chip-like products. The product contained dried potatoes, gelatinized maize, gluten, and oil. The dough prepared from these ingredients is shaped into disks, dried to 12% moisture, and deep-fat fried. Various products of such types have been commercially prepared.[132-134] Such products exhibit homogeneity or uniformity. However, the flavor of the fried product does not resemble that prepared from raw potatoes. Several methods have been suggested to improve flavor,[135] texture,[136] crispness,[137] and shelf life of the fabricated products.[138]

G. DEHYDRATED PRODUCTS
1. Potato Granules

Granules are one of the important products prepared from dehydrated mashed potatoes. This product contains 6 to 7% moisture. The granules are reconstituted to a texture that is either dry and mealy or moist and creamy, according to individual preference. Many reviews[139-141] outline the process of production of granules from potatoes.

Even though several processes have been developed for the production of potato granules, the add-back process is commercially employed in the U.S. (Figure 9). The manufacture of potato granules by this process has been recently described.[142] In this process, cooked potatoes are partially dried by adding back previously dried granules to give a moist mix which, after holding, can be satisfactorily granulated to a fine powder.[142] A method for the production of dehydrated potato granules by a straight-through freeze-thaw process without add back of a large proportion of the product has been described.[143] This type of granule has been found superior in nutritional quality[144] and in flavor, color, and texture.[143] The freeze-thaw granules exhibited higher water-binding capacity, lower bulk density, and larger particle size than those from the add-back process[145] (Table 9). While preparing granules, it is necessary to minimize rupturing of the cells. The excess rupture of cells releases starch and the products become unduly sticky or pasty if this is excessive. Granulation is significantly improved by conditioning the moist mix. This includes adjustment of temperature and moisture. Potter[146] reported a decrease in soluble starch during conditioning. Changes in the physical properties of starch play an important role during conditioning in the preparation of potato granules.

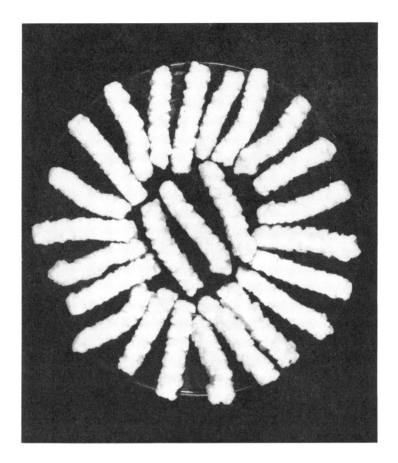

A

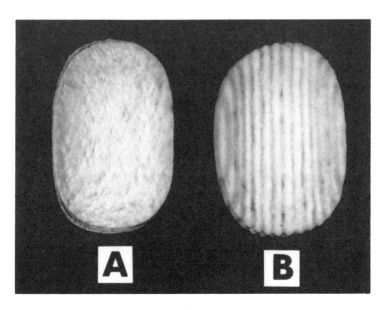

B

FIGURE 8. (A) Preformed French fries. Courtesy of Dr. B. N. Wankier, Ore-Ida Potato Processor, Ontario, OR. (B) Preformed chips (A = regular plain potato chip; B = preformed, crinkled potato chips).

Nonenzymatic browning is a common problem with dehydrated products of potatoes. Hence, the use of potatoes with low sugar content, sulfiting, drying the product to low moisture content, and avoiding high storage temperature are common methods of controlling nonenzymatic browning.

Oxidative deterioration is markedly influenced by storage temperature. Fats present in potatoes contribute to oxidative deterioration. Packing in nitrogen greatly retards oxidative deterioration, but has little effect on the rate of browning. Although lowering moisture content retards nonenzymatic browning, it accelerates oxidative deterioration.[147] The use of antioxidants such as BHA and BHT has been found to be helpful in retarding oxidation in potato granules.

2. Potato Flakes

These are dehydrated mashed potatoes prepared by applying cooked mashed potatoes to the surface of a drum drier fitted with applicator rolls, drying the deposited layer of potato solids rapidly to the desired final moisture content, and breaking the sheet of dehydrated potato solids into a suitable size for packaging.[148] In this process (Figure 10), cells are ruptured to a considerable extent. However, the reconstituted product is acceptably mealy because of the precooking and cooling treatments to which the potatoes are subjected during processing and the addition of an emulsifier. Since flakes are dried quickly, the potato cells are rehydrated easily and the potato starch retains its high absorption power. The flakes can be reconstituted with cold water.

The flakes rehydrate rapidly with boiling water, resulting in excessive cell rupture with a pasty texture. When potato flakes are broken into smaller particles for economical packaging, a certain amount of cell rupture occurs along the edges of the flakes. The gelatinized starch released from these cells would cause a pasty or rubbery texture if it had not been subjected to retrogradation in the precooking and cooling steps. An emulsifier added to the potatoes reacts with the released amylose molecules, forming a starch emulsifier complex with less solubility and reduced stickiness. Potato flakes made with the precooking/cooling process and containing added monoglyceride emulsifier are more mealy because the water used for reconstitution is not as strongly held by the intercellular material. As a result, more water penetrates the cell wall of intact cells, expanding them, thus creating more mealy, less cohesive mashed potatoes. If more of the reconstitution liquid is held between intact cells, the texture of the mash will be pasty or rubbery or gummy.

Many additives are employed in flake making from potatoes with a view to improve texture and extend the shelf life of the product. These include sodium bisulfite (to retard nonenzymatic browning), monoglyceride emulsifier, antioxidants, and chelating agents such as sodium acid pyrophosphate and citric acid.[148] Dehydrated potato flakes can be fortified at the mash stage prior to dehydration.[149] However, ascorbate reacts with potato protein, causing a pink Schiff-base compound. Pinking occurs erratically and takes time to appear after dehydration. Flake fortification can be done by mixing potato flakes with vitamined flakes of approximately the same size and shape. The vitamin flakes are actually 50 to 75% fat, containing water-soluble vitamins and minerals.

The shelf-life of flakes depends upon the chemical composition, the variety, the extent of cooking, the drying conditions, the amount of water used in processing, and the residual level of antioxidants (especially SO_2) to prevent off-flavor development. Bitterness was related to phenolic compounds. The off-flavors in stored flakes are associated with hexanal and other aldehydes produced by oxidation of the natural fat of potatoes and branched aldehydes such as 2- and 3-methyl butanal, resulting from reactions of amino acids giving the browning an off-flavor. Sapers et al.[150] reported that the haylike off-flavor found in potato flakes results from oxidation rather than from nonenzymatic browning. It is reported

GRANULES

Add Back Process

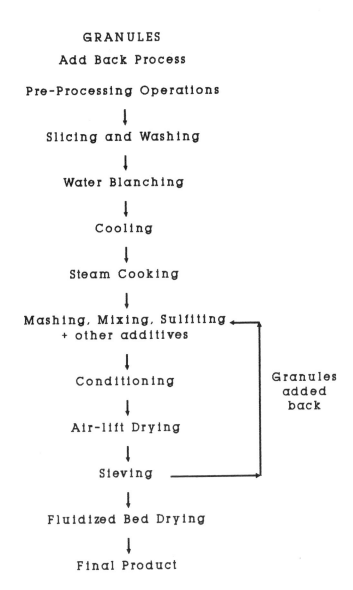

FIGURE 9.   Processes for producing dehydrated granules.

that flakes treated with emulsifier (0.66%), BHA (150 ppm), BHT (150 ppm), and SO$_2$ (40 ppm) had the best after-storage quality.

Augustin et al.[151] reported that the retention of thiamine in flake processing is higher than that in granule processing even though sulfites are used. Ascorbic acid content decreases gradually when flakes are placed in storage. They further reported nutrient losses during preparation of potato flakes and holding of the reconstituted product.[152] Retention of vitamin C and folic acid were 63 and 56%, respectively, after 1 h on the steam table.

Attempts have been made to fortify potato flakes with iron[153] and proteins.[154] However, fortification with seven ion compounds resulted in after-cooking darkening and objectionable off-flavor during storage.

Potato flakes have a problem of reconstitution with boiling water. Hence, these cannot be served as a hot product. They cannot be mixed with milk. Potato flakes can be used to form French fries for restaurant feeding using a dry mix combined with other dry binding

TABLE 9

The Relationship Between Water-Binding Capacity (WBC), Size, and Bulk Density of Two Types of Dehydrated Potato Granules

Fisher sieve size mesh (per in.)	Freeze-thaw granules[a]			Add-back granules[a]		
	% of total	Bulk density (g/cc)	WBC[b] (%)	% of total	Bulk density (g/cc)	WBC[b] (%)
Whole mixture	100	0.726	394	100	0.945	284
+60	4	—	421	0	—	—
−60	40	0.705	402	5	—	379
−80	36	0.758	375	13	0.846	340
−100	20	0.786	348	82	0.952	272

[a] Moisture content: freeze-thaw granules, 5.4%; add-back granules, 7.2%.

[b] Water-binding capacity (% bound water); least significant difference ($p < 0.05$): 16 for WBC of whole mixture; 19 for WBC of the freeze-thaw granules; 12 for WBC of the add-back granules.

From Jadhav, S. J., Berry, L. M., and Clegg, L. F. L., *J. Food Sci.*, 41, 852, 1976. With permission.

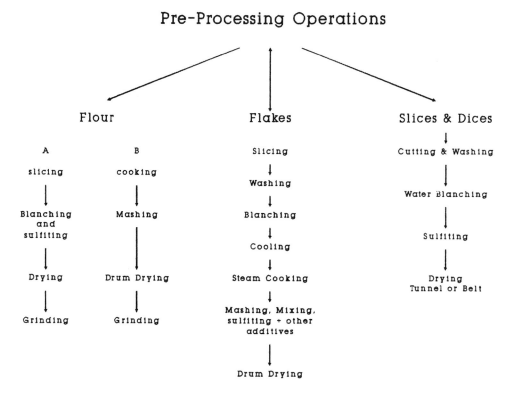

Pre-Processing Operations

Flour   Flakes   Slices & Dices

A	B	
slicing	cooking	Slicing

Flour:

A: slicing → Blanching and sulfiting → Drying → Grinding

B: cooking → Mashing → Drum Drying → Grinding

Flakes: Slicing → Washing → Blanching → Cooling → Steam Cooking → Mashing, Mixing, sulfiting + other additives → Drum Drying

Slices & Dices: Cutting & Washing → Water Blanching → Sulfiting → Drying Tunnel or Belt

FIGURE 10.   Processes for producing dehydrated and flour products.

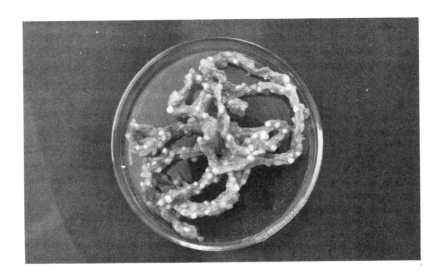

FIGURE 11.   Chakali.

agents. These can be ground to produce flour, which is used as an ingredient in soups, baby foods, and baked goods.

3. Diced Potatoes

Diced potatoes are prepared from whole potatoes by slicing, followed by blanching and dehydration (Figure 11). These are used in canned meat. One of the important problems associated with preparation of diced potatoes is graying or darkening of the products. This occurs in dry or reconstituted products. The presence of tyrosine has been implicated in discoloration. The enzyme tyrosinase catalyzes the conversion of tyrosine into quinones, resulting in the development of brown pigment. Iron salt is known to increase discoloration. Sodium pyrophosphate treatment can prevent after-cooking darkening in dehydrated potatoes. Certain commercial cultivars exhibit this phenomenon, whereas others do not. The color of the product in either dry or reconstituted form after processing or storage is influenced by a nonenzymatic discoloration caused by a nonenzymatic browning reaction. During dehydration and/or storage, amino acids react with sugars. In storage, dehydrated diced potatoes tend to develop a reddish-brown discoloration with the production of CO_2 and bitter off-flavors. Hence, potatoes used for dehydrated products should contain low amounts of reducing sugars.

Potatoes with high total solid content tend to slough more excessively on reconstitution unless processing procedures are adjusted to suit these conditions. Potatoes with low solid content tend to be firmer and do not lend themselves well to end-use requirements where mealy characteristics are desirable.

4. Potato Flour

Potato flour is commonly used in the baking industry. The flour is prepared (Figure 11) by dehydration of peeled, cooked potatoes on a single drum drier equipped with applicator rolls. The thin, dried sheet of potato solids is then ground to the desired fineness. The single drum drier is one of the most efficient means of dehydrating potatoes. By spreading the mash into a thin sheet, extremely rapid evaporation of water can be achieved. Potato flours are available in two forms. These are granular and fine flours. Granular potato flour is normally used in operations where freedom from lumping is desired.

In general, potato flour is used in bread to the extent of 6%. At 2 to 3% it helps materially to preserve freshness due to increased water absorption afforded by potato flour. Potato flour is used in cookie pastries, doughnuts, cakes and cake mixes, and snack food.[155-161] The use of flour in such products results in a substantial increase in quality when added up to 5%. Potato flour can be used as an ingredient in many dehydrated soup mixes. Bushway et al.[162] showed that 1.5% potato flour and 1.5% potato starch can be used in the manufacture of frankfurters to increase the tenderness and juiciness.

H. MISCELLANEOUS PRODUCTS

Potatoes serve as raw material for several food and industrial products. The industrial waste from potato processing may be utilized for cattle feed or for isolation of some chemicals. Potatoes have been used for the production of several miscellaneous food products such as canned potato salad, canned beef hash, canned beef stew, canned French fries, pancakes, soups, chip bars, chip confections, potato nuts, potato puffs, sponge dehydrated potatoes, potato snack items, etc.

1. Canned Potato Salad

Canned potato salad is prepared from potatoes, bacon, onions, and sauce. It is served as a hot dish. The potatoes are peeled, sliced, and blanched in water. Cut onions, fried bacon, and chopped parsley are mixed with the blanched potatoes. A sauce is prepared by adding flour or starch to bacon fat. To this sauce, a mixture of vinegar, water, sugar, salt, and pepper is added. The sauce is heated to the desired consistency. The mixed solid ingredients and sauce are mixed and filled in cans. The cans are heat processed and cooled. The pH of the product is about 3.5 or less.

2. Canned Beef Hash and Beef Stew

Potatoes are used as an ingredient in some meat products. Canned beef hash is prepared from meat, potatoes, onions, fat, sugar, salt, pepper, and sodium nitrite. Canned beef stew, in addition, contains carrots, tomato puree, and cereal flour.[163]

3. Potato Soup

Diced potatoes, chopped onions, milk, salt, pepper, butter, vegetable oil, and flour are generally used as ingredients in the preparation of potato soup.

4. Canned French-Fried and Shoestring Potatoes

French-fried or shoestring potatoes prepared for canning are fried in oil until the moisture is reduced to about 5 to 8%. After frying, the excess fat is drained off and the product is salted. The product may be packed in air, nitrogen, or a vacuum. The shelf-life is quite long if it is packed in nitrogen or a vacuum.

5. Potato Pancakes and Pancake Mixes

Potato pancakes are prepared from grated raw potatoes mixed with eggs, flour, onion, baking powder, salt, and bacon fat. The batter is placed in hot fat and fried until golden brown. Dehydrated pancake mixes are manufactured commercially.

6. Potato Chip Bars

Treadway et al.[164] developed a potato hip bar suitable for military use by crushing potato chips and compressing them at 500 to 3000 p.s.i. to about 1/20th of their original volume. The potato chip bars are about 7 cm long, 2.5 cm wide, and 1.5 cm thick.

7. Potato Chip Confections

The preparation of various candy-coated and flavored potato chip confections has been described by Townsley and Dixon.[165] These products are prepared by dusting or glazing the chips with the desired coating. Chocolate-coated potato chips are prepared by dipping the chips in semisweet baker's chocolate.

8. Potato Nuts

For preparation of potato nuts, fresh potato pieces about $^3/_8 \times {}^3/_8 \times {}^1/_4$ in in size are dehydrated to about 12% moisture and fried in deep fat to give an end product containing about 20% fat.[166,167] Willard[168] patented a process for making potato nuts. Diced potato pieces dehydrated to 20 to 40% moisture were fried at 177 to 218°C for less than 60 s. The product has low bulk density, tender texture, and fat content of about 15 to 25%.

9. Potato Puffs

For preparation of potato puffs, the potatoes are peeled and cut into pieces of about $^3/_8 \times {}^3/_8 \times {}^1/_8$ in size. These pieces are blanched in boiling salt water and dried. Flavoring agents may be added to the blanching water.[169] After drying, the puffed pieces become rigid and take on a light brown color. They are approximately $1^1/_2 \times {}^3/_4$ in in size.[170] Preparation of potato puffs using egg and vegetable shortening has been described.[171] A mixture containing 80% cooked, peeled potatoes, 4% butter or margarine, 9% cream, 3.5% egg yolk, and 3.5% egg white is extruded in the form of croquettes, allowed to cool, and fried in deep fat.

10. Sponge-Dehydrated Potatoes

Potatoes with high solid content are washed, peeled, cut to half dice, and cooked in a steam blancher. The pieces are cooled and frozen to $-23°C$. They are placed at 4.4°C and allowed to thaw slowly. The thawed product is dried to about 8% moisture.[172] The dehydrated product can be converted to mashed potatoes in a few minutes by the addition of hot water. The product can be used in potato cakes, soups, casserole dishes, hash chowder, etc.

11. Potato Snack Item

A potato snack item is prepared from the following ingredients:[173]

Steamed potatoes	—	200 parts
Potato starch	—	100 parts
Potato flour	—	20 parts
Salt	—	5 parts
Shortening	—	19 parts and
Karaya gum	—	1 part

The steamed potatoes are mashed in a mechanical mixer. The starch, flour, salt, and gum are mixed. Shortening is added and mixing is continued for another 5 min. The dough is rolled, cut into pieces, and dried at 149°C. The product is hollow, crisp, and light golden brown.

12. Lactic Acid

Potato starch has been used as a raw material for the manufacture of lactic acid. The starch is first hydrolyzed by using malt, acid, or fungal amylase.[174] The sugars formed are fermented using *Lactobacilli* to produce lactic acid. Cardon et al.[175] have described the production of lactic acid from potatoes. The yield of lactic acid is about 80 to 90% of the original starch content of potatoes.

13. Potato Pulp

Although potatoes are mainly considered as human food, more potatoes are fed to

TABLE 10
Analyses of Dried Potato Pulp from Two Types of Plants

Constituent	Table type plant		Vat type plant	
	Without Ca(OH)$_2$	0.3% Ca(OH)$_2$ added	Without Ca(OH)$_2$	0.3% Ca(OH)$_2$ added
Protein	6.7	3.2	8.3	5.1
Fat	0.9	0.1	0.3	0.3
Fiber	12.0	10.8	9.2	9.0
Ash	2.7	4.9	3.6	10.1
Nitrogen-free extract	77.9	81.1	78.5	75.1

Note: Percent on dry-weight basis.

From Highlands, M. E., in *Potato Processing,* Talburt, W. F. and Smith, O., Eds., AVI Publishing, Westport, CT, 1967, 540. With permission.

livestock than are consumed by humans. Potato pulp is obtained as a by-product during starch plant operations. The chemical composition of dried potato pulp is presented in Table 10. The amount of pulp produced depends upon the method of starch extraction employed. The moisture from the pulp can be removed by natural drainage or by mechanical pressing. Treatment with coagulants like calcium hydroxide (0.3 to 0.5%) for 15 to 30 min prior to pressing improves the removal of water. The drying of pulp can be accomplished by the conventional steam tube drier[176] or by a direct-fired rotary drier.[177] The dried pulp is fed to the cattle.[178]

IV. TRADITIONAL/DOMESTIC PROCESSING

Potatoes form a staple food in many countries of the world. These include countries such as Rwanda, Nepal, Tibet, China, Peru, Bolivia, Ecuador, and many countries in Eastern Europe. In addition, potatoes serve as a major side dish to principal staples. This pattern is evident in Central America where corn, tortillas, and beans form the mainstay of the diet. Potatoes are consumed as a vegetable side dish on a regular basis, but in comparatively smaller quantities and not every day. This is very common in India, Pakistan, and Bangladesh. In some countries, like Indonesia, Malaysia, and the Philippines, potatoes are consumed as a small item among many other dishes prepared to complement the basic staple foods.

Potatoes are generally cooked before consumption. There are various methods of cooking potatoes. These include boiling with or without peels, baking, frying, and microwave cooking. Similarly, potatoes are boiled with or without the peel and then macerated into potato mash which is used for the preparation of various traditional products. These products are fried before consumption. This section summarizes in brief various traditional methods of processing potatoes and various traditional products prepared from potatoes in different parts of the world.

A. PEELING

Potato peeling is one of the important traditional processes commonly employed before direct consumption of boiled potatoes or processing of peeled potatoes into various products. The method of peeling influences significantly the quantity of loss through peeling. The average weight loss from peeling has been reported to range from 5 to 24%.[178,179] Augustin et al.[180] found that peel fractions of whole tubers when the peel was removed with a domestic peeling knife either before or after cooking were 6, 2, and 10% in the case of peeled raw

TABLE 11
Nutrient Losses (%) in
Boiled and Steamed
Potatoes

| | Cooking method | |
Constituent	Boiled	Steamed
Dry matter	9.4	4.0
Protein	—	—
Calcium	16.8	9.6
Magnesium	18.8	14.0
Phosphorus	18.3	14.0
Iron	18.3	11.7

From Domah, A. A. M. B., Davidek, J., and Valisek, J., *Z. Lebensm. Unters.-Forsch.*, 154, 272, 1974. With permission.

potatoes, those peeled after boiling, and those peeled after baking, respectively. It was suggested that potatoes should be peeled after and not before boiling when cooked domestically to reduce losses. Peeling results in loss of some minerals and trace elements,[181-186] crude fibers,[180] riboflavin,[180] and amino acids.[185] The loss of nutrients through peeling removed from boiled potatoes, expressed as a percentage of total tuber nutrients, were lower in all cases than those from raw potatoes. The weight loss varies from 5 to 24% with the average weight loss 8 to 12%.[88] Damage in the form of penetrating cracks followed by superficial crushing and bruising results in a significant loss of weight.

B. METHODS OF COOKING
1. Boiling

Boiling is the most common method of preparing potatoes domestically in all continents of the world. In developing countries, potatoes are usually boiled intact by the consumer, whereas in developed countries potatoes are peeled and cooked. The peeling of potatoes before boiling results in a higher loss of nutrients than boiling of potatoes intact.[87,187,188] It has been shown that cooking potatoes by boiling increases the digestibility of the starch.[189] However, the presence of starch that is resistant to enzymatic hydrolysis *in vitro* has been reported.[190,191] Cooking results in the loss of nitrogen. Peeled potatoes when cooked exhibited more loss of nitrogen than unpeeled potatoes.[192] Jaswal[193] observed negligible losses of free amino acids. These losses were attributed to carbohydrate/amino acid interactions. Changes in dietary fiber content due to boiling of potatoes have been reported by several workers.[88,194,195] Significant losses in ascorbic acid have been reported due to boiling of peeled or unpeeled potatoes.[196] Changes in B group vitamin contents varies according to boiling conditions.[197] Losses of minerals through boiling, except in the case of iron, are probably smaller than losses due to careless peeling before cooking.[198] Nutrient losses in boiled potatoes were less than in steamed potatoes (Table 11).

2. Frying

Potatoes are generally boiled or blanched before frying. The boiled potatoes are peeled, or peeled potatoes are boiled, and then fried in oil or margarine. The fried products are used for consumption. As frying results in a decrease in moisture content, nutrients are concentrated. Hence, changes occurring due to frying are difficult to assess when nutrient contents are assessed on a fresh-weight basis. Fenton[199] and Richardson et al.[200] observed ascorbic

TABLE 12
Effect of Frying on Stability of Ascorbic Acid (AA) and Dehydroascorbic Acid (DAA) in Potatoes

Sample	Dry matter	DAA (mg/100 g dry matter)	AA (mg/100 g dry matter)	Total content of vitamin C (mg/100 g dry matter)
Raw, peeled potatoes (before frying)	26.20	7.4	44.6	52.0
Fried potatoes				
140°C/10 min	83.01	29.7	20.6	50.3
140°C/20 min	84.00	33.7	7.3	41.0
140°C/30 min	88.00	42.7	0.0	42.7
180°C/5 min	89.10	42.8	0.0	42.8

From Domah, A. A. M. B., Davidek, J., and Velisek, J., *Z. Lebensm. Unters. Forsch.*, 154, 272, 1974. With permission.

acid retentions of 67% in fried potatoes. Domah et al.[201] studied the effects of frying potatoes at 140°C for 10, 20, and 30 min and at 180°C for 5 min on the ascorbic acid content in potatoes (Table 12). The retention of total vitamin C was good and in fact better than that in boiled potatoes. However, ascorbic acid is oxidized to dehydroascorbic acid (DAA) more rapidly in boiling than frying, but hydrolysis of DAA is slowed by the dehydration of the product during frying, and therefore DAA accumulated in fried potatoes. During boiling, DAA is converted to 2-3 diketogulonic acid. Frying decreases mineral content significantly in both the cortical and the pith areas, with most of the loss occurring in the cortical areas (10 to 45%).[184]

3. Baking

Baking of potatoes is one of the common methods of potato processing. Potatoes are generally baked in their skin. Several studies have been conducted to study the effects of baking on nutrient retention in potatoes.[202]

Kahn and Halliday[203] compared baked (in skin), parabaked, and French-fried potatoes and observed ascorbic acid retention of 80, 80, and 77%, respectively. After standing approximately 1 h, the retention rates were 41, 52, and 71%, respectively. Page and Hanning[202] studied 58 samples of boiled and baked potatoes for vitamin B_6 and niacin retention (Table 13). Baking losses (9% for B_6 and 4% for niacin) were less than for boiling (20 and 18%, respectively). The difference was essentially found in the cooking liquid. Pelletier et al.[204] found that the cooking method used on potatoes has a significant effect on the retention of vitamin C, with boiled potatoes retaining about 80% compared to only about 30% for hashed browns. (Table 14). During the baking of potatoes the movement of minerals such as potassium, phosphorus, and iron toward the interior tissues has been demonstrated by Mondy and Ponnampalam.[184] Baking increased the content of potassium (14 to 23%), phosphorus (2 to 9%), and iron (2 to 8%) in the interior pith tissue.

4. Microwave Cooking

The use of microwaves in cooking has been on the rise, and it is estimated that by the year 2000, microwave ranges will be common in food-service units.[205] In conventional cooking heat is applied to the outside of food by convection, radiation, or by conduction and then heat is conducted to the interior of the food. Heat generated from within the food by a series of molecular vibrations is the basis for microwave cooking. Advantages of microwave cooking include higher energy efficiencies, greater time saving, convenience, and easy clean up. Its disadvantages include greater cooking losses and less palatability.[205]

TABLE 13
Niacin and Vitamin B$_6$ Retention (%) in Boiled and Baked Potatoes

Location and cultivar	Number of samples	Retention in boiled potatoes		Loss in cooking liquid		Retention in baked potatoes	
		Niacin	Vitamin B$_6$	Niacin	Vitamin B$_6$	Niacin	Vitamin B$_6$
Wisconsin 'Cobbler'	7	80.7 ± 2.73	81.0 ± 3.03	17.1 ± 1.47	15.9 ± 1.32	91.2 ± 6.25	90.9 ± 9.17
'Triumph'	14	82.4 ± 4.45	78.5 ± 6.24	15.5 ± 2.32	14.5 ± 2.14	99.1 ± 4.60	92.8 ± 9.44
Minnesota 'Cobbler'	12	82.6 ± 6.76	81.8 ± 8.03	16.1 ± 6.11	15.8 ± 5.67	93.6 ± 3.05	93.2 ± 8.52
Kentucky 'Cobbler'	9	82.6 ± 3.14	78.2 ± 10.70	17.6 ± 2.04	15.2 ± 1.84	91.1 ± 4.66	88.6 ± 12.10
Indiana 'Chippewa'	10	79.0 ± 2.36	79.8 ± 8.10	17.3 ± 1.15	14.0 ± 1.11	94.1 ± 4.93	88.8 ± 5.32
Colorado 'McClure'	3	84.8 ± 5.01	79.0 ± 4.31	16.8 ± 1.91	15.4 ± 1.35	105.4 ± 5.36	91.8 ± 8.06
Idaho 'Russet Burbank'	3	84.7 ± 2.92	84.7 ± 8.26	16.2 ± 3.57	18.4 ± 5.40	90.6 ± 0.85	91.3 ± 2.85
Overall mean		81.9	80.0	16.6	15.2	95.8	91.2

From Page, E. and Hanning, F. M., *J. Am. Diet. Assoc.*, 42, 42, 1963. With permission.

TABLE 14
Percentages of Weight Changes and Vitamin C Retained in Canadian Potatoes after Different Types of Cooking

Preparation	% Weight changes 4 seasons	% Vitamin C retained 4 seasons
Boiled in skin	94 ± 2	81 ± 10
Boiled pared before cooking	99 ± 2	73 ± 8
Boiled, mashed + milk	120 ± 2	77 ± 10
Boiled, mashed + margarine + milk	123 ± 2	72 ± 12
Boiled, browned	72 ± 2	28 ± 11
Fried	57 ± 5	72 ± 14
Baked	85 ± 2	78 ± 9
Raw, browned	68 ± 4	60 ± 13
Scalloped + cheese	123 ± 6	80 ± 15
Scalloped	113 ± 6	68 ± 10

From Pelletier, O., Nantel, C., Leduc, R., Tremblay, L., and Bressard, R., *Can. Inst. Food Sci. Technol. J., 10*, 138, 1977. With permission.

TABLE 15
Changes in Nutrient Composition of Potatoes Baked by Microwave and Conventional Means

Nutrient	Tissue	Retention (%) Conventional (60 min/400°F)	Microwave 3 min
Protein	Cortex	87	94
	Pith	87	84
Total amino acids	Cortex	97	102
	Pith	109	96
Free amino acids	Cortex	85	108
	Pith	108	83
Potassium	Cortex	85	101
	Pith	122	113
Phosphorus	Cortex	88	98
	Pith	110	100
Calcium	Cortex	98	102
	Pith	104	117
Magnesium	Cortex	90	104
	Pith	104	99
Manganese	Cortex	89	91
	Pith	104	109
Iron	Cortex	88	98
	Pith	122	90
Copper	Cortex	93	96
	Pith	93	105
Zinc	Cortex	91	100
	Pith	136	100

From Klein, L. B. and Mondy, N. I., *J. Food Sci., 46*, 1874, 1981. With permission.

A comparison between conventional and microwave baking of potatoes was studied by Klein and Mondy.[206] Conventional cooking primarily resulted in the migration of some nutrients from the cortex to the pith, whereas microwave cooking resulted in the loss of volatiles from the interior pith tissue (Table 15). It appears that baking potatoes conventionally is less nutritious with respect to nitrogenous and mineral constituents, especially when the skin and adhering cortical tissue are not consumed.

C. TRADITIONAL PRODUCTS

1. Chuno

In the high Andean areas of Peru and Bolivia, potatoes are processed by traditional methods. One product, called *chuno,* is very popular in these areas. There are two forms of *chuno:* one is *chuno blanco* (white) and the other is *chuno negro* (black).

a. Chuno Blanco

On a night when a particularly heavy frost is expected, potatoes are spread evenly on the ground over a selected site. The following morning, a careful examination of the tubers determines whether or not they have been frozen. They are exposed to additional nights of frost if freezing is incomplete. The frozen tubers are then thawed during the day as the temperature rises. They may be covered with straw to protect them from blackening by exposure to the sun. The freezing and thawing process results in the separation of tuber cells and the destruction of the differential permeability of the cell membrane.[207] This allows cell sap to diffuse from the cells into the intercellular spaces. The released liquid is squeezed out of the tubers by trampling, a procedure which also removes the tuber skin. The trampled tubers are transferred to a running stream and immersed, covered with protective layers of straw, for 1 to 3 weeks. After removal of the water, they are spread in the sun to dry. The white crust which forms on the drying tubers makes *chuno blanco.* Chuno blanco normally forms an ingredient in soups or cheese. Mixed with fruit and molasses, it is consumed as a sweet dessert.[208]

b. Chuno Negro

The process of preparation of *Chuno negro* is similar to that of *Chuno blanco,* except that the tuber skins are not removed during trampling to squeeze out the juice and the trampled tubers are not soaked in water. After trampling, they are immediately sun dried and the product is dark brown to black in color. *Chuno negro* is soaked in water for 1 or 2 d prior to cooking to remove strong, undesirable flavor. It is mainly used in soups.

2. Papa Seca

This is mainly produced by boiling and hand peeling potatoes which are sliced or broken into small pieces and sun dried. When dry, they are ground into finer pieces with a hand meat grinder. *Papa seca* is used for the preparation of a special dish called *Carapulca,* which is a mixture of *Papa seca* meat, tomatoes, onions, and garlic. It can also be prepared as a soup.

3. Chakali

This is a popular product in Southeast Asia. It is prepared by using either whole potato flour or composite flour in which potato flour is substituted by chickpea or peanut meal (10 to 20%). Water is added to the whole meal or composite flour to form a soft dough. Salt and chili powder are added to taste. The dough is passed through an orifice under pressure by using a kitchen press. The product is sun dried (Figure 11) and fried in vegetable oil. Yahya[209] prepared this product by using composite flour. It was noted that the substitution of potato flour with defatted groundnut flour at a 20% level produced acceptable *chakalies.* The *chakalies* prepared with composite flour had a higher protein content than in *chakalies* prepared from whole flour.

4. Papad

Potatoes are boiled, hand peeled, and macerated into a dough. The dough is made into

FIGURE 12.   Papad.

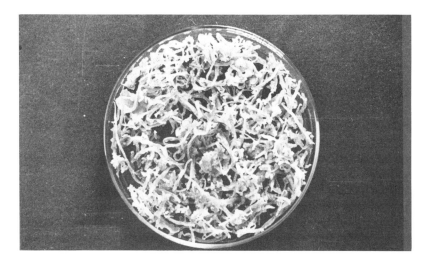

FIGURE 13.   Shreds.

small pieces and rolled into thin round sheets with hand rollers and then dried. The dried product (Figure 12) is fried in oil before it is consumed. Papads exhibit excellent keeping quality for several months. These are popular in Southeast Asia.

5. Shreds

Potatoes are boiled with skin, hand peeled, and then made into thin shreds with kitchen shredders. The shreds (Figure 13) are sun dried and the dried product is fried in oil before it is consumed.

6. Snack Foods

a. Vada

The potatoes are boiled, peeled, and cut into pieces. These are spiced with onion and other spices and fried in oil. These pieces are made into balls and dipped into a thick slurry (dough) of chickpea flour. The balls are then deep fried in vegetable oil. The product is used as a snack item in Southeast Asia.

b. *Alu-Bhajiya*

Potato slices prepared from peeled potatoes are dipped into a thick slurry (dough) of chickpea flour and deep fried in vegetable oil. This product is a popular snack item in the Indian subcontinent.

c. *Other Products*

Anand and Maini[210] suggested several possible outlets of utilizing potatoes economically under glut conditions, including potatoes for human food, starch, alcohol, and cattle feed. Processing of potatoes into potato sticks, chips, buds, mashed potatoes, and several other human foods, such as snacks like "samosa", *"tikki"*, and *"stuffed chapaties"* using potato flour. Nankar and Nankar[211] similarly reported a number of processed potato products such as *"sev"*, *"chakli"*, *"papad"*, and *samosa*, which could be made from potatoes.

7. Weaning Foods

Potatoes either in the fresh state or after processing, mixed with other foods or made denser calorically through the addition of oil, provide an excellent weaning food for infants. Potatoes are used as a weaning food in Central America and in Southeast Asia.

8. Other Traditional Foods

As potato cultivation increases in developing countries and in rice fields, potato products will be developed that are suitable and acceptable to the cultures of the nations.

REFERENCES

1. **Howard, H. W.,** The production of new varieties, in *The Potato Crop: The Scientific Basis for Improvement,* Harris, P. M., Ed., Chapman and Hall, New York, 1978, 607.
2. **Harris, J. R.,** Potato processing industry—what the market requires, in *Agriculture Group Symp.: Factors Influencing Storage Characteristics and Cooking Quality of Potatoes, J. Sci. Food Agric.,* 32(Abstr.), 104, 1981.
3. **French, W. M.,** Varietal factors influencing the quality of potatoes, in *Agriculture Group Symp.: Factors Influencing Storage Characteristics and Cooking Quality of Potatoes, J. Sci. Food Agric.,* 32(Abstr.), 126, 1981.
4. **Talburt, W. F. and Smith, O.,** *Potato Processing,* AVI Publishing, Westport, CT, 1967.
5. **Gray, D. and Hughes, J. C.,** Tuber quality, in *The Potato Crop: The Scientific Basis for Improvement,* Harris, P. M., Ed., Chapman and Hall, New York, 1978, 504.
6. **McComber, D. R., Osman, E. M., and Robert, R. A.,** Factors related to potato mealiness, *J. Food Sci.,* 53, 1423, 1988.
7. **Sterling, C. and Bettelheim, F. A.,** Factors associated with potato texture. III. Physical attributes and general conclusions, *Food Res.,* 20, 130, 1955.
8. **Reeve, R. M.,** Pectin, starch and texture of potatoes: some practical and theoretical implications, *J. Texture Stud.,* 8, 1, 1977.
9. **Nonaka, M. and Timm, H.,** Textural quality of cooked potatoes. II. Relationship of steam cooking time to cellular strength of cultivars with similar and differing solids, *Am. Potato J.,* 60, 685, 1983.
10. **Loh, J., Breene, W. M., and Davis, E. A.,** Between species differences in fracturability loss: microscopic and chemical comparison of potato and Chinese waterchestnut, *J. Texture Stud.,* 13, 325, 1982.
11. **Bretzloff, C. W.,** Some aspects of cooked potato texture and appearance. II. Potato cell size stability during cooking and freezing, *Am. Potato J.,* 47, 176, 1970.
12. **Briant, A. M., Personius, C. J., and Cassel, E. G.,** Physical properties of starches from potato of different culinary quality, *Food Res.,* 10, 437, 1945.
13. **Barrios, E. P., Newsom, D. N., and Miller, J. C.,** Some factors influencing the culinary quality of Irish potatoes: physical character, *Am. Potato J.,* 40, 200, 1963.
14. **Unrau, A. and Nytund, R.,** The relation of physical properties and chemical composition to mealiness in the potato. I. Physical properties, *Am. Potato J.,* 34, 245, 1957.

15. **Heinze, P. H., Kirkpatrick, M. E., and Dochtermann, E. F.,** Cooking quality and compositional factors of potatoes of different varieties from several commercial locations, *U.S. Dept. Agric. Tech. Bull.*, Washington, D.C., 1106, 1955.

16. **McClendon, J. H.,** Evidence for the pectic nature of middle lamella of potato tuber cell walls based on chromatography of macerating enzymes, *Am. J. Bot.*, 51, 628, 633.

17. **Warren, D. S., Gray, D., and Woodman, J. S.,** Relationship between chemical composition and breakdown in cooked potato tissues, *J. Sci. Food Agric.*, 26, 1689, 1975.

18. **Bettelheim, F. A. and Sterling, C.,** Factors associated with potato texture. 1. Specific gravity and starch content, *Food Res.*, 20, 71, 1955.

19. **Bettelheim, F. A. and Sterling, C.,** Factors associated with potato texture. 2. Pectic substances, *Food Res.*, 20, 118, 1955.

20. **Gray, D.,** Some effects of variety, harvest date and plant spacing on tuber breakdown on canning tuber dry matter content and cell surface area in the potato, *Potato Res.*, 15, 317, 1972.

21. **Warren, D. S. and Woodman, J. S.,** Distribution of cell wall components in potato tubers: a new titrimetric procedure for the estimation of total polyuronide (pectic substances) and its degree of esterification, *J. Sci. Food Agric.*, 25, 129, 1974.

22. **Hughes, J. C., Faulks, R. M., and Grant, A.,** Texture of cooked potatoes: relationship between the compressive strength of cooked potato discs and release of pectic substances, *J. Sci. Food Agric.*, 26, 731, 1975.

23. **Hughes, J. C. and Faulks, R. M.,** Texture of cooked potatoes of different maturity in relation to pectic substances and cell size (Abstract), *Proc. Fifth Triennial Conf. Eur. Assoc. Potato* (Norwich, 1972), 1963, 1973.

24. **Reeve, R. M.,** A review of cellular structure, starch and texture qualities of processed potatoes, *Econ. Bot.*, 21, 294, 1967.

25. **Reeve, R. M.,** Histological survey of conditions influencing texture in potatoes. II. Observations on starch in treated cells, *Food Res.*, 19, 333, 1954.

26. **Ophuis, B. G., Hesen, J. C., and Krosbergen, E.,** The influence of the temperature during handling on the occurrence of blue discolorations inside potato tubers, *Eur. Potato J.*, 1, 48, 1958.

27. **Salunkhe, D. K., Wheeler, E. J., and Dexter, S. T.,** The effects of environmental factors on suitability of potatoes for chip making, *Agron. J.*, 46, 195, 1954.

28. **Salunkhe, D. K.,** The Influence of Certain Environmental Factors on the Production and Quality of Potatoes for Potato Chips Industry, Ph.D. thesis, Michigan State University, East Lansing, 1953.

29. **Salunkhe, D. K., Wittwer, S. H., Wheeler, E. J., and Dexter, S. T.,** The influence of a pre-harvest foliar spray of maleic hydrazide on the specific gravity of potato chips, *J. Food Sci.*, 18, 191, 1953.

30. **Burton, W. G.,** The potato, in *A Survey of Its History and of the Factors Influencing Its Yield, Nutritive Value, Quality and Storage,* Veenman, H. and Zonen, N. V., Eds., Wageningen European Association of Potato Research, The Netherlands, 1966.

31. **Smith, O.,** Environmental factors, in *Potatoes: Production, Storing, Processing,* AVI Publishing, Westport, CT, 1968, 259.

32. **Killick, R. J. and Simmonds, N. W.,** Specific gravity of potato tubers as character showing small genotype-environmental interactions, *Heredity,* 32, 109, 1974.

33. **Kunkel, R. and Holstad, N.,** Potato chip color, specific gravity and fertilization of potatoes with N, P and K, *Am. Potato J.,* 49, 43, 1972.

34. **Mazza, G., Hung, J., and Dench, M. J.,** Processing/nutritional quality changes in potato tubers during growth and long term storage, *Can. Inst. Food Sci. Technol. J.,* 16(1), 39, 1983.

35. **Mazza, G.,** Correlations between quality parameters of potato during growth and long-term storage, *Am. Potato J.,* 60, 145, 1983.

36. **Mazza, G.,** Selected Factors Affecting Frying Quality of Potatoes, paper presented at the Eleventh Annual Meeting of the Prairie Potato Council, Winnipeg, Manitoba, February 12 to 16, 1983.

37. **Salunkhe, D. K. and Pollard, L. H.,** Microscopic examination of starch grains in relation to maturity of potatoes, *Proc. Am. Soc. Hortic. Sci.,* 64, 331, 1954.

38. **Reeve, R. M.,** Suggested improvements for microscopic measurement of tissue cells and starch granules in fresh potatoes, *Am. Potato J.,* 44, 41, 1967.

39. **Sharma, K. N. and Thompson, N. R.,** Relationship of starch grain size to specific gravity of potato tubers, *Mich. Agric. Exp. Stn. Q. Bull.,* 38, 559, 1955.

40. **Barrios, E. P., Newsom, D. W., and Miller, J. C.,** Some factors influencing the culinary quality of southern- and northern-grown Irish potatoes. I. Chemical composition, *Am. Potato J.,* 38, 182, 1961.

41. **Barrios, E. P., Newsom, D. W., and Miller, J. C.,** Some factors influencing the culinary quality of Irish potatoes. II. Physical characters, *Am. Potato J.,* 40, 200, 1963.

42. **Unrau, A. M. and Nylund, R. E.,** The relation of physical properties and chemical composition of mealiness in the potato. I. Physical properties, *Am. Potato J.,* 34, 245, 1957.

43. **Unrau, A. M. and Nylund, R. E.,** The relation of physical properties and chemical composition of mealiness in the potato. II. Chemical composition, *Am. Potato J.,* 34, 303, 1957.

44. **Esau, K.,** *Plant Anatomy,* John Wiley & Sons, New York, 1953.

45. **Reeve, R. M.,** Histological survey of conditions influencing texture in potatoes. I. Effects of heat treatments on structure, *Food Res.,* 19, 323, 1954.

46. **Shewfelt, A. L., Brown, D. R., and Troop, K. D.,** The relationship of mealiness in cooked potatoes to certain microscopic observations of the raw and cooked product, *Can. J. Agric. Sci.,* 35, 513, 1955.

47. **Dastur, R. H. and Agnihotri, S. D.,** Study of the pectic changes in the potato tubers at different stages of growth and in storage, *Indian J. Agric. Sci.,* 4, 430, 1934.

48. **Sharma, K. N., Isleib, D. R., and Dexter, S. T.,** The influence of specific gravity and chemical composition on hardness of potato tubers after cooking, *Am. Potato J.,* 36, 105, 1959.

49. **Whittenberger, R. T. and Nutting, G. C.,** Observations on sloughing of potatoes, *Food Res.,* 15, 331, 1950.

50. **Mapson, L. W. and Swain, T.,** *Production and Application of Enzyme Preparation in Food Manufacture,* SCI Monograph II, Society of Chemical Industry, London, 1961, 121.

51. **Mapson, L. W., Swain, T., and Tomalin, A. W.,** Influence of variety, cultural conditions and temperature of storage on enzymatic browning of potato tubers, *J. Sci. Food Agric.,* 14, 673, 1963.

52. **Clark, W., Mondy, N., Bedrosan, K., Ferrari, R., and Michon, C.,** Polyphenolic content and enzymatic activity of two varieties of potatoes, *Food Technol.,* 11, 297, 1957.

53. **Mulder, E. G.,** Mineral nutrition in relation to the biochemistry and physiology of potatoes. I. Effect of nitrogen, phosphate, potassium, magnesium, and copper nutrition on the tyrosine content and tyrosinase activity with particular reference to blackening of the tubers, *Plant Soil,* 2, 59, 1949.

54. **Welte, E. and Muller, K.,** The influence of potassium manuring on the darkening of raw potato pulp, *Eur. Potato J.,* 9, 38, 1966.

55. **Mondy, N. I., Mobley, E. O., and Gedde-Dahl, S. B.,** Influence of potassium fertilization on enzymatic activity, phenolic content and discoloration of potatoes, *J. Food Sci.,* 32, 378, 1967.

56. **Hoff, J. E., Jones, C. M., Wilcox, G. E., and Castro, M. D.,** The effects of nitrogen fertilization on the composition of the free amino acid pool of potato tubers, *Am. Potato J.,* 48, 390, 1971.

57. **Mulder, E. G. and Bakema, K.,** Effect of the nitrogen phosphorus, potassium, and magnesium on the content of free amino acids and on the amino acid composition of the protein of the tubers, *Plant Soil,* 7, 135, 1956.

58. **Grewal, S. S. and Sukumaran, N. P.,** Effect of maleic hydrazide (MH) sprout suppressant on various biochemical constituents, *Annu. Sci. Rep. 1979, of the Central Potato Res. Inst., Simla, India,* 40, 1980.

59. **Hodge, J. E.,** Chemistry of browning reactions in model systems, *J. Agric. Food Chem.,* 1, 928, 1953.

60. **Schallenberger, R. S., Smith, O., and Treadway, R. H.,** Role of the sugars in the browning reaction in potato chips, *J. Agric. Food Chem.,* 7, 274, 1959.

61. **Schwimmer, S., Hendel, C. E., Harrington, W. O., and Olson, R. L.,** Interrelation among measurements of browning of processed potatoes and sugar components, *Am. Potato J.,* 34, 119, 1957.

62. **Denny, F. E. and Thornton, N. C.,** Factors for color in the production of potato chips, *Contrib. Boyce Thompson Inst.,* 11, 291, 1940.

63. **Burton, W. F. and Wilson, A. R.,** The apparent effect of the latitude of the place of cultivation upon the sugar content of potatoes grown in Great Britain, *Potato Res.,* 13, 269, 1970.

64. **Sowokinos, J. R.,** Maturation of *Solanum tuberosum.* 1. Coomparative sucrose and sucrose synthetase levels between several good and poor processing varieties, *Am. Potato J.,* 50, 234, 1973.

65. **Carlsson, H.,** Production of potatoes for chipping, *Vaxtodling,* 26, 9, 1970.

66. **Pressey, R.,** Role of invertase in the accumulation of sugars in cold-storage potatoes, *Am. Potato J.,* 46, 291, 1969.

67. **Schwimmer, S., Makower, R. W., and Rorem, E. S.,** Invertase and invertase inhibitor in potato, *Plant Physiol.,* 36, 313, 1961.

68. **Hughes, J. C. and Swain, T.,** After-cooking blackening in potatoes. II. Core experiments, *J. Sci. Food Agric.,* 13, 229, 1962.

69. **Hughes, J. C. and Swain, T.,** After-cooking blackening in potatoes. III. Examination of the interaction of factors by *in vitro* experiments, *J. Sci. Food Agric.,* 13, 358, 1962.

70. **Hughes, J. C. and Evans, J. L.,** Studies on after-cooking blackening in potatoes. 4. Field experiments, *Eur. Potato J.,* 10, 16, 1967.

71. **Smith, O.,** Potato chip research in 1959, *Proc. Prod. Tech. Div. Meeting, Potato Chip Inst. Intern.,* Ithaca, New York, 15, 1960.

72. **Smith, O.,** Chip color tester, *Am. Potato J.,* 37, 308, 1960.

73. **Smith, O.,** Choosing potatoes for chipping, *Natl. Potato Chip Inst. Release, Potatoes,* 4, 1, 1950.

74. **Smith, O.,** Use of antioxidants in making potato chips, *Natl. Potato Chip Inst. Fats and Oils,* Article 2, 1, 1950.

75. **Davis, C. O. and Smith, O.,** Potato quality. XVII. Objective measurement of chip color. *Potato Chipper,* 21, 72, 1962.

76. **Smith, O. and Davis, C. O.,** Preventing discoloration in cooked and French fry potatoes, *Am. Potato J.,* 37, 352, 1960.

77. **Smith, O. and Davis, C. O.,** Potato quality. XIV. Prevention of graying in dehydrated potato products, *Potato Chipper,* 21(3), 84, 1961.

78. **Isleib, D. R.,** Objective method for potato chip color determination, *Am. Potato J.,* 40, 58, 1968.

79. **Brady, H. V.,** A simple reliable procedure for moisture determination in potato chips, *Proc. Prod. Tech. Div. Meetings, Potato Chip Inst. Intern.,* Ithaca, New York, 23, 1964.

80. **Smith, O. and Davis, C. L.,** Potato processing, in *Potatoes: Production, Storing and Processing,* Smith, O., Ed., AVI Publishing, Westport, CT, 1968, 558.

81. **Smith, T. J. and Huxsoll, C. C.,** Peeling potatoes for processing, in *Potato Processing,* Talburt, W. F. and Smith, O., Eds, Van Nostrand Reinhold, New York, 1987, 333.

82. **Anon.,** Commercial infrared peeling process, *Food Process,* 31, 28, 1970.

83. **Graham, R. P., Huxsoll, C. C., Hart, M. R., Weaver, M. L., and Morgan, A. I., Jr.,** Dry caustic peeling of potatoes, *Food Technol.,* 23, 61, 1969.

84. **Huxsoll, C. C., Weaver, M. L., and Ng, K. C.,** Double-dip caustic peeling of potatoes. I. Laboratory scale development, *Am. Potato J.,* 58, 327, 1980.

85. **Weaver, M. L., Timm, H., Nonaka, M., Sayre, R. N., Ng, K. C., and Whitehand, L. C.,** Potato composition. III. Tissue selection and its effects on total nitrogen, free amino acid nitrogen, and enzyme activity, *Am. Potato J.,* 55, 319, 1978.

86. **Burton, W. G.,** Postharvest behavior and storage of potatoes, in *Applied Biology,* Vol. 2, Coaker, T. H, Ed., Academic Press, New York, 1978.

87. **Weaver, M. L., Ng, K. C., and Huxsoll, C. C.,** Sampling potato tubers to determine peel loss, *Am. Potato J.,* 56, 217, 1979.

88. **Finglas, D. M. and Faulks, R. M.,** Nutritional composition of UK retail potatoes both raw and cooked, *J. Sci. Food Agric.,* 35, 1347, 1984.

89. **True, R. H., Hogan, J. M., Augustin, J., Johnson, S. R., Teitzel, C., Toma, R., and Orr, P.,** Changes in the nutrient composition of potatoes during home preparation III. Minerals, *Am. Potato J.,* 56, 339, 1979.

90. **Zarneger, L. and Bender, A. E.,** The stability of vitamin C in machine-peeled potatoes, *Proc. Nutr. Soc.,* 30, 94, 1971.

91. **Gorun, E. G.,** Effect of mode of potato peeling on contents of B group vitamins (in Russian), *Izv. Vyssh. Uchebn. Zaved. Pishch. Tekhnol.,* 6, 154, 1978.

92. **Mondy, N. I. and Barry, G.,** Effect of peeling on total phenols, total glycoalkaloids, discoloration and flavor of cooked potatoes, *J. Food Sci.,* 53, 756, 1988.

93. **Marquez, G. and Anon, M. C.,** Influence of sugars and amino acids in color development in fried products, *J. Food Sci.,* 51, 157, 1986.

94. **Sullivan, J. F., Kozempel, M. F., Egoville, M. J., and Talley, E. A.,** Loss of amino acids and water soluble vitamins during potato processing, *J. Food Sci.,* 50, 1249, 1986.

95. **Mishkin, M., Saguy, I., and Karel, M.,** A dynamic test for kinetic models for chemical changes during processing: ascorbic acid degradation in dehydration of potatoes, *J. Food Sci.,* 49, 1267, 1984.

96. **Warner, K., Evans, C. D., List, G. R., Boundy, B. K., and Kwolek, W. F.,** Pentane formation and rancidity in vegetable oils and in potato chips, *J. Food Sci.,* 39, 761, 1974.

97. **Moore, M. D., Van Blaricom, L. D., and Senn, T. L.,** The effect of storage temperature of Irish potatoes on the resultant chip color, *Clemson Univ. Coll. For. Recreat. Resour. Dep. For. For. Res. Ser.,* 43, 1, 1963.

98. **Agle, W. M. and Woodbury, G. W.,** Specific gravity dry matter relationship and reducing sugar changes affected by potato variety, production area and storage, *Am. Potato J.,* 45, 119, 1968.

99. **Clegg, M. D. and Chapman, H. W.,** Postharvest discoloration of chips from early summer potatoes, *Am. Potato J.,* 39, 176, 1962.

100. **Dexter, S. T. and Salunkhe, D. K.,** Improvement of potato chip color by hot water treatment of slices, *Mich. State Coll. Exp. Stn. Q. Bull.,* 34, 399, 1952.

101. **Dexter, S. T. and Salunkhe, D. K.,** Chemical treatment of potato slices in relation to the extraction of sugars and other dry matter and quality of potato chips, *Mich. State Coll. Exp. Stn. Q. Bull.,* 35, 102, 1952.

102. **Dexter, S. T. and Salunkhe, D. K.,** Control of potato chip color by treatment of slices with glucose solutions following an acid treatment, *Mich. State Coll. Agric. Exp. Stn. Q. Bull.,* 35, 156, 1952.

103. **Smith, O.,** Improving the color of potato chips, *Natl. Potato Chip Inst. Potatoes,* Article 10, 1, 1950.

104. **Smith, O.,** Factors affecting the methods of determining potato chip quality, *Am. Potato J.,* 38, 265, 1961.

105. **Smith, O.,** Potato chip research in 1962, *Proc. Prod. Tech. Div. Meetings, Potato Chip Inst. Intern.,* Ithaca, New York, 1963, 2.

106. **Smith, O.,** Changes in the manufacture of snack foods, *Cereal Sci. Today,* 19, 306, 1974.

107. **Smith, O.,** *Potatoes: Production, Storing, Processing,* 2nd ed., AVI Publishing, Westport, CT, 1977.

108. **Quast, D. G. and Karel, M.,** Effects of environmental factors on the oxidation of potato chips, *J. Food Sci.,* 37, 584, 1972.

109. **Reeves, A. F.,** Potato chip color ratings of advanced selections from Maine potato breeding program, *Am. Potato J.,* 59, 389, 1982.

110. **Sherman, M. and Ewing, E. E.,** Temperature, cyanide and oxygen effects on the respiration, chip color, sugars and organic acids of stored tubers, *Am. Potato J.,* 59, 165, 1982.

111. **Singh, R. P., Heldman, D. R., and Cargill, B. F.,** The influence of storage time and environmentals on potato chip quality, in *The Potato Storage: Design, Construction, Handling and Environmental Control,* Cargill, B. F., Ed., Michigan State University, East Lansing, 1976.

112. **Whiteman, T. M.,** Improvement in the color of potato chips and French fries by certain precooking treatments, *Potato Chipper,* 11, 24, 1951.

113. **Young, N. A.,** Potato products: production and markets in European communities, Information on Agriculture No. 75, Office for Official Publications of the European Communities, Luxembourg, 1981.

114. **Fitzpatric, T. J., Talley, E. A., and Porter, W. L.,** Preliminary studies on the fate of sugars and amino acids in chips made from fresh and stored potatoes, *J. Agric. Food Chem.,* 13, 10, 1965.

115. **Fitzpatric, T. J. and Porter, W. L.,** Changes in sugars and amino acids in chips made from fresh, stored and reconditioned potatoes, *Am. Potato J.,* 43, 238, 1966.

116. **Jaswal, A. S.,** Effects of various processing methods on free and bound amino acid content of potatoes, *Am. Potato J.,* 50, 86, 1973.

117. **Bucko, A., Obonova, K., and Ambrova, P.,** Effects of storage and culinary processing on vitamin C losses in vegetables and potatoes, *Nahrung,* 21, 107, 1977.

118. **Pelletier, O., Nantel, C., Leduc, R., Tremblay, L., and Brassard, R.,** Vitamin C in potatoes prepared in various ways, *J. Inst. Can. Sci. Technol. Aliment.,* 10, 138, 1977.

119. **Deutsch, R. M.,** Science looks at potato chips, *Chipper Snacker,* 35, 15, 1978.

120. **Shaw, R., Evans, C. D., Munson, S., List, G. R., and Warner, K.,** Potato chips from unpeeled potatoes, *Am. Potato J.,* 50, 424, 1973.

121. **Feustel, I. C. and Kueneman, R. W.,** Frozen French fries and other frozen products, in *Potato Processing,* 2nd ed., Talburt, W. F. and Smith, O., Eds., AVI Publishing, Westport, CT, 1967.

122. **Weaver, M. L., Reeve, R. M., and Kueneman, R. W.,** Frozen French fries and other frozen potato products, in *Potato Processing,* 3rd ed., Talburt, W. F. and Smith, O., Eds., AVI Publishing, Westport, CT, 1975.

123. **Zobel, M.,** Nutritional aspects of potato peeling in the DDR and from the international viewpoint, *Ernaehrungsforschung,* 24, 74, 1979.

124. **Murphy, E. W., Marsh, A. C., White, K. E., and Hagan, S. N.,** Proximate composition of ready to serve potato products, *J. Am. Diet. Assoc.,* 49, 122, 1966.

125. **Augustin, J., Swanson, B. G., Teitzel, C., Johnson, S. R., Pometto, S. F., Artz, W. E., Huang, C. P., and Shoemaker, C.,** Changes in the nutrient composition during commercial processing of frozen potato products, *J. Food Sci.,* 44, 807, 1979.

126. **Oguntuna, T. E. and Bender, A. E.,** Loss of thiamin from potatoes, *J. Food Technol.,* 11, 347, 1976.

127. **Gorun, E. G.,** Changes in the vitamin activity of potatoes during production of quick frozen fresh fries (in Russian), *Kholod. Tekh.,* 10, 15, 1978.

128. **Feustel, I. C. and Kueman, R. W.,** Frozen French fries and other frozen potato products, in *Potato Processing,* 2nd ed., Talburt, W. F. and Smith, O., Eds., AVI Publishing, Westport, CT, 1967.

129. **Tressler, D. K., Van Arsdel, W. B., and Copley, M. J.,** *The Freezing Preservation of Foods,* Vol. 4, 4th ed., AVI Publishing, Westport, CT, 1968.

130. **Talburt, W. F.,** Canned white potatoes, in *Potato Processing,* 2nd ed., Talburt, W. F. and Smith, O., Eds., AVI Publishing, Westport CT, 1967.

131. **Markakis, P., Freeman, T. M., and Harte, W. H.,** Dough containing vital gluten, amylopectin and inert starches, U.S. Patent 3027258, 1962.

132. **Fast, R. B., Spotts, C. E., and Morck, R. A.,** Process for making a puffable chip-type snack food product. U.S. Patent 3451822, 1969.

133. **Reinertsen, B. J.,** Method of Making a Chip-Type Food Product, U.S. Patent 3361573, 1968.

134. **Singer, N. S. and Beltran, E. G.,** Process of Making a Snack Product, U.S. Patent 3502479, 1970.

135. **Liepa, A. L.,** Potato Food Product, U.S. Patent 3396036, 1968.

136. **Hilton, B. W.,** Potato Products and Process for Making Same, U.S. Patent 3109739, 1963.

137. **Murray, D. G., Marota, N. G., and Boettger, R. M.,** Novel Amylose Coating for Deep Fried Potato Products, U.S. Patent 3597227, 1971.

138. **Slakis, A. J., Kubr, W. K., Hughes, R. L., and Neilson, A. J.,** Cyclohexylsulfamic Acid as a Taste Improver of Raw Potato Products, U.S. Patent 3353962, 1970.

139. **Olson, R. L. and Harrington, W. O.,** Dehydrated Mashed Potatoes—a Review, U.S. Department of Agriculture, Western Regional Research Laboratory, Albany, California, AIC-297, 1951.

140. **Kueneman, R. W.,** Dehydrated mashed potatoes, *Proc. Potato Utiliz. Conf.,* 8, 64, 1957.

141. **Feustel, I. C., Hendel, C. F., and Juilly, M. E.,** Potatoes, in *Food Dehydration—Vol. II, Processes and Products,* AVI Publishing, Westport, CT, 1964.

142. **Talburt, W. E., Boyle, F. P., and Hendel, C. F.,** Dehydrated mashed potatoes—potato granules, in *Potato Processing,* Talburt, W. F. and Smith, O., Eds., Van Nostrand Reinhold Company, New York, 1987, 12.

143. **Ooraikul, B.,** Processing of Potato Granules with the Aid of Freeze-Thaw Technique, Ph.D. thesis, University of Alberta, Edmonton, Canada, 1973.

144. **Jadhav, S., Steele, L., and Hadziyev, D.,** Vitamin C losses during production of dehydrated mashed potatoes, *Food Sci. Technol.,* 8, 225, 1975.

145. **Jadhav, S., Berry, L. M., and Clegg, L. F. L.,** Extruded French fries from dehydrated potato granules processed by a freeze-thaw technique, *J. Food Sci.,* 41, 852, 1976.

146. **Potter, A. L., Jr.,** Changes in physical properties of starch in potato granules during processing, *J. Agric. Food Chem.,* 2, 516, 1954.

147. **Burton, W. G.,** Mashed potato powder. IV. Deterioration due to oxidative changes, *J. Soc. Chem. Ind.,* 68, 119, 1949.

148. **Willard, M. J., Hix, V. M., and Kluge, G.,** Dehydrated mashed potatoes—potato flakes, in *Potato Processing,* Talburt, W. F. and Smith, O., Eds., Van Nostrand Reinhold Company, New York, 1987, 13.

149. **Pedersen, D. C. and Sautier, P. M.,** Vitamin Enriched Potato Flakes, U.S. Patent 3,833,739, Sept. 3, 1974.

150. **Sapers, G. M., Panasuik, O., Talley, F. B., Osman, S. F., and Shaw, R. L.,** Flavor quality and stability of potato flakes: volatile components associated with storage changes, *J. Food Sci.,* 37, 579, 1972.

151. **Augustin, J., Swanson, B. G., Pometto, S. F., Teitzel, C., Artz, W. F., and Huang, C. P.,** Changes in nutrient composition of dehydrated potato products during commercial processing, *J. Food Sci.,* 44, 216, 1979.

152. **Augustin, J., Marousek, G. A., Artz, W. F., and Swanson, B. G.,** Vitamin retention during preparation and holding of mashed potatoes made from commercially dehydrated flakes and granules, *J. Food Sci.,* 47, 274, 1982.

153. **Kluge, G., Appoldt, F. S. Y., Seiler, G., and Petutschnig, K.,** Production of Dried Potato Flakes, British Patent 1473036, 1977.

154. **Sapers, G. M., Panasuik, O., Jones, S. B., Kalan, E. B., Talley, F. B., and Shaw, R. L.,** Iron fortification of dehydrated mashed potatoes, *J. Food Sci.,* 39, 552, 1974.

155. **Glabau, C. A.,** Cookie production with potato flour as an ingredient, *Bakers Wkly.,* 178, 40, 1958.

156. **Kim, H. S. and Lee, H. J.,** Development of composite flours and their products utilizing domestic raw materials. IV. Effect of additives on the bread making quality of composite flours, *Korean J. Food Sci. Technol.,* 9, 106, 1977.

157. **Rivoche, E. J.,** Food Products and Method of Making the Same, U.S. Patent 2791508, 1957.

158. **El-Samahy, S. K., Elias, A. M., and Morad, M. M.,** Effect of potato flour on the rheological properties of the dough, bread quality and staling, *Cereal Microbiol. Technol. Lebensm.,* 4, 186, 1976.

159. **Ceh, M.,** Baking bread with potato flour addition, *Hrana Ishrana* (Yugoslavian), 15, 113, 1974.

160. **Yanez, E., Ballester, D., Wuth, H., Orrego, W., Gattas, V., and Estay, S.,** Potato flour as a partial replacement of wheat flour in bread: baking studies and nutritional value of bread containing graded levels of potato flour, *J. Food Technol.,* 16, 291, 1981.

161. **Zahana, T., Sutescu, P., Popescu, F., and Gontea, L.,** The quality of bread supplemented with maize flour and other ingredients, *Igiena* (Romanian), 22, 11, 1972.

162. **Bushway, A. A., Belyea, P. R., True, R. H., Work, T. M., Russell, D. O., and McGann, D. F.,** Potato starch and flour in frankfurters: effect on chemical and sensory properties and total plate counts, *J. Food Sci.,* 47, 402, 1982.

163. **Feustel, I. C.,** Miscellaneous products from potatoes, in *Potato Processing,* Talburt, W. F. and Smith, O., Eds., Van Nostrand Reinhold, New York, 1987, 727.

164. **Treadway, R. H., Wagner, J. R., Woodward, C. F., Heisler, E. G., and Hopkins, R. M.,** Development and evaluation of potato chip bars, *Food Technol.,* 12, 479, 1958.

165. **Townsley, P. M. and Dixon, E.,** Potato chip confections, *Can. Food Ind.,* 23, 21, 1952.

166. **Highlands, M. E. and Getchell, J. S.,** A new snack item—French fried potato dice, *Maine Farm Res.,* 4, 10, 1956.

167. **Siciliano, J., Woodward, C. F., Treadway, R. H., and Heisler, E. G.,** Potato nuts—a new type of snack, U.S. Department of Agriculture, ARS-73-15, 1956.

168. **Willard, M. J.,** Preparing a potato snack product, U.S. Patent 3,634,094, 1972.

169. **Harrington, W. O. and Griffiths, F. P.,** WRRL "Puffs" potatoes, *Food Ind.,* 22, 1872, 1950.

170. **Talburt, W. F., Weaver, M. L., Reeve, R. M., and Kueneman, R. W.,** Frozen French fries and other frozen potato products, in *Potato Processing,* Talburt, W. F. and Smith, O., Eds., Van Nostrand Reinhold, New York, 1987, 491.

171. **Anon.,** Potato puff formula, *Food Eng.,* 27, 194, 1955.

172. **Harrington, W. O., Olson, R. L., and McCready, R. M.,** Quick cooking dehydrated potatoes, *Food Technol.,* 5, 311, 1957.

173. **Potter, A. L. and Belote, M. L.,** New potato snack item, *Bakers Wkly.,* 197, 42, 1963.

174. **Baczkowicz, M. and Tomasik, P.,** A novel method of utilization of potato juice, *Starch,* 37, 241, 1985.

175. **Cordon, T. C., Treadway, R. H., Walsh, M. D., and Osborne, M. F.,** Lactic acid from potatoes, *Ind. Eng. Chem.,* 42, 1833, 1950.

176. **Eskew, R. K., Edwards, P. W., and Redfield, C. S.,** Recovery and utilization of pulp from white potato starch factories, U.S. Department of Agriculture, Eastern Regional Research Laboratories, AIC-204, 1948.

177. **Highlands, M. E.,** Potatoes and potato pulp for livestock feed, in *Potato Processing,* Talburt, W. F. and Smith, O., Eds., AVI Publishing, Westport, CT, 1967, 540.

178. **Treadway, R. H.,** Potato starch, in *Potato Processing,* Talburt, W. F. and Smith, O., Eds., Van Nostrand Reinhold, New York, 1987, 647.

179. **Szkilladziowa, W., Secomska, B., Nadolna, I., Trzebska-Jeska, I., Wartanowicz, M., and Rakowska, M.,** Results of studies on nutrient content in selected varieties of edible potatoes, *Acta Aliment. Acad. Sci. Hung.,* 3, 87, 1977.

180. **Augustin, J., Toma, R. B., True, R. H., Shaw, R. L., Teitzel, C., Johnson, S. R., and Orr, P.,** Composition of raw and cooked potato peel and flesh: proximate and vitamin composition, *J. Food Sci.,* 44, 805, 1979.

181. **Bretzloff, C. W.,** Calcium and magnesium distribution in potato tubers, *Am. Potato J.,* 48, 97, 1971.

182. **Johnston, F. B., Hoffonan, I., and Petrosovits, A.,** Distribution of mineral constituents and dry matter in potato tuber, *Am. Potato J.,* 45, 287, 1968.

183. **Kubisk, A., Tomkowiak, J., and Andrzejewska, M.,** The content of some trace elements in different parts of the tuber in five potato varieties, *Hodowla Rosl. Aklim. Nasienn.,* 22, 81, 1978.

184. **Mondy, N. I. and Ponnampalam, R.,** Effect of baking and frying on nutritive value of potato minerals, *J. Food Sci.,* 48, 1475, 1983.

185. **Chick, H. and Slack, E. B.,** Distribution and nutritive value of the nitrogenous substances in the potato, *Biochem. J.,* 45, 211, 1949.

186. **Talley, E. A., Toma, R. B., and Orr, P. H.,** Composition of raw and cooked potato peel and flesh: amino acid content, *J. Food Sci.,* 48, 60, 1983.

187. **Herrera, H.,** Potato Protein: Nutritional Evaluation and Utilization, Ph.D. thesis, Michigan State University, East Lansing, 1979.

188. **Toma, R. B., Augustin, J., Orr, P., True, R. H., Hogan, J. M., and Shaw, R. L.,** Changes in the nutrient composition of potatoes during home preparation. I. Proximate composition, *Am. Potato J.,* 55, 639, 1978.

189. **Hellendoorn, E. W., Vanden Top, M., and Van der Weide, J. E. M.,** Digestibility *in vitro* of dry mashed potato products, *J. Sci. Food Agric.,* 21, 71, 1970.

190. **Englyst, H., Wiggins, H. S., and Cummings, J. H.,** Determination of non-starch polysaccharide in plant foods by gas-liquid chromatography of constituent sugars as alditol acetates, *Analyst,* 107, 307, 1982.

191. **Jone, G. P., Briggs, D. R., Wahlquist, M. L., and Flentge, L. M.,** Dietary fiber content of Australia foods. I. Potatoes, *Food Technol. Aust.,* 37, 81, 1985.

192. **Choudhari, R. N., Joseph, A. A., Ambrose, D., Narayana Rao, V., Swaminathan, M., Srenivasan, A., and Subramanyan, V.,** Effect of cooking, frying, baking and canning on the nutritive value of potato, *Food Sci.* (Mysore), 12, 253, 1963.

193. **Jaswal, A. S.,** Effect of various processing methods on free and bound amino acid content of potatoes, *Am. Potato J.,* 50, 86, 1973.

194. **Johnston, D. E. and Oliver, W. T.,** The influence of cooking technique on dietary fiber of boiled potato, *J. Food Technol.,* 17, 99, 1982.

195. **Paul, A. A., Southgate, D. A. T., McCance, and Widdowson,** 's, the Composition of Foods, 4th ed., MRC Serial report No. 297, London, 1978.

196. **Swaminathan, K. and Gangwar, B. M. L.,** Cooking losses of vitamin C in Indian potato varieties, *Indian Potato J.,* 3, 86, 1961.

197. **Leichsering, J. M., Norris, L. M., and Pilcher, H. L.,** Ascorbic acid contents of potatoes. I. Effect of storage and of boiling on the ascorbic, dehydroascorbic and diketogluconic acid contents of potatoes, *Food Res.,* 22, 37, 1957.

198. **Woolfe, J. A.,** Effects of storage, cooking and processing on nutritive value of potato, in *The Potato in Human Diet,* Cambridge University Press, London, 1987, 4.

199. **Fenton, F.,** Vitamin C retention as a criterion of quality and nutritive value in vegetables, *J. Am. Diet. Assoc.,* 16, 524, 1940.

200. **Richardson, J. E., Davis, R., and Mayfield, H. L.,** Vitamin C content of potatoes prepared for table use by various methods of cooking, *Food Res.,* 2, 85, 1937.

201. **Domah A., Davidek, J., and Velisek, J.,** Changes of L-ascorbic acid and L-dehydroascorbic acids during cooking and frying of potatoes, *Z. Lebensm. Unters. Forsch.,* 154, 272, 1974.

202. **Page, E. and Hanning, F. M.,** Vitamin B_6 and niacin in potatoes, *J. Am. Diet. Assoc.,* 42, 42, 1963.

203. **Kahn, R. M. and Halliday, E. G.,** Ascorbic acid content of white potatoes as affected by cooking and standing on steam table, *J. Am. Diet. Assoc.,* 20, 220, 1944.

204. **Pelletier, O., Nantel, C., Leduc, R., Tremblay, L., and Brassard, R.,** Vitamin C in potatoes prepared in various ways, *Can Inst. Food Sci. Technol. J.,* 10, 138, 1977.

205. **Korschgen, B., Baldwin, R., and Snider, S.,** Quality factors in beef, pork and lamb cooked by microwaves, *J. Am. Diet. Assoc.,* 69, 635, 1976.

206. **Klein, L. B. and Mondy, N. I.,** Comparison of microwave and conventional baking of potatoes in relation to nitrogenous constituents and mineral composition, *J. Food Sci.,* 46, 1874, 1981.

207. **Treadway, R. H., Heister, E. G., Whittenberger, R. T., Highlands, M. E., and Getchell, J. S.,** Natural dehydration of cull potatoes by alternate freezing and thawing, *Am. Potato J.,* 32, 293, 1955.

208. **Werge, R. W.,** Potato processing in the Central highlands of Peru, *Ecol. Food Nutr.,* 7, 229, 1979.

209. **Yahya, M.,** Studies on the Utilization of Potato for Processing into Convenience and Nutritive Food, M.Sc. thesis, Mahatma Phule Agricultural University, Rahuri, India, 1988.

210. **Anand, J. C. and Maini, S. B.,** Utilization of potato under glut condition possible outlets, *Proceedings of the International Symposium on Postharvest Technology & Utilization of Potatoes,* Kishore, H., Ed., International Potato Center, Region VI, New Delhi, 1979.

211. **Nankar, J. T. and Nankar, V. J.,** Studies on the processing and utilization of potato in rural India, in *Proceeding of International Symposium on Postharvest Technology and Utilization of Potatoes,* Kishore, H., Ed., International Potato Center, Region VI, New Delhi, 1979.

Chapter 6

STARCH AND ITS DERIVATIVES

S. P. Phadnis and S. J. Jadhav

TABLE OF CONTENTS

I. INTRODUCTION

Starch is an abundant and well-known source of renewable raw material. It is used as a feed or food, and about 1 to 1.5% is processed further so as to make it suitable for other applications.[1] Maize or corn is the most important source of this raw material, but potatoes are also of significant importance. Because of its unique properties,[2] potato starch (sometimes termed *farina*) has maintained a position in certain applications in the face of lower-priced maize starch (Table 1). About 3% of the world crop of potatoes is used for the production of starch. The world production of potato starch is approximately 2 million tons.[2] Apart from the U.S., the Netherlands, Poland, and West Germany are the major producers of potato starch. Some important manufacturers of potato starch are Avebe (the Netherlands), Henkel (West Germany), Colby Starch Co., A.E. Staley Mfg. Co., Boise Cascade Corp., and J. R. Simplot Co. (U.S.). In the U.S., potato starch[3] accounted for 4.4% of the total of 2.3 billion kg of starch consumed in the year 1979. Maize starch held 91% of the market and the remainder was mostly wheat and tapioca starches. If potato starch becomes available in sufficient quantity at corn starch prices, it would certainly be preferred in most applications. Its most important characteristics are (1) high consistency on pasting followed by a decrease in viscosity on further heating and agitation, (2) excellent flexible film formation, (3) good binding power, and (4) low gelatinization temperature. Thus, in specialty applications where performance is needed, potato starch justifies its premium over maize starch.

The acceptance of high fructose corn syrup (HFCS) in soft drinks and the growing support of gasohol are likely to improve the position of potato starch in the years ahead. Substantial quantities of maize starch are likely to get diverted for the production of HFCS and gasohol and may limit its expansion, thus putting an upward pressure on its price. If that happens, potato starch may become more competitive and some expansion of it may result. Hence, the future of potato starch appears to be reasonably secure and it is likely to continue its role as an important material for food and industrial applications.[3]

II. PRODUCTION OF POTATO STARCH

Before undertaking the manufacture of potato starch,[4] certain economic considerations must be taken into account. The market for all kinds of starch has been affected by overproduction. The capital requirement of the industry is high and the working period for the factory is short. In temperature zones where potatoes can be grown satisfactorily, it becomes impossible to run a factory in frosty weather. It is also necessary to use only those varieties of potatoes that have very high starch content, and these should be preferably grown for captive consumption. Alternatively, these should be contracted for long years to ensure the unbroken supply of the raw material. Since the water requirement of this industry is quite large (100 m³/ton of starch), the supply of water should be assured and a corresponding outlet for the dilute waste and its treatment has to be provided.

In comparison with the cereal starches, the manufacture of potato starch is a relatively easy process. The factory scheme for the isolation and purification of starch is confined to a succession of simple subprocesses, viz., milling, screening, settling, and drying. A product of reasonably good quality can be made in a rather simple arrangement of machinery around one good rasp. A typical scheme of the potato starch factory involves milling, screening (coarse), settling, screening (fine), settling, filtering (vacuum or centrifuge), drying, and packaging. The choice of machinery depends, among other things, upon the yield of starch per hour, the cost of machinery and utilities, and the quality specifications.

Factory potatoes usually harvested by machinery are first cleaned. The arrangement and the sequence of the cleaning machinery depend upon local circumstances. On unloading, the potatoes are wetted, stored for a while, and washed with jets of water. Transportation

TABLE 1
Properties of Potato Starch

Shape of granules	Spherical, oval, pronounced oyster shell-like striations around an eccentrically placed hilum
Size (diameter) of granules	
Range (μm)	5 to 10
Number average	33
Chemical composition of granules (%)	
Moisture	19 (at 65% RH and 20°C)
Lipids	0.05 (on dry substance)
Proteins	0.06 (on dry substance)
Ash	0.4 (on dry substance)
Phosphorus	0.08 (on dry substance)
Amylose content	21% (w/w); DP 3000
Amylopectin content	79% (w/w); DP 2 million
Gel temp range (Kofler)	58 to 68°C
Brabender viscosity (8% paste)	
Pasting temp	60 to 65°C
Peak viscosity	3000 Brabender units
Swelling power at 95°C	1153
Properties of starch paste	
Viscosity	Very high
Texture	Long
Clarity	Translucent
Resistance to shear	Medium or low
Rate of retrogradation	Medium

Note: DP, degree of polymerization; w/w, weight per weight.

in flumes to the cleaning machinery then follows. The surface of the flumes is skimmed off to eliminate floating material like straw and haulms. The potatoes are then swirled up by a strong flow of water to eliminate stones, usually in cylindrical vats with a horizontal axis. Finally, they are picked up from a moving belt by knives or pointed spikes, leaving behind hard substances. The cleaned potatoes are dried superficially by passing them over a sieve.

The next step is milling, which aims at opening each cell of the parenchyma and at pressing the starch out of the remaining sack. Milling is done mostly in rotary saw-blade rasps, a number of variations of which are known. The performance of rasps depends upon such variables as the type of saw blades, the rotation velocity, and the number of silts and their width. Milling is usually a two-stage process, although everything is done to obtain high efficiency in the first stage itself. This is followed by screening which is done mainly to remove the last traces of large particles from the final starch suspension or the so-called mill flow. There are a multitude of machines available for this purpose and most of these are fitted with a fine-weave textile. In general, sieves which vibrate freely and are enclosed totally can be dismantled easily for maintenance and are leakproof. Normally, a minimum of four stages of sieving are used in sequence with one set of second rasps. A common sequence is two rotary sieves, a second rasp, and again two rotary sieves. A fifth sieve may be provided if necessary at the delivery point. A first stage of concentration of the starch milk in centrifugal or separatory hydrocyclone is interposed between crude and fine sieving as an amount of cell wall material and many other impurities are removed in this stage.

The further purification of starch has two aspects: (1) removal of soluble impurities and (2) removal of insoluble impurities. The soluble impurities are removed by alternate cycles of filtration and redispersion of the starch cake in water. After three cycles each, most of the soluble impurities are removed. As the washing efficiency depends upon the temperature of the water, a temperature of around 30°C is recommended. The insoluble impurities are removed by the settling method. This can be done in settling vats, tables, nozzle separators, hydrocyclones, or a basket centrifuge with a scraper.

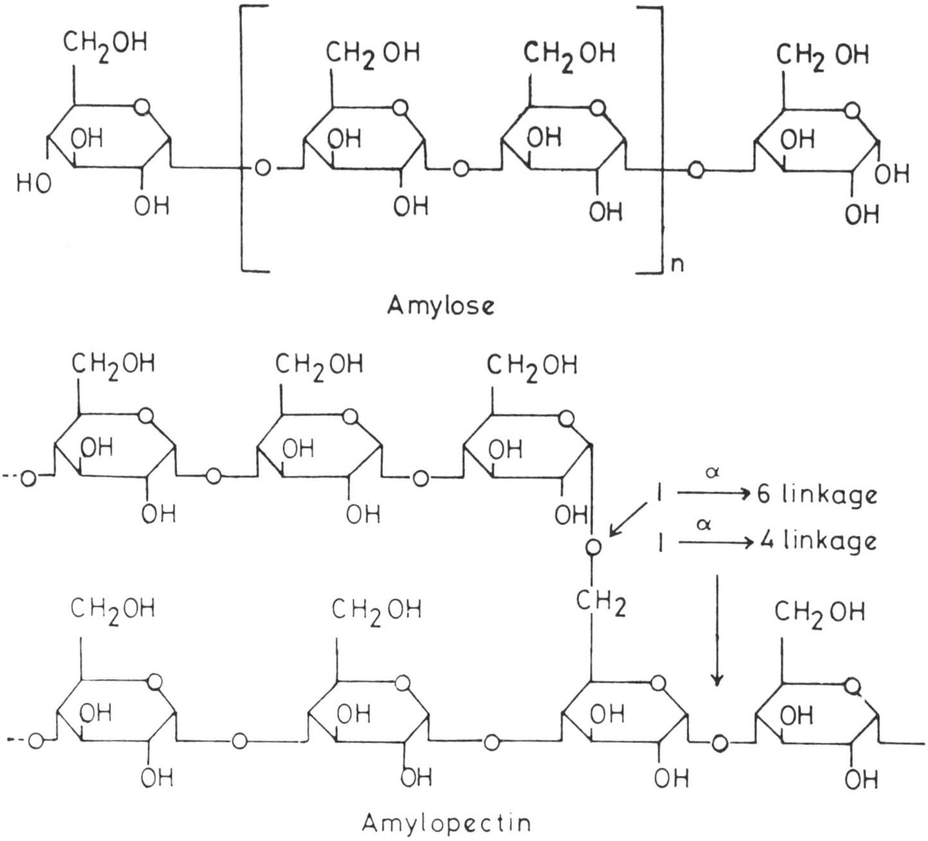

FIGURE 1.   Amylose, the linear polymer of glucose (above). Amylopectin, the branched polymer of glucose (below).

The clean starch from the last settling stage is dewatered, usually on a rotatory suction filter or in an open basket centrifuge. A dry matter content of around 60% is obtained. As a rule, a suspension of relatively low starch content is fed to the apparatus to facilitate the even spread of the layer.

The last stage of dewatering is a hot-air drying which is done with a turbulent airstream heated to 140°C. In the ordinary commercial product, a water content of 20% is left. The process is very fast and is so organized that the gelatinization of starch, if any, is negligible. Final screening may be resorted and the product is packed in jute bales or paper sacks.

III. STARCH FRACTIONS AND PROPERTIES

Most starches[5-9] consist of a mixture of two polysaccharide types, viz., amylose and amylopectin. The former is essentially a linear polymer, while the latter is a highly branched polymer (Figure 1). The relative amounts of these two fractions in a particular starch are a major factor in determining the properties of that starch. Generally, amylopectin in starch granules is about 75% and the rest is amylose. By genetic manipulation, however, varieties of plants have been developed which give starches containing almost 100% amylopectin or those containing up to 70% amylose. Since growth conditions are different in each plant, the starch from different plant sources varies somewhat in appearance, composition, and properties. The starch is therefore described by its plant source, e.g., potato starch, corn starch, wheat starch, and rice starch. The starch granules are a highly organized complex formation of macromolecules of D-glucose units and they exhibit interesting properties when dispersed in water and heated. The pastes thus formed exhibit excellent adhesive properties and act as binders for cellulosic and other hydrophilic surfaces. They also act as dispersants, stabilizers, and thickening agents. Because of these properties they are used in the textile industry[5,10,11] for sizing of warps, as thickeners for print pastes, and as finishing agents. In the paper industry[5,12-14] starches find application as wet-end additives to pulp for paper formation, as surface-sizing agents, and as coating components. They are also used for paper laminations in the manufacture of corrugated boards and boxes and as sealants for envelope flaps and stamps. Other uses[15-17] of starch are as thickeners for foods, suspending agents for colloidal sols and medicinal formulations, and as binders in tablet making.

There are four prime factors which largely determine the properties of starch.

1. The starch granules contain two types of glucose polymer: linear (amylose) and branched (amylopectin).
2. The polymeric molecules are organized and packed into granules which are insoluble in water at ambient temperature.
3. Disruption of granule structure is required to render the starch polymers dispersible in water.
4. In aqueous starch paste, amylose molecules can associate with each other to render starch insoluble in water or form a gel.

The salient properties of amylose and amylopectin are summarized below.

Property	Amylose	Amylopectin
Molecular shape	Linear	Branched
Molecular weight	10,000 to 60,000	50,000 to 100,000 or more
Tendency to retrograde	Strong, rapid	Weak or lacking
Action of β-amylase	100% hydrolysis	Up to 60% hydrolysis
Cooking with water	Solid and liquid phases separate	Forms gel or paste
Color with iodine	Deep blue	Reddish brown

The properties of starch granules in water are the major factors in the commercial utility of starch. Since the granule is insoluble in water below about 50°C, the starch can be extracted from its plant source in aqueous system, purified and modified in suspension in water, and recovered by filtration and drying. It can be handled as an aqueous slurry (30 to 40%) which is pumpable until its thickening, adhesive, binding, and film-forming properties are needed. When the aqueous slurry is heated to a temperature of 55 to 80°C, the intermolecular hydrogen bonds holding the granules together are weakened and the granule undergoes a rapid, irreversible swelling. The critical temperature at which this occurs is known as the pasting or gelatinization temperature. The granules take up the water, swelling to many times their original volume, rupturing and collapsing as the heating and agitation of the mixture continues, and releasing concurrently some of the starch molecules, particularly amylose, into solution. The viscosity increases to a maximum that corresponds to the largest hydrated swollen volume of the granule before it bursts apart and then the viscosity declines as the swollen granules disintegrate. The result is a viscous colloidal dispersion which is a complex mixture of residual swollen granule masses, hydrated molecular aggregates, and dissolved molecules.

One of the most important properties of starch is the formation of a viscous paste on cooking in water. Thus, the measurement of changes in viscosity or consistency as an aqueous suspension of starch is heated is of practical value in predicting the useful properties of a particular starch or modified starch.[5,18-22]

In addition to the type of modification, the properties of the starch paste are determined by the cooking procedure and the presence of other materials. Thus, careful attention must be given to controlling the conditions of cooking in order to obtain the desired properties consistently. The factors involved are concentration, temperature, time, intensity of agitation, pH, and the type of additive or impurity present.

The most important of all the tests that are employed for characterizing starches is the paste viscosity.[5,18-22] To obtain a measure of viscosity of a hot starch paste, it must be determined at low concentration. At higher concentration, pastes deviate from true Newtonian properties of viscous flow. For most industrial purposes, however, it is the latter type of pastes that are employed and which normally possess anomalous viscosities. Since the observed viscosity of starch paste depends upon so many variables, the entire procedure for determining viscosity should be standardized with great care in order to obtain satisfactory results. The variables involved are the origin, history, and source of the sample, the concentration of the paste, the initial temperature of water used for making the paste, the rate of increase of temperature, the rate and mode of stirring, the pH, the presence of extraneous material, the technique, and the instrument used for the determination of viscosity. The Brabender® viscograph is used most widely and commonly to characterize starch pastes.[19]

IV. STARCH AND ITS DERIVATIVES

Considering the various uses of starch, it is to be expected that not all starches will satisfy the property requirements for all end uses. For certain end uses, no starch is suitable in its natural form. It is therefore necessary that starch has to be modified by suitable means such as physical, chemical, or enzymic to suit the specific requirements of a particular use. The primary objective of the modification of starch is to enhance or repress its intrinsic properties or to impart new ones. This amenability to modification has been an important factor in developing new uses of starch. The commercial, degraded, or depolymerized starches are referred to as the converted starches and usually comprise the acid-converted, thin-boiling starches, the dextrins as well as the oxidized starches. The term "modified starch" usually refers to any starch the hydroxyl group of which has been altered by a chemical reaction like esterification, etherification, oxidation, and the like.

A. THIN BOILING STARCH AND DEXTRINS

The term "thin boiling starch" refers to any starch depolymerized under the influence of acid catalysis. Alternatively, the products are called as fluidity starch, acid-cut starch, acid-modified starch, acid-converted starch, or acid-hydrolyzed starch, and these terms are used interchangeably.[5,23-26] These products are made for making a Newtonian or almost Newtonian solution of a much higher concentration than would be possible with native starch.

In making these products, the starch is suspended in dilute sulfuric or hydrochloric acid held at around 50°C under pressure for a period of up to 24 h. Depending upon the desired degree of conversion, actual conditions of the treatment are suitably modified. The reaction is controlled by monitoring the viscosity during the reaction and the action is stopped by the addition of a suitable reagent, viz., sodium hydroxide, sodium carbonate, or calcium hydroxide. Subsequently, the product is filtered, washed, and dried. It is necessary to ensure that the hydrolysis is not extensive enough to make the process economically unviable in terms of product recovery and pollution in effluents.

The rate of hydrolysis is determined by the acid concentration and temperature, and these factors are taken into account in predicting the time to stop the hydrolysis. Since the product is recovered by filtration, the degree of hydrolysis needs to be limited so that the starch granule is not degraded to a point where it becomes solubilized. During hydrolysis, starch fragments of various sizes are split off from the molecule which dissolve in the reaction medium. They are lost when the acid-converted starches are filtered and washed.

When an aqueous suspension of acid-converted starch is heated to gelatinize the starch, the modified granules swell less than those of the native starch and tend to fragment after limited swelling. More starch is dissolved and dispersed and the peak viscosity is lower than that of the unmodified starch in the initial cooking cycle. The fluidity starch disperses to a clear fluid sol. On cooling, these sols retrograde to form firm gels, particularly at the lower range of conversion. The degree of conversion of starch with respect to the conditions of acid treatment can be ascertained with the help of parameters such as dextrose equivalent, alkali number, reducing value, copper number, and fluidity number.[5,9,22,27,28]

The textile industry forms the main outlet for acid-modified starch. Large tonnages are used annually for sizing and finishing mainly of cotton-synthetic blend goods. Usually, the finer the cloth or yarn, the higher the fluidity that is required.

Although the meaning of the term ''dextrin'' is very broad, covering almost any starch-degradation product obtained by physical, chemical, or enzymic methods, for all practical purposes it refers to products derived by the pyroconversion of starch, i.e., the treatment of a dried, acidified starch with heat.[29-31] These products are needed when a high solid starch dispersion with a pumpable and workable viscosity is required. Their solutions are Newtonian or nearly Newtonian at very high concentrations. These are almost free of structure and are therefore able to penetrate fibrous materials like paper and textiles and can act as glue.

Depending upon the extent of modification, the pyrodextrins are classified into three groups. The British gums are prepared by heating starch with agitation for 10 to 20 h at 170 to 195°C. The products have a buff or brown color. Sometimes alkaline materials such as sodium bicarbonate or ammonia are added before pyrolysis. White dextrins are prepared by spraying a powdered starch with acid (usually dilute hydrochloric acid) before it is heated at 95 to 120°C for 3 to 8 h. Yellow or canary dextrins are produced by using less acid, but more strenuous conditions, viz., 150 to 180°C for 6 to 18 h.

British gums and dextrins are used in a great variety of adhesive applications. Because their pastes give lower viscosities than those from raw starches, they can be used at higher concentrations, giving films which can dry rapidly with fast tack. They are mostly used for envelope and labeling adhesives, as adhesives for postage stamps, gummed tapes, and cardboard boxes, and in many other binding and sizing applications.

B. OXIDIZED STARCH

Oxidized starches[23,26,32] are made for the explicit purpose of making a Newtonian or almost Newtonian solution of a much higher concentration than would be possible with native starch. In general appearance, oxidized starch resembles common starch in most respects. The individual granules retain their characteristic shapes and their polarization crosses. The granules are insoluble in cold water and give the characteristic color reaction with iodine. However, the difference between unmodified and oxidized starches becomes apparent when the starches are made into pastes.

There are many ways of oxidizing starch, but the classic method is to treat starch with sodium hypochlorite. Oxidation can be effected with ammonium persulfate, potassium permanganate, and hydrogen peroxide. However, the resulting products are characterized by low reagent treatment compared with conventional oxidation using sodium hypochlorite.

The manufacture of oxidized starch consists of the addition of a predetermined quantity of sodium hypochlorite to the aqueous starch slurry. Alkali is added to maintain the pH at 8 to 10 throughout the reaction. A cooling system is employed to dissipate the heat involved in the reaction and the temperature is maintained around 30°C. The amount of hypochlorite added is usually equivalent to between 0.5 and 6% of available chlorine based on starch and is determined by the fluidity grade of starch to be made. Usually, the more reagent added, the less will be the viscosity of the starch when gelatinized. The rate of oxidation of starch by hypochlorite can be increased by the presence of catalysts such as Co^{+++}, Br^-, or nickel sulfate. After a reaction period of 5 to 20 h, the slurry is neutralized and any free chloride present in it is destroyed with sodium bisulfite. The starch is washed on a suitable device to remove soluble byproducts before being collected for drying.

Oxidation increases the clarity of the paste, but reduces the strength of the resulting gel to a much greater extent than acid treatment. Hence, these are of interest for very tender gums or jell confections of high clarity. Oxidation of starch with calcium peroxide is claimed to produce a starch which will give a strong gel as well as a high degree of clarity. The degree of oxidation of starch governs the viscosity which is tailored to the particular use to which the starch is to be put. The number of carboxyl groups and carbonyl groups determines the way in which the oxidized starch will disperse when cooked. A well-regulated oxidation gives a product that disperses satisfactorily while being cooked. After cooling down the solution, an immediate retrogradation will not occur. There are still very many hydroxyl groups in oxidized starch, but the relatively small number of COOH, CO, and CHO groups limits the formation of hydrogen bonds as these groups prevent the chains from coming together. However, a sufficient stability level also means a rather extensive degradation, and this implies the loss of binding power.

When the starch has been oxidized to a sufficient degree, the product is soluble in hot water, dissolving by fragmentation and not by swelling to give mobile solutions which contain a high proportion of oxidized starch, and the products show an enhanced reducing power. Although the film produced from such solutions have a somewhat lower tensile strength than those from unmodified starch, they possess an advantage in that they are more transparent and have better power of penetration that make them of great value in the adhesive, paper, and textile industries. They are whiter in color and have the additional advantage of limited setting up (or retrogradation) on standing, thus providing products that give permanently clear solutions. These properties make oxidized starches different from acid-treated starches of similar fluidity. Commercially, these are available over a wide range of viscosity. Products based on maize starch and potato starch are available.

C. DIALDEHYDE STARCH

The periodic oxidation of starch is a specific reaction involving the scission of the carbon-carbon link in the 2-3 position of glucose units and oxidation of the adjacent hydroxyl groups on these carbon atoms to aldehyde groups. It is possible to prepare dialdehyde starch[26,33,34] at all levels of oxidation up to 10% of the anhydroglucose units in the starch.

Although the periodic acid oxidation of starch poses no particular problem, the high cost of the reagent requires recycling of the iodate and its reoxidation. Both chemical and electrochemical methods can be used for this operation, but its economic feasibility is questionable. Starch is oxidized with periodic acid and the spent oxidant solution is recycled through an electrolytic cell to regenerate the periodic acid which is then pumped back to the reaction tank.

Dialdehyde starch is insoluble in water, but swells to some extent on cooking at 90°C. The higher the degree of oxidation, the more difficult it is for the dialdehyde starch to disperse by cooking in water. Mild alkalinity or the presence of sodium bisulfite in small amounts helps its dispersion by controlled degradation and yields clear fluid sols of the degraded polymer. Under strongly alkaline conditions, however, dialdehyde starch undergoes rapid depolymerization.

Many applications have been described for this reactive polymeric polyaldehyde. It can cross link with substrates containing amino, hydroxyl, and imino groups and is useful in hardening gelatin and polyvinyl alcohol in tanning leather, for imparting water resistance to adhesives and photographic films, in developing wet strength in paper, and for the production of glyoxal and erythrose by molecular hydrolysis.

D. PREGELATINIZED STARCH

These are the derivatives of starch in which starch has been cooked and dried prior to use. Whenever a cold water-soluble or swellable starch is required these products are employed. The preparation of these products involves disruption of the granule structure of starch. The starch manufacturer simply cooks the starch for the user. The use of these products is convenient when the user has either no cooking facility or when it is intended not to install one.

Pregelatinized starch[9,35] can be made from almost any native or derivatized starch. However, the best products are obtained from potato and tapioca starches. The most versatile and commonly used method involves the use of a double drum dryer with a feed into the center valley. A concentrated aqueous slurry of starch is gradually poured over the rotating drums preheated by steam. The main drawback of this unit is its low rpm and certain inherent mechanical problems. The process may be simple in theory, but becomes unpredictable and elusive to control in actual practice. Hence, commercial operation of a drum dryer can be an exasperating and frustrating experience. Minor drifts in operating conditions seem to produce magnified changes in the finished product, and output of finished product is often not commensurate with operating and maintenance costs.

The nature of the final product depends upon the processing technology. If the drum dryer is adjusted to produce a thick and dense film which results in a horny product, the particles after grinding will not hydrate rapidly on dispersing and may give grainy paste with low viscosity, which may or may not be desirable depending upon the intended use. On the other hand, a thin film with rapid drying on the drum will give a thin flake that may hydrate quickly, giving a high viscosity without appreciable grainy nature. The paste between the rolls should have consistent viscosity along with the valley and should be distributed evenly on the drums. The depth of the starch paste pool in the valley should also be constant. Control of the gap at the nip of two rolls is a critical factor, since a variation of ± 0.0075 cm in gap width is enough to give performance less than optimum. The thickness of the walls of drums and the temperature and condition of the roll surface are also important. The supply of starch can be modified prior to the drum-drying process in accordance with the intended use. Gelatinization aids like salts and alkalies, surface-active agents to control rehydration or to minimize sticking to rolls, sweeteners, flavor-improvers, and other ingredients may be included in the feed stream. The final product may be mixed with a dispersing agent like borax, which not only helps redispersion of the dry product, but also helps prevent retrogradation.

The way in which a pregelled starch behaves on addition to water is important. The hydration rate may be slow, or if initially fast the particles may form lumps because the initial wetting produces a sticky surface on the particles which then lump together. Sometimes on initial wetting a particle forms a swollen shell which prevents further diffusion or causes very slow penetration of water into the particles so that it is not completely wetted and dispersed. This leads to the so-called fisheyes in paper manufacture.

Retrogradation shown by starch is one of the factors militating against the successful use of precooked and redried starches, because when gelled the amylose portion may precipitate on aging and cooling and cannot be redispersed except by means of autoclaving or by the action of an alkali. Thus, precooked starches redispersed in water give a lumpy, granular paste quite dissimilar to freshly prepared starch gels. This may lead to better retention, but gives poor dispersion throughout the web, resulting in poor overall efficiency. Overgrinding the dry product may reduce its retention in the web at the wire, so the particle size of pregelled, redried product must be closely controlled.

In Europe, the use of pregelatinized starch is very common. The addition of cold swelling starch in amounts up to 5% to the paper is made at any stage of paper manufacture. The market for the pregelatinized starches in prepared foods is expanding rapidly, particularly in the development of convenience foods such as instant soups, puddings, beverages, cake mixes, and salad dressing mixes. Pregelatinized tapioca starches are of particular value. The dry mixes are usually added to water or milk at room temperature and the pregelatinized starch ingredient is required to hydrate and thicken the mixture to give a palatable texture and to hold the other components in a uniform suspension.

Pregelatinized starches have been used as a binder for charcoal briquettes and as foundry core binders. They are useful as adhesives, such as in cold-water dispersible wallpaper pastes. The use of soluble, pregelatinized starch, apart from obviating the need for cooking, also eliminates a fault sometimes encountered, viz., that caused by a local concentration of granular starch which on the drying cylinders gives the so-called windows or shiners.

Starch-based adhesives for corrugated paper boards are usually prepared by suspending unmodified starch in an aqueous solution of a pregelatinized starch. The former acts as a binder and the latter as a carrier. The aqueous solution of a gelatinized starch should have a sufficient viscosity to keep the raw material in suspension and to cause the composition to be picked up and transferred properly on the applicator roll of the machine. Sodium hydroxide is added in these compositions because it lowers the gelatinization temperature. Borax is also added to increase the degree of viscosity developed by starch on gelatinization and as an effective buffer for sodium hydroxide.

E. CROSS-LINKED STARCH

Starch can react with multifunctional reagents like phosphorus oxychloride or epichlorohydrin so that interconnection of starch molecules, more commonly known as cross linking, occurs.[23,26,36,37] Because of the multiplicity of hydroxyl groups in starch molecules, some intramolecular cross-linking may also take place. This reaction brings about reinforcement of hydrogen bonds holding the granule together, as a result of which significant changes occur in the gelatinization and swelling properties of starch. Usually, the granules become toughened and their swelling during gelatinization is restricted. Hence, such products are also referred to as the inhibited starches. Intermolecular cross-linking obviously leads to an increase in the molecular weight of starch. Significant changes in the properties of starch granules may occur on cross-linking. The cross-linking reagent present in amounts as low as 0.005% on weight of starch is sometimes enough to completely change the properties of the starting material. Very low degrees of cross-linking reaction are difficult to determine directly. Hence, characterization of products as well as quality control during their production are dependent upon the measurement of their viscosity, solubility pattern, and resistance to shear.

Cross-linked starches are used when a stable, high-viscosity starch paste is needed and particularly when the dispersion is to be subjected to high temperature, high shear, or low pH. For resistance to rigorous conditions, a high degree of cross linking is required. In the food industry, cross-linked starches are used to provide appropriate gelatinization, viscosity, and textural properties as cross linking provides optimum thickening and rheological properties in food systems. Cross-linked starch ethers containing carboxymethyl and/or hydroxyalkyl groups are suitable absorbents for personal sanitary applications. Cross-linked starches are used in textile printing pastes and in adhesives used for corrugated paper boards where they give high viscosity in alkaline pH. Other application areas are oil well-drilling muds, printing inks, binders for coal/charcoal briquettes, fiberglass, and textile sizing.

F. STARCH ETHERS

Many starch ethers have been described in the literature,[5,37-42] but a great majority of them are only of academic interest as they have contributed to clarification of the structure of the starch molecule without being developed commercially. The only starch ethers of industrial importance are perhaps the cationic starch, hydroxyalkyl starch, and carboxyalkyl starch.

1. Cationic Starch

The distinctive characteristics imparted to starches by the introduction of a small number of cationic groups have attracted sustained attention from the time of the introduction of the first of this group of compounds which were tertiary aminoalkyl starches primarily for the use in the paper trade. A wide range of tertiary aminoalkyl, imino, and quaternary ammonium products have been produced and also quaternary phosphonium and tertiary sulfonium products.

The term "cationic starch" implies a derivative of starch which carries a cationic group. Cationic starches are under constant development to satisfy the changing needs of the paper industry. More efficient products which enhance the strength properties of paper, maintain performance under a variety of mill conditions, increase strength and retention over a wide range of pH, and reduce load on effluent-treatment plants are in use today in the developed countries.[43]

A large number of reactions have been cited mostly in patent literature[39,40] for the preparation of cationic starches. The substrate can be any starch such as maize, tapioca, potato, or wheat starch and the substituent can be any cationic group. The two most commonly encountered groups, however, are the tertiary aminoalkyl group and the quaternary ammoniumalkyl group. Most of the reactions are commonly carried out in aqueous slurries using alkali as a catalyst or by a semidry method. The resulting starch derivative is often finished at an acidic pH in the form of a tertiary amine salt. The reactions are to be carefully controlled to prevent any cross linking that would impair subsequent dispersion of the granules when cooked. It is desirable to wash the product to remove the by-products of reaction which may interfere with the paper-making process. Chemically, cationic starch can be classified as an ether. Usually, a low degree of substitution (i.e., 0.01) is required to make starch cationic. They are characterized by low gelatinization temperature, stable paste viscosity, and high substantiveness for cellulosic fibers. Cationic starches are widely used in the paper industry, mainly to improve the strength of paper and to increase drainage on the paper machine. Cationic maize starch is common in the U.S., while cationic potato starch is more widely used in Europe and Scandinavian countries, and cationic wheat starch is more common in Australia.

The utility of cationic starch in the paper industry as a wet-end additive, as a size press starch, and as a binder in pigment coating is based upon its attraction to the fiber surface and its reaction with the anionic sites of cellulose, pigments, and other anionic materials in the paper-making pulp slurry (furnish). Initially, cationic starch found its largest acceptance in fine paper as a dual strength and retention additive. Later, it found expanded uses in unbleached grades of kraft and ground-wood furnishes. It can be used in combination with almost all paper-making fibers including reclaimed fiber and post-consumer waste. Cationic starch has also gained acceptance in such special applications as emulsification and retention of highly reactive internal sizing agents, e.g., ketene dimers and alkenyl succinic anhydrides.

Starch, like cellulose, is a polymer of D-glucopyranose units and binds well to cellulose through hydrogen bonding, which is believed to be the main cause of higher strength. Normal starch solutions, however, are anionic as is cellulose when in suspension with water. The repulsion due to like charges on the two polymers leads to a lowering of freeness and drainage and a consequent reduction in machine speed. Poor retention on pulp leads to a higher concentration of starch in the back water. As a result, there is a slime buildup on the machine and biological oxygen demand (BOD) goes up. The mutual repulsion must be overcome to allow maximum bonding to occur. Cationic starch does just that as it is directly attracted to the negative cellulose.

Cationic starches, because of their attraction to the anionic cellulosic fiber, are virtually 100% retained on the pulp at the levels of addition usually employed (0.2 to 2% on the fiber). The strongly absorbed cationic starch is not removed from the fiber by the usual paper-making processes and repulping. Thus, it is a very effective agent and does not contribute to pollution in the paper-making process. Unmodified starches or anionic starch derivatives are not retained as well. Hence, they must be used at higher levels of addition (i.e., 2 to 3% on dry pulp) and do contribute to the buildup of solubles in the recycling water system as well as to pollution problems directly and on repulping.

Fillers or pigments such as clay, calcium carbonate, or titanium dioxide are used to increase capacity and to improve the brightness and printing qualities of the paper sheet. It is important to retain these fillers in the wet sheet as it is formed on the wire to realize fully the potential of the added material and to minimize the economic loss of not holding the filler in the sheet.[44] Further, the cleanliness of the water system in the paper-making process and the pollution load of the effluent waste water from the mill are affected by the efficiency of filler retention. Filler and pigment retention is the percent of pigment/filler added to the pulp slurry (furnish) that is retained in the finished paper. The filler particles generally have a negative surface charge in water suspension, and positively charged cationic polymers which promote flocculation of the pigment particles are effective retention aids.

Cationic starches act as retention aids besides increasing the strength of the sheet. Retention of fillers decreases the internal strength of the sheet, but cationic starches have the unique property of maintaining the internal bonding strength in the presence of fillers. Further, cationic starches help in retaining the fiber fines in the wet sheet, thus reducing the amount carried away in the back water. The fiber fines retained on the paper mat contribute to the strength of the sheet. Cationic starch graft copolymers are also of interest for filler retention.

When used as a combined strength and retention aid, cationic starch is generally added at 0.25 to 1% (w/w) on the pulp. The level of addition depends upon the system and the character of the pulp being used. The cationic demand can be assessed by a titration with cationic polymer as measured by the zeta potential, which is the electrical potential between the particle surface and the bulk of the medium (i.e., water) and is related to the charge on the particle.[45] A test for the cationic efficiency to predict the retention efficiency of cationic starch involves optical measurement of the amount of anionic blue pigment remaining in suspension after the addition of a cationic starch-pulp complex.[46]

If an excess of cationic starch is added to the pulp-filter-water system, the negative surface potential of the pigment particles will again repel each other, thereby reducing retention. The cationic starches should be fully cooked (dispersed) for maximum retention efficacy. The drainage of the water from the wet sheet on the moving wire of the paper-making machine is improved by the addition of cationic starch and polymers which act as retention aids and also through flocculation and redistribution of the fines throughout the wet sheet or the web, which increase the permeability of the wet web.

As concern for reducing pollution in mill effluent becomes a major factor in planning and operations, the paper mills are recycling the drained water to minimize solids and solubles in waste-water discharge. This leads to increased solubles in the water used to make up the furnish and more problems in the process and the quality of the product that must be overcome. Additives such as cationic starches and other polymers are employed in the advanced countries to solve these problems in the paper-making process. Similarly, with the cost of energy rising precipitously, any means that will minimize energy requirements is in demand. The use of additives to remove water more rapidly and easily from the wet sheet is of interest. Cationic starches make up a major portion of the starch wet-end additives used at present. Oxidized starches, on the contrary, have a detrimental effect on filler retention even in the presence of alum. Since oxidized starch, which is used as a wet-end additive and in sizing and coating, may return to the paper-making process through the broke (i.e., trimmings and other wastepaper that recycled through pulping), a means of nullifying its dispersant effect on pigments must be used. Cationic starches are effective in the presence of small amounts of oxidized starch.

Cationic starches are tenaciously held on the fibers by electrostatic attraction, and, when used as surface sizes, remain substantially with the cellulosic fiber and the filler during the repulping of the broke, thereby reducing pollution measured by the BOD and providing more effective use of the broke. Depending upon the type of paper mill, the finished paper can contain from 15 to 35% broke. Some reports have indicated that as much as 50 to 60% of the starch that comes back with the broke goes through the paper-making process and comes out as effluent.[5]

The cationic starches have other advantages in surface sizing. Since there is an electro-chemical attraction between the positively charged starch and the negatively charged fiber, the size is attracted to the surface fibers and retained on or near the surface of the sheet. Consequently, less starch is needed to maintain surface strength and quality. Less starch is therefore used, which improves opacity and often increases drying rates and machine speeds, thereby improving economics. The printing quality of the paper sized with the cationic starches is also improved.

2. HYDROXYALKYL STARCH

Hydroxyalkyl starches are starch ethers readily prepared by the action of starch with alkylene oxide in the presence of a base. Reactions have also been carried out using such reagents as ethylene chlorohydrin. The presence of a small amount of sodium ethylenediamine tetra-acetate is claimed to increase the efficiency of the utilization of alkylene oxide. In many wet-milling factories throughout the world, the reaction is usually carried out on the final starch milk which has been concentrated to 40 to 45% solid suspension and made alkaline with either an alkali or an alkaline earth metal hydroxide. Still other processes involve the reaction of swollen, but ungelatinized starch in a substantially dry state with the etherifying agents. The substituent group most commonly encountered in these products is either hydroxyethyl or hydroxypropyl.

Being ethers, these compounds are resistant to cleavage by acids, alkalies, and mild oxidizing agents.[47,48] Hence, these products can be subjected to further conversion to obtain products of desired properties. Hydroxyalkyl starches are nonionic in nature and their dispersions are not subjected to the solubility and viscosity effects that dissolved electrolytes normally have on polyelectrolyte polymers. The hydroxyalkyl groups attached to the starch molecules interfere with the associative hydrogen bonding that holds the granule together and thus bring down the gelatinization temperature. The gelatinization temperature is progressively lowered as the level of substitution is increased, and, if enough substituents are introduced, it comes down to room temperature and the starch derivative becomes cold water soluble. The substituent groups effectively inhibit the tendency of starch molecules to re-associate or retrograde and this results in excellent viscosity stability of starch dispersion even at high concentration. The starch granules on cook-out absorb and carry a greater amount of water, which results in a continuous film of high clarity and toughness.

Since ethylene oxide and propylene oxide are highly flammable and form explosive mixtures with air over a broad range of composition, care must be taken in running large-scale reactions. Preferably, these reactions are performed in a closed-system pressure reactor under a nitrogen blanket. It should be ensured that there is no residual epoxide before further processing.

The low DS-hydroxyalkylated products have received considerable attention in the textile industry[5,9-11] since about 1950. Considerable tonnages are now being produced for the textile and paper industries in the western world. These derivatives show reduced gelatinization temperature, increased rate of swelling and dispersion when cooked with water, increased paste clarity and adhesiveness, and less tendency to retrograde on standing with greatly improved film clarity, flexibility, smoothness, and solubility. The higher substituted starches become increasingly soluble in cold water and cannot be prepared in a granular state. They have merit in that they resist microbial attack, and effluents from desizing plants would have low BOD. A reduction of 85% in BOD is reported to be obtained in the commercial products of most firms.

Hydroxyalkyl starches find application in the paper industry[5,12-14] for wet-end addition and surface sizing, as well as for paper coating. Their use as a wet-end additive is very limited, however, since they are useful only in heavy-grade paper where their retention occurs by physical entrapment as the wet sheet is formed. When used as a surface-sizing agent for paper, they improve the surface property of paper, namely, appearance, surface strength, scuff resistance, fluids, inks, and coating into the sheet. The size is employed to anchor the surface fibers and particles so that less dusting and linting occur. The degree of conversion and the viscosity level needed for these purposes vary, and hence usually a range of converted products is used. The less-converted starch has a higher bonding strength and will also penetrate to a lesser degree into the sheet. The lower-viscosity products penetrate further into the sheet, giving a greater effect in improving internal strength.

Low substituted hydroxyethyl starches are good adhesives in paper-coating and surface-sizing applications. As a result of improved flow properties, the colored pastes have good leveling characteristics. Reduced penetration into the paper stock due to the increased water-holding properties minimizes coating failures, and the improved adhesive strength results in uniformly high pick values. Ink receptivity and printing properties of coatings containing this derivative are improved. Hydroxyethyl starches are said to be superior to ordinary starch for preparing starch-clay glyoxal paper coatings. Owing to the lower swelling temperatures of these products, they can be used as beater additives to increase fiber bonding which takes place readily under the conditions of temperature and moisture present on the drying rolls of the paper machine. In corrugated boards, when used as adhesives, they stop the adhesive migrating from the glue line due to their high water-holding capacity.

In the food industry, hydroxyethyl starches are used for pie fillings, salad dressing, and food thickening since they have water-binding capacity, do not retrograde, and possess clarity and resistance to acid treatment. Hydroxyethyl starches are used as blood plasma volume expanders in the pharmaceutical industry.

3. Carboxymethyl Starch

Although many sodium O-carboxyalkyl starches are known, sodium O-carboxymethyl starch (CMS) is the most important from the viewpoint of industrial applications. By hooking carboxymethyl anions onto the starch backbone, the properties of the native starch can be changed radically. Thus, starch which is neutral in its native form becomes anionic after the modification. It gelatinizes at a lower temperature and its water solubility increases. The very anionic character of the carboxymethyl group gives the starch a polyelectrolyte behavior along with high viscosity. Hence, it is used as a thickening agent in various fields. The anionic polymer also assists in detergency similar to that of carboxymethyl cellulose, and is used in detergent formulations.

Carboxymethylation[5,38-40,42] of starch can be carried out in solid phase, solvent medium, or homogenous paste. It involves the reaction of sodium chloroacetate with starch in the presence of an alkali. Carboxymethylation can lead to cold water-soluble products which give clear, nonretrograding viscous gels; and although these are not stable to high-speed stirring, their stability can be increased by cross linking with polyfunctional agents like $POCl_3$ and epichlorohydrin. This in return allows the introduction of more carboxyl groups to increase the viscosity of the final product.

The α-hydroxycarboxyalkyl starch ethers are useful as food additives and as filler retention and sizing aids in paper manufacture. If the starch product is depolymerized, it can be used as a sizing agent for cotton warps in the textile industry. CMS is used as a thickening agent in various fields, e.g., paints, oil well-drilling muds, wallpaper adhesives, and others.

It has been recommended as a tablet disintegrant in pharmaceutical applications and as an antidirt-deposition agent in detergents. CMS can be insolubilized by reaction with polyvalent ions such as aluminum, ferric, chromic, and cupric ions leading to precipitation or gelling of dispersions or insolubilization of films. CMS has the dispersion stability and film-forming capabilities of particular value in the surface sizing of paper. The anionic starch is retained almost completely by means of aluminum ions. Hence CMS is useful only in an acid paper-making system. Its retention in neutral or alkaline paper-making systems is poor because of the lack of aluminum ions.[5,10,42]

The printing of colored patterns on cloth is carried out by using pastes of dyes in certain hydrophilic, polymeric substances called thickeners. Special starch derivatives like CMS can compete against natural gums and other materials used for this purpose. The thickener provides good dispersion of the expensive dye stuff and helps give sharp transfer of the dye pattern to the cloth by restricting print-paste boundaries as well as good color value. CMS has been recommended for this purpose either alone or in combination with other thickeners. Its use as a printing thickener for disperse dyeing of polyester fabrics is relatively less well known. The use of CMS in printing paste seems to improve dye fixation.

G. STARCH ACETATES

Although a great variety of starch esters have been described, only a few have been marketed, mainly due to the nonspecific properties arising from the substituent groups. Among the starch esters now available commercially, the typical is perhaps the starch acetate which has applications in many industries. The starch acetates are manufactured by treatment of the aqueous starch suspension with acetic anhydride or vinyl acetate. During the reaction, there is a competition between hydrolysis of the anhydride, reaction on the starch, and deacetylation of the product. The acetic anhydride reaction is carried out with the gradual addition of the reagent at about 8.4 pH and 20°C. The vinyl acetate reaction is performed at around 40°C and at a pH between 9 and 10. Under these conditions, approximately 75% of the reactive part is fixed onto the starch molecule. Although the esterification is symbolized by the introduction of an acetyl group on the starch backbone, the nature of the substituent may influence the final product. The products are recovered by neutralization with dilute mineral acids to pH 5 followed by filtration, washing, and drying.

The steric disturbance of the acetyl group makes starch solubilization easier by weakening the energy level of the starch granule network created by the hydrogen bonds and also slow down the retrogradation. The solubility of the starch acetates is influenced by the degree of polymerization (or depolymerization) and the type of treatment given for acetylation. Starch acetates up to about 15% acetyl content tend to be soluble in hot water. Degradation usually increases the water solubility. The stability of starch acetate sols is good. This improves their compatibility with hydrophilic colloids used in size formulations and makes application of the size possible at low temperature. Starch acetates are used for sizing textiles, particularly fancy cotton and cotton synthetic blends.[5,6,10] Since the acetate groups generate a hydrophobic type of structure, the starch acetates are used for certain applications like adhesives and grease resistance in paper sizing. They are also employed in acid pH-resistant binders in the food industry. Due to their low gelatinization temperature and better water retention, they are added in foundry moulds.

H. SUPER SLURPERS

Among the new starch-transformation technologies, grafting is certainly original in that it allows the association of a synthetic structure with a natural molecular organization. The more common derivatives of starch mentioned here have been extensively studied. Grafting synthetic polymers to starch is, however, a relatively new area of research.

Water-absorbing polymers or the super slurpers, made by chemically grafting acrylic monomers to starch followed by hydrolysis, are of much current interest due to their versatility of use. Several water-gelling absorbents made by chemically grafting acrylic monomers to cereal grain starch followed by hydrolysis were introduced as super slurpers over a decade ago by scientists at the Northern Regional Research Laboratory, Peoria.[5,49,50] These products absorbed up to 5000 times their own weight of pure water in the laboratory and established new standards of absorbancy as well as the name for the standard products and the industry. The super slurpers are now produced by many companies on a commercial scale.

A graft copolymer is a high polymer, the molecules of which consist of two or more polymeric parts of different compositions chemically united together. A graft copolymer may be produced by polymerizing a given monomer with subsequent polymerization of another kind of monomer onto the product of the first polymerization. The union of the two different polymers by chemical reaction between their molecular end groups or by a reaction producing cross links between the different materials would also produce a graft copolymer.

Starch-graft copolymers can be prepared by generating free radicals on the starch backbone and then allowing these materials to react with the desired monomers (Figure 2). Any starch or flour can be used as the substrate for grafting. These already possess some hydration characteristics and their modification by grafting further increases their water-absorbing capacities. A large number of reagents can be employed for the purpose of polymerization. These may be nonionic, cationic, or anionic in nature. The presence of a vinyl group in a given reagent is desirable. The more common reagents which are used for this purpose are acrylonitrile and acrylic acid.

A number of initiating methods have been used to prepare starch-graft copolymers. These may be divided into the following broad categories:

1. Initiation by chemical methods generally involve the use of ceric ammonium nitrate or the ferrous ion-peroxide redox system.
2. Initiation by irradiation involves the use of UV or gamma (CO^{60}) irradiation.
3. Initiation by mastication, physically ruptures starch molecules to give free radical sites where cleavage of the molecule occurred. The products of polymerization here can be referred to as block rather than graft copolymers.

Many polymers obtained by graft copolymerization display their water-absorbing properties only when they are saponified; e.g., starch-PAN copolymers. These polymers may have different starch-PAN ratios dependent upon the conditions of their synthesis. When they are treated with KOH, varying amounts of carboxylic (COOH) and amido ($CONH_2$) groups are formed by the hydrolysis of nitrile groups. The degree of hydrolysis determines the properties of these polymers. The degree of saponification can be varied by using different saponifying agents. Thus, mineral acids, enzymes, and combination of methanolic alkali and periodate can be used to obtain products of different properties.

Starch-graft copolymers can be characterized by three parameters, namely (1) the percent add on, i.e., the weight percent of synthetic polymer incorporated into the graft copolymer; (2) the average molecular weight of the grafted branches; and (3) the grafting frequency.

Grafted branches of synthetic polymers are easily separated for molecular weight determination by degradation of the starch portion of the copolymer to glucose and other small carbohydrate fragments through its treatment with hot mineral acids or enzymes.

The term "grafting efficiency" often used in describing graft polymerizations is defined as the percentage of the total synthetic polymer formed that has been grafted to starch. High grafting efficiencies are desirable, since a polymerization of low grafting efficiency gives mainly a physical mixture of starch and homopolymer.

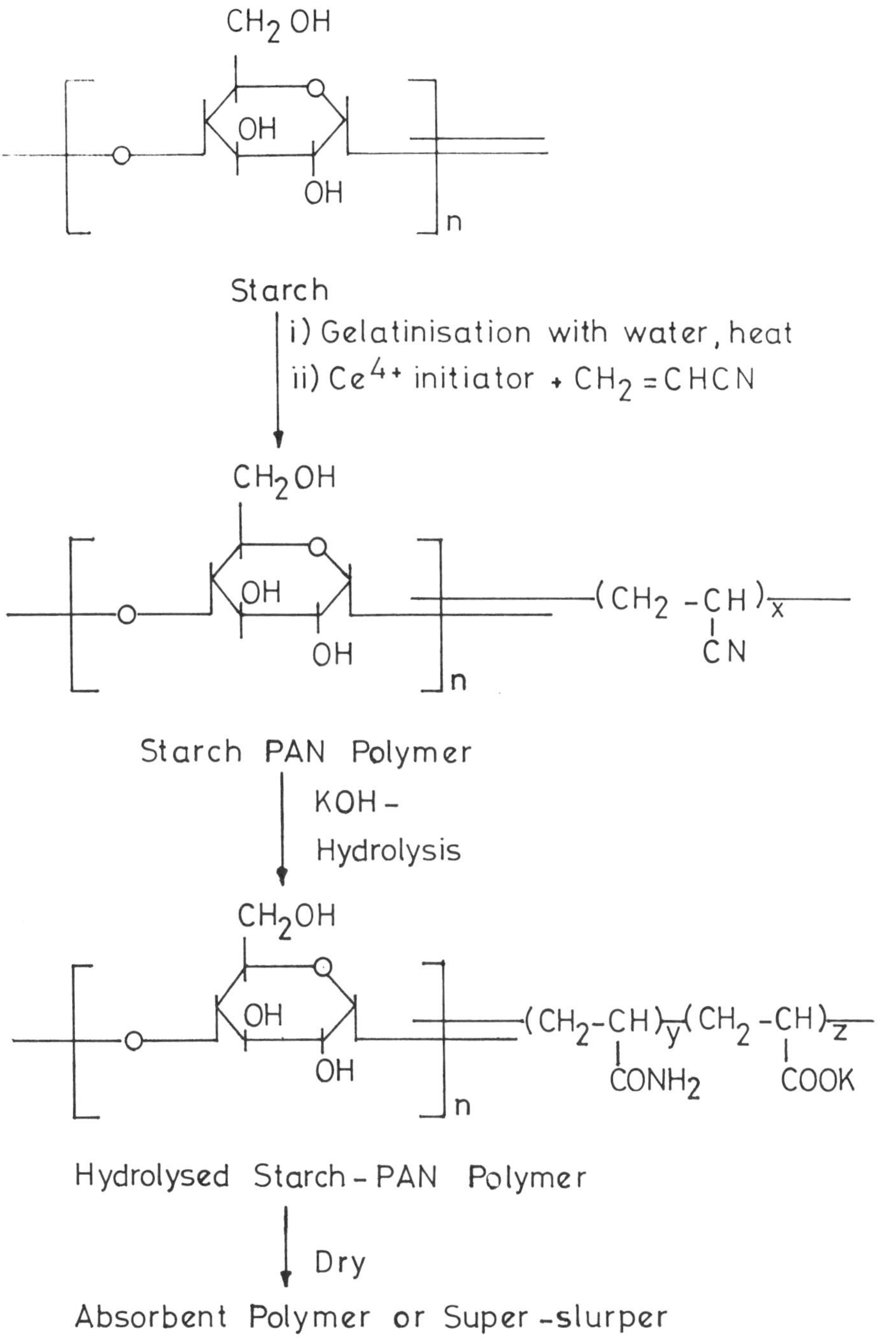

FIGURE 2. Preparation of super slurper from starch.

Grafting frequency expressed as the average number of glucose units per grafted branch is calculated from the percent add on and the molecular weight of the graft. Thus,

$$\% \text{ Grafting efficiency} = \frac{100 \times \text{total wt of polymer} - \text{wt of polymer extractable}}{\text{total wt of polymer}}$$

Gel-permeation chromatography is also used to calculate the molecular weight distribution of grafted chains. The average molecular weight ($\overline{MW}$), the number of average molecular weight ($\overline{Mn}$), and the polydispersity ratio ($\overline{MW}$ + $\overline{Mn}$) can be calculated from the gel-permeation chromatogram.

The polymers described above exhibit certain unusual properties, and a number of potential applications have been suggested for them. The most remarkable property of these products is their ability to absorb several hundred times their own weight of water while still remaining insoluble. Films prepared by air drying the water dispersions of these polymers retain their integrity when they imbibe water and swell to form continuous sheets of gel. The suggested uses of these products range from incorporation into personal care products to forest fire-fighting aid and from wet-end additives for paper to thickeners for textile printing.

The synthesis of block and graft copolymers of starch has been the area of much current interest in several laboratories in the world. The research in this field is mainly aimed at the simplification of synthetic methods, the correlation of the structures of polymers with their properties and improvement of absorbent properties, and salt tolerance of these products.

The current and potential applications of super slurpers are listed below.

1. Medical: mainly for the absorption of body fluids, viz., blood, urine, saliva, perspiration.
   - Disposable diapers
   - Bandages
   - Bedpans
   - Surgical gloves and gowns
   - Incontinent pads
2. Agricultural and horticultural
   - Seed coating and root coating
   - Soil additive
   - Pesticide formulations
   - Agricultural sprays and drip irrigation
   - Seed pelletizing
3. Thickening, dispersing, solidifying agent
   - Solidification of radioactive waste, sludge
   - Removal of excess water from athletic fields, construction sites, basements, etc.
   - Waxes, polishes, paints
   - Fire-fighting fluids
4. Paper and textile
   - Absorbent paper towels and tissues
   - Moisture retention in synthetic textiles (antistats)
   - Printing of textiles
5. Miscellaneous
   - Binder for foundry core, charcoal briquettes
   - Foods, feeds, toys, novelties, etc.
   - Water removal from emulsions, solvents, oils

REFERENCES

1. **Van Beynum, G. M. A. and Roels, J. A., Eds.,** *Starch Conversion Technology,* Marcel Dekker, New York, 1985, 3.
2. **Swinkels, J. J. M.,** Sources of starch, its chemistry and physics, in *Starch Conversion Technology,* Van Beynum, G. M. A. and Roels, J. A., Eds., Marcel Dekker, New York, 1985, 15.
3. **Mitch, E. L.,** Potato starch: production and uses, in *Starch: Chemistry and Technology,* Whistler, R. L., BeMiller, J. N., and Paschall, E. F., Eds., Academic Press, London, 1984, 479.
4. **De Willigen, A. H. A.,** The manufacture of potato starch, in *Starch Production Technology,* Radley, J. A., Ed., Applied Science Publishers, London, 1976, 135.
5. **Rutenberg, M. W.,** Starch and its modifications, in *Handbook of Water Soluble Gums and Resins,* Davidson, R. L., Ed., McGraw-Hill, New York, 1980, 22.
6. **Whistler, R. L. and Paschall, E. F., Eds.,** *Starch Chemistry and Technology,* Vol. I., Fundamental Aspects; Vol. II., Industrial Aspects, Academic Press, New York, 1965.
7. *Kirk-Othmer Encyclopedia of Chemical Technology,* John Wiley & Sons, New York, First edition, Vol. 12, 764; Second edition, Vol. 18, 672; 3rd ed., Vol. 21, 492, 1975.
8. **Whistler, R. L., BeMiller, J. N., and Paschall, E. F., Eds.,** *Starch: Chemistry and Technology,* 2nd ed., Academic Press, London, 1984.
9. **Kerr, R. W., Ed.,** *Chemistry and Industry of Starch,* Academic Press, New York, 1950.
10. **Radley, J. A.,** The textile industry, in *Industrial Uses of Starch and Its Derivatives,* Radley, J. A., Ed., Applied Science Publishers, London, 1976, 149.
11. **Compton, J. and Martin, W. H.,** Starch in the textile industry, in *Starch Chemistry and Technology,* Vol. II, Whistler, R. L. and Paschall, E. F., Eds., Academic Press, New York, 1965, 147.
12. **Zuderveld, A. H. and Stoutjesdijek, P. G.,** The paper industry, in *Industrial Uses of Starch and Its Derivatives,* Radley, J. A., Ed., Applied Science Publishers, London, 1976, 199.
13. **Mentzer, M. J.,** Starch in the paper industry, in *Starch: Chemistry and Technology,* Vol. II, Whistler, R. L., BeMiller, J. N., and Paschall, E. F., Eds., Academic Press, New York, 1965, 543.
14. **Nissen, E. K.,** Starch in the paper industry, in *Starch: Chemistry and Technology,* Vol. II, Whistler, R. L., BeMiller, J. N., and Paschall, E. F., Eds., Academic Press, New York, 1965, 121.
15. **Osman, E. M.,** Starch in the food industry, in *Starch: Chemistry and Technology,* Vol. II, Whistler, R. L., BeMiller, J. N., and Paschall, E. F., Eds., Academic Press, New York, 1965, 163.
16. **Moore, C. O., Tuschhoff, J. V., Hastings, C. W., and Schanefelt, R. V.,** Application of starches in food, in *Starch: Chemistry and Technology,* Vol. II, Whistler, R. L., BeMiller, J. N., and Paschall, E. F., Eds., Academic Press, New York, 1965, 163.
17. **Radley, J. A.,** The food industry, in *Industrial Uses of Starch and Its Derivatives,* Radley, J. A., Ed., Applied Science Publishers, London, 1976, 51.
18. **De Willigen, A. H. A.,** The rheology of starch, in *Examination and Analysis of Starch and Starch Products,* Radley, J. A., Ed., Applied Science Publishers, London, 1976, 61.
19. **Radley, J. A.,** Physical methods of characterising starch, in *Examination and Analysis of Starch and Starch Products,* Radley, J. A., Ed., Applied Science Publishers, London, 1976, 91.
20. **Leach, H. W.,** Gelatinization of starch, in *Starch: Chemistry and Technology,* Vol. II, Whistler, R. L., BeMiller, J. N., and Paschall, E. F., Eds., Academic Press, New York, 1965, 289.
21. **Zobel, H. F.,** Gelatinization of starch and mechanical properties of starch pastes, in *Starch: Chemistry and Technology,* Vol. II, Whistler, R. L., BeMiller, J. N., and Paschall, E. F., Eds., Academic Press, New York, 1965, 285.
22. **Whistler, R. L., Ed.,** *Methods in Carbohydrate Chemistry,* Vol. IV, Academic Press, New York, 1964.
23. **Radley, J. A.,** The manufacture of modified starches, in *Starch Production Technology,* Radley, J. A., Ed., Applied Science Publishers, London, 1976, 449.
24. **Rohwer, R. G. and Klem, R. E.,** Acid-modified starch: production and uses, in *Starch: Chemistry and Technology,* 2nd ed., Whistler, R. L., BeMiller, J. N., and Paschall, E. F., Eds., Academic Press, London, 1984, 529.
25. **Shildneck, P. and Smith, C. F.,** Production and uses of acid-modified starch, in *Starch Chemistry and Technology,* Vol. II, Academic Press, New York, 1965, 217.
26. **Fleche, G.,** Chemical modification and degradation of starch, in *Starch Conversion Technology,* Marcel Dekker, New York, 1985, 273.
27. **Simms, R. L.,** The technology of corn wet-milling, in *Starch Conversion Technology,* Marcel Dekker, New York, 1985, 47.
28. **Lyne, F. A.,** Chemical analysis of raw and modified starches, in *Examination and Analysis of Starch and Starch Products,* Radley, J. A., Ed., Applied Science Publishers, London, 1976, 133.
29. **Evans, R. B. and Wurzburg, O. B.,** Production and use of starch dextrins, in *Starch Chemistry and Technology,* Vol. II, Academic Press, New York, 1965, 254.

30. **Action, W.,** The manufacture of dextrins and British gums, in *Starch Production Technology,* Radley, J. A., Ed., Applied Science Publishers, London, 1976, 273.
31. **Kennedy, H. M.,** Starch and dextrins in prepared adhesives, in *Starch: Chemistry and Technology,* 2nd ed., Whistler, R. L. and Paschall, E. F., Eds., Academic Press, London, 1984, 593.
32. **Scallet, B. L. and Sowell, E. A.,** Production and use of hypochlorite-oxidized starches, in *Starch Chemistry and Technology,* Vol. II, Whistler, R. L. and Paschall, E. F., Eds., Academic Press, New York, 1965, 237.
33. **Radley, J. A.,** The manufacture and chemistry of dialdehyde starch, in *Starch Production Technology,* Radley, J. A., Ed., Applied Science Publishers, London, 1976, 423.
34. **Mehltretter, C. L.,** Production and uses of dialdehyde starch, in *Starch Chemistry and Technology,* Vol. II, Whistler, R. L. and Paschall, E. F., Eds., Academic Press, New York, 1965, 433.
35. **Powell, E. L.,** Production and use of pre-gelatinised starch, in *Starch Chemistry and Technology,* Vol. II, Whistler, R. L. and Paschall, E. F., Eds., Academic Press, New York, 1965, 523.
36. **Hullinger, C. H.,** Production and use of crosslinked starch, in *Starch Chemistry and Technology,* Vol. II, Whistler, R. L. and Paschall, E. F., Eds., Academic Press, New York, 1965, 445.
37. **Rutenberg, M. W. and Solarek, D.,** Starch derivatives: production and uses, in *Starch: Chemistry and Technology,* 2nd ed., Whistler, R. L. and Paschall, E. F., Eds., Academic Press, London, 1984, 312.
38. **Roberts, H. J.,** (a) Non degradative reactions of starch in Ref. 6 (a), 439; (b) Starch derivatives in *Starch Chemistry and Technology,* Vol. II, Academic Press, Inc., New York, 1965, 293.
39. **James, R. W., Ed.,** *Industrial Starches,* Noyes Data Corporation, Park Ridge, NJ, 1974.
40. **Johnson, J. C., Ed.,** *Industrial Starch Technology: Recent Developments,* Noyes Data Corporation, Park Ridge, NJ, 1979.
41. **Paschall, E. F.,** Production and uses of cationic starches, in *Starch Chemistry and Technology,* Vol. II, Academic Press, New York, 1965, 403.
42. **Radley, J. A.,** The manufacture of esters and ethers of starch, in *Starch Production Technology,* Radley, J. A., Ed., Applied Science Publishers, London, 1976, 481.
43. **Reynolds, W. F., Ed.,** *Dry Strength Additives,* Tappi Press, Atlanta, 1980.
44. **Foster, W. A.,** Water-soluble polymers as flocculants in paper-making, in *Water Soluble Polymers,* Bikales, N. M., Ed., Plenum Press, New York, 1973, 3.
45. **Penniman, J. G.,** The importance of eliminating anionic trash, *Paper Trade J.,* 162(7), 1978.
46. **Mehltretter, C. L., Weakhy, F. B., Ashby, M. L., Herlocker, D. W., and Rist, C. E.,** Spectrophotometric method for determining cationic efficiency of pulp-cationic starch complexes, *Tappi,* 46, 506, 1963.
47. **Van Der Bij, J.,** The analysis of starch derivatives, in *Examination and Analysis of Starch and Starch Products,* Radley, J. A., Ed., Applied Science Publishers, London, 1976, 189.
48. **Lortz, H. Z.,** Determination of hydroxyalkyl groups in low substituted starch ethers, *Anal. Chem.,* 28(5), 892, 1956.
49. **Fanta, G. F. and Bagley, E. B.,** Starch graft copolymers, in *Encyclopedia of Polymer Science and Technology,* Suppl. 2, John Wiley & Sons, 1977.
50. **Otey, F. H. and Doane, W. M.,** Chemicals from starch, in *Starch: Chemistry and Technology,* 2nd ed., Academic Press, London, 1984, 389.

Chapter 7

WASTE UTILIZATION

R. B. Natu, G. Mazza, and S. J. Jadhav

TABLE OF CONTENTS

I. INTRODUCTION

Losses of potatoes occur in the field as well as during harvesting, handling, transporting, storing, and processing. During potato processing, waste is generated as a result of peeling, trimming, slicing, washing, grading, screening, blanching, and disintegration or homogenization. The factors responsible for the increase in potato processing wastes in French fry and potato chip plants are mechanical injury, photo-induced greening, abnormal size, age of tubers, and physiological disorders. Potato processing waste is of two types: liquid waste due to soluble solids and solid waste from potato tissue. The use of a large quantity of water in French fry, chip, and starch manufacturing plants increases the volume of the liquid waste. Green, immature, and cull potatoes can also be considered potato processing waste. In this chapter, available information on the utilization and treatments of potato processing waste is summarized.

II. PROTEIN RECOVERY

Potatoes annually provide the world with about 6 million t of protein. Potato protein is rich in lysine and contains methionine and cystine as the limiting amino acids.[1] However, the nutritional quality of potato protein is comparable to that of whole egg. The crop productivity, in terms of protein per hectare per day, is considerably higher than that of cassava, sweet potatoes, yams, beans, wheat, and rice. In view of the quantity of potato protein and the need of reducing waste effluent, waste from potato processing plants should be utilized for feed, food, and useful products.[2] It has been pointed out that commercially processed potato protein concentrates, recovered from the waste effluents of European starch mills, are extensively used as animal feed.[1] These products are equal to or better than casein in quality and can be improved by supplementation with methionine. In North America, the upward trend in processing frozen fries, chips, etc., has resulted in potato waste suitable for value-added products.[3] On the other hand, much emphasis has been given to recovery of proteins from potato starch waste effluents because of their prominence in most European countries. De Noord[4] reported that 25,000 tons of protein could be produced annually from the potato starch mills of the Netherlands. The volume of protein water from potato starch factories ranges from about 0.7 to 7.0 m^3/t processed potatoes. About 1.5% solids out of 5% can be precipitated by heat treatment.

The protein recovery from potato processing water as reviewed by Knorr[2] is carried out in three stages: coagulation (precipitation), separation (dewatering), and drying. Concentration may be included if necessary. Other soluble compounds which remain in the protein water after the coagulation step can also be recovered.

Heat coagulation of potato protein is shown in Figure 1. The steam-injection technique is normally used to heat the protein water. Heating (80 to 90°C) at pH values of 4 to 4.5 is most effective for protein recovery.[5] Heat coagulation has been reported by several other authors.[7,8] Protein precipitation by polyphosphoric acid,[6] $FeCl_3$ with HCl and HCl with H_3PO_4 have been attempted.[5] Finley and Hautala[6] used 1% polyphosphoric acid to precipitate protein at various pH values. Calcium chloride was employed to remove precipitated phosphate by centrifugation. Meister and Thompson[5] treated the protein water with lime, followed by the acidic coagulants including $FeCl_3$, HCl, and H_3PO_4. The results indicated that HCl and H_3PO_4 showed identical results, while $FeCl_3$ compared favorably with HCl. In general, $FeCl_3$ offers several advantages. It is relatively inexpensive, has acidic properties, and the ferric ion is a good nucleating site for a large floc formation. Also, it requires no heat treatment and has nutritional significance along with the precipitated protein. Although protein recovery is satisfactory with lime-H_3PO_4 or -$FeCl_3$ treatment, a high amount of calcium in the product is a drawback. Lime-H_3PO_4 treatment requires neutralization of water

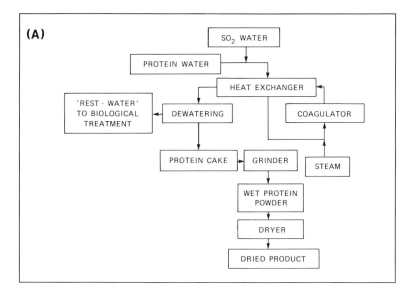

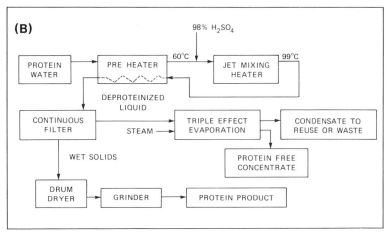

FIGURE 1. (A) Scheme of potato protein plant and (B) protein recovery from protein water and concentration of protein-free waste from processing potato starch plant. (From (A) Peters, H., *Pure Appl. Chem.*, 29, 1 & 129, 1972. With permission.[151] (B) Stabile, R. L., Turkot, V. A., and Aceto, N. C., Second National Symp. on Food Processing Wastes, Denver, 1971. With permission.[10])

before it is discharged. Moreover, phosphate solubility increases below pH 9, although it can be precipitated with lime above pH 11.8. The rate of sedimentation of proteins as enhanced by heat was effective up to 60 min.[5] The rate of initial sedimentation in low protein concentrates was faster, but with a lower percentage of protein that settled over the entire period. Similar results were reported by Strolle et al.[8] by varying the pH and the concentration of solids. These authors[8] concluded that the efficiency of protein recovery can be increased if higher waste strength can be obtained. Less water use and recycling of waste effluent were recommended. The working conditions for obtaining a protein-free potato juice were acidification to pH 4.5 and the addition of bentonite[9] to achieve a weight ratio of soluble protein:bentonite of 0.9. Stabile et al.[10] compared various methods for processing potato starch effluents for presenting an economic analysis (Table 1).

Separation of the coagulated proteins can be accomplished by using a continuous rotary

TABLE 1
Comparison of Alternative Methods for Processing Potato Starch Effluents

Alternative method	Resulting product	Use	Fixed capital cost (U.S.$)	Total operating cost (U.S.$/year)	Sales (U.S.$/year)
Concentration by evaporation	Concentrate with protein	Feed	514,000	148,500	186,700
Protein recovery + biological treatment	Protein	Feed/food	807,000	198,100	44,000
Protein recovery and concentration by evaporation	Protein Concentrate without protein	Feed/food —	881,000	281,500	165,000
Protein recovery + ion exchange + biological treatment	Protein Amino acid mixtures Organic acid mixture K_2SO_4-$(NH_4)_2SO_4$ $(NH_4)_2 SO_4$	Feed Feed/Food Beverages Fertilizer Fertilizer	2,550,000	755,750	444,000

From Stabile, R. L., Turkot, V. A., and Aceto, N. C., Second National Symp. on Food Processing Wastes, Denver, 1971. With permission.[10]

filter, a plate and frame-type filter press, by gravity settling, or by centrifugation in continuous conveyor-discharge centrifuges. Strolle et al.[8] recommended the use of the plate frame filter press. Knorr[2] concluded that protein recovery could possibly be economical in combination with starch recovery and/or recovery of other compounds such as amino acids, potassium, organic acids, and phosphates.

III. SINGLE-CELL PROTEIN

Single-cell protein (SCP) refers to unicellular organisms grown specifically for their protein content under controlled fermentation conditions on an industrial scale.[11] A wide range of microorganisms as well as substrates can be employed. The details on SCP developments are summarized in Table 2. The potential interest in SCP has been mainly in the animal feed sector. This section deals with the utilization of potato waste for this protein-rich product.

A. YEAST

Media from potato starch production wastes were evaluated for accumulation of food yeast biomass.[12] In this study, biomycin stimulated yeast growth and nonhydrolyzed residues in the media did not depress yeast yield. Starchy waste from potato processing, pretreated with an amylolytic preparation and green malt, was found to be suitable as a carbon source for the biosynthesis of feed proteins by *Candida humicola* strains on a commercial scale.[13] The wastes from food and beverage processing industries including the potato processing industry were converted to SCPs[14] using such organisms as *Rhodotorula glutinis, Geotrichum candidum, Trichoderma viride, Bliocladium deliquescens, Endomycopsis fibuliger, Saccharomyces fragilis,* and *C. utilis.* Thus, it was possible to utilize 70 to 90% of organic carbon from these wastes by microbial conversion.

In a systematic study,[15] SCPs were produced from potatoes after enzymatic hydrolysis using *C. utilis.* Fermentation was conducted at pH 3.5 and 30°C with aeration and stirring. Ammonium sulfate and potassium dihydrogen phosphate were added as nutrients. The retention time was 2 h 15 min with eluate containing 21 g dry matter resulting from 1 kg potatoes. The yeasts, *C. utilis* and *S. fibuligera,* were propagated as a source of SCP in a continuously mixed, anaerobic, single-stage cultivation on blancher water produced from potato processing.[16] Various organisms were inoculated on a dilute mixture (1:1) of potato pulp hydrolyzates and distillery slop containing 1% reducing substances.[17] Protein concentration was highest, in decreasing order, in *T. wittenbergi* D27, *T. utilis* D25, *C. utilis* D11, *C. tropicalis* CK4, *C. humicola* D12, and *C. curvata* D66 (55.7 to 52.2%). However, the percent utilization of the reducing substances was highest for *C. tropicalis* CK4 (88.2%) and *T. candida* D4 (81.8%) and lowest for *T. wittenbergi* D27 (18.7%). The maximum relative growth rate of the yeasts during 72-h cultivation was 0.385 g/h. Fully aerated cultures of *Schwanniomyces alluvius* were grown on 4% soluble potato starch in a defined minimal medium at a doubling time of 1.5 h at 30°C for the production of SCP.[18] The recovery was 51%.

B. FUNGI

Cultivation of mycelial fungi on potato processing industry waste is useful as they contain a complex of hydrolytic enzymes[19] which eliminate the necessity of hydrolyzing potato substrates before culturing.[20] *Penicillium digitatum* 24 P grown on potato processing wastes increased the biological value of the protein preparations because of their high contents of unsaturated fatty acids and vitamin F.[21] Further, *P. notatum* and *P. digitatum* biomass contained 62% protein, 5 to 7.5% fat, and also B vitamins and vitamin D.[20] The biological value of protein was 58 to 80%. It was also observed that mixed cultures of *Penicillium*

TABLE 2
Organizations Involved in SCP Development

Type of substrate	Organism	Organization/location	Production	Use[a]		Commercial status
				A	H	
Starches, sugars/yeast						
Molasses	*Candida utilis*	Cuban sugar	70,000	+	?	Product being developed for H application
Lactose/whey	*Klugveromyces fragilis*	Bellyeast/France	6,000	+	+	High-value application sought
Sulfite liquor	*C. utilis*	Pekilo/Finland	3,000	+		
Starches, sugars/fungi						
Hydrolyzed starch	*Fusarium graminarium*	RHM/U.K.	Pilot plant	+		Awaiting scale up
Alcohols/yeast						
Ethanol	*C. utilis*	Amocol/U.S.	5,000	+	+	High-value markets
Methanol	*Pichia pastoris*	IFP/France	Pilot plant	+		Research phase
Methanol	*Pichia* spp.	Philips/U.S.	Pilot plant		+	Research phase
Alcohols/bacteria						
Methanol	*Methylopophus*	ICI/U.K.	55,000	+	?	Successful large-scale development
Methanol	*Methylomonas clara*	Hoescht/Germany	Pilot plant	?	+	Research phase
Hydrocarbons/yeast						
Gas, oil	*C. tropicalis*	BP/France	15,000	+		Project closed
	Candida spp.	VEB/East Germany	100,000	+		Commercial development
n-Alkanes	*C. lipolytica*	BP/U.K.	4,000	+	?	Project closed
	C. lipolytica	ANIC/Italy	100,000	+		Project closed
	C. maltosa	Liquichemical/Italy	100,000	+		Project closed
	C. tropicalis	IFP/France	Pilot plant	+		Project closed

[a]   A = animal; H = human.

From Stringer, D. A., *Comprehensive Biotechnology*, Vol. 4, Robinson, C. W. and Howell, J. A., Eds., Pergamon Press, New York, 1985, 685. With permission.

and yeast did not have improved protein biological value over *Penicillium* alone, but protein yield was greater. *T. album* cultured on media which is largely agricultural residues or wastes or by-products provided 57 to 65% proteins on a dry weight basis.[22]

The solid state fermentation of starchy materials such as cassava flour, banana refuse, and potato waste with *Aspergillus niger* at 40°C and 55% moisture in the presence of nitrogen and minerals increased the protein content of these materials from 2.5 to 5.0% to 17 to 20%.[23] The carbohydrates decreased from 65 to 90% to 25 to 35%, which represents a 40 to 50% biomass conversion. Protein enrichment of these materials can also be carried out by fermenting with amylolytic fungi.[24] Selected fungi, TK2, TK41, and TK42, were found to be effective in enriching the protein content of the potato waste by solid-state fermentation.[25] The use of 1% potato peels in the medium for the growth of *A. foetidus* increased the protein content to 29 to 70% after 48 h of incubation. The protein and glycoamylase yields were improved by the addition of corn steep liquor and wort.[26] The fungus, *Cephalosporium lichorniae* 152, was capable of converting the fresh potato processing waste into microbial proteins at 45°C and pH 3.75.[27]

Production of yeast protein by Symba Process involving *E. fibuligera* and *Candida utilis* was reviewed by Bergkvist.[28] Later on, Skogman[29] described the process for producing high-quality food yeast by the symbiotic cultivation of these microorganisms on starch containing food-processing wastewater. The former microorganism breaks down starch to glucose and the latter utilizes it for growth.[30]

The potato distillery by-products obtained from French fry industry waste after alcoholic fermentation had fair amounts of crude protein (17.6 to 25.3%), but a large portion of it was nonprotein nitrogen.[31] Although the products were good sources of lysine and methionine, high ash and glycoalkaloid contents would limit their potential feeding value.

IV. ANIMAL FEED

Potato peel, cull potatoes, and products derived from potato processing waste have been considered for animal feed. For cultivating fodder yeast on potato starch production waste, best results were obtained with 1.18% reducing compounds and 3.73% inoculating volume.[32] Out of 20 fodder yeasts grown in deproteinized potato juice, 2 strains were capable of producing biomass containing nearly 56% protein.[33] The effectiveness of *Rhizopus oligosporus* and *E. fibuligera* in the production of animal feed or bacterial media from potato waste was shown by substantial saccharification of starch yielding 93 and 87% reducing sugars, respectively.[34] The potato processing waste was fermented into a crude protein feed[35] supplement by *Lactobacilli* at a temperature of 43°C and a pH of 5.5. The fermented ammoniated potato waste proved to be an excellent nitrogen source for ruminant animals. It is estimated that nearly 1.3 million t of potato processing waste could be available annually in the U.S. for such products. The potato peelings were used for growing hyphae of molds, *T. viride* B83, B115, B116, B117, and B118, *Myrothecium verrucaria* B91, and *A. nigera* A60 and a high protein content indicating possible use as animal feed.[36] Also, the possibilities of wastewater treatment with molds are not neglected because of improvement in the filtration time of wastewater. The fermentation of a potato juice-potato pulp medium with *Penicillium terlikowskii* at the dilution rate of 0.20 and 0.66/h during the first and second stages of continuous fermentation, respectively, was adequate enough to utilize 86.6% dry matter, giving rise to a biomass containing 29.0 to 39.6% crude protein having a biological value of 54 to 57.9.[37]

Using a composite feed for animals has been an age-old practice. Potato flakes made out of wheat starch-processing waste, potato waste, and refuse starch sludge (1:1.5:1) has more than 97% digestibility coefficient of protein content.[38] The waste potato pressing after extraction of starch mixed with grain dust (20 to 30 parts); malt (2 to 5); urea (0.2 to 2);

ammonium sulfate; salt and mineral additives (1 to 2.5) increased the weight increments from 0.48 to 1.07 kg per cow/day compared with classical feeds.[39] Waste from the manufacture of potato or bean flakes was processed into feed together with protein from the activated sludge and dried *Lemna*. This feed[40] contained moisture 73 to 78%, proteins 17 to 20%, phosphorus 1.6%, ash 0.37%, fat 0.75%, cellulose 1.19%, vitamin B_1 0.25, mg/100 g, and vitamin B_2 0.6 mg/100 g.

A feed high in proteins and minerals has been prepared by blending potato juice with waste solids from potato starch processing followed by heating for starch gelatinization and cooling.[41] The use of caustic potato peel waste plus acidic wastewater from fish processing operations was suggested as animal feed.[42] Another way of neutralizing dry caustic potato peeling waste useful as feed stuff is by an inexpensive lactic fermentation.[43] The soluble potato solids, isolated from wastes from potato starch processing, contain substantial amounts of protein and sugar. The product was used for broiler rations after heat treatment to reduce the trypsin inhibitors and CaO to reduce the hygroscopicity.[44]

Escher Wyss® drum driers were used to dry the disintegrated waste from manufacturing of dry potato products.[45] The waste contained carbohydrates, proteins, and cellulose. The product resembled potato flour. Similarly, the potato proteins from effluents of potato starch mills can be drum dried and converted to useful animal fodder. It was observed that the drum drying saves energy, keeps loss of useful ingredients to a minimum, and produces a less hygroscopic product. The potato wastewater slurry was processed to form a gel which, on stirring with anhydrous calcium chloride followed by pressing, produced solids useful as a feed for ruminants.[46] Potato processing residue can replace barley as an energy source for beef cattle.[47] Potato starch waste, when treated with phosphoric acid and then dried to 13% moisture, was found to be more palatable to cows than untreated starch waste.[48] The potato waste meal can be considered a good substitute ingredient for a proportion of the ground corn in a practical diet for broiler chickens[49] and livestock.[50] It is produced by drying and grinding whole potatoes, potato pulp and peelings, etc. It contained slightly less crude protein, but more lipid, crude fiber, and ash when compared with corn. It is not devoid of any of the amino acids than corns and seems limiting with respect to methionine, cystine, arginine, and the aromatic amino acids. Potato meal was also ensiled with a grass legume forage and evaluated as a feedstuff for lactating cows.[51] Milk production for cows receiving this feed was slightly better than that for cows consuming untreated silage. Bushway et al.[52] produced a new potato meal by low-temperature dehydration as a by-product of starch manufacturing. The new potato meal contained 15.53 mg α-chaconine and 4.73 mg α-solanine per 100 g meal. It contained 2% more protein and 1.9-, 3.4-, and 1.2-fold more niacin, riboflavin, and thiamine, respectively. Potassium and phosphorus concentrations were about 2 and 0.2%, respectively. The meal was recommended for fish and chicken. Under the Federal Food, Drug and Cosmetic Act, the tolerance limits for Diquat[53] pesticide and Metalaxyl[54] fungicide in or on dried potato waste used for animal feed are 1 and 4 ppm, respectively.

V. ALCOHOL PRODUCTION

A. BACKGROUND

Since the incredible rise in oil prices on two occasions, one in 1973 and the other in 1979, many countries worldwide initiated programs to develop alternate energy sources. Consequently, the concept of fermentative ethanol occupied importance due to its potential fuel/chemical uses. Basically, any biomass can be converted readily to alcohol. A number of potential raw materials of plant origin are considered for fermentative conversion to ethanol. These materials include sugar-bearing products and wastes (molasses, cane juice, beet juice, sulfite liquor, whey, fodder beets, sweet sorghum, Jerusalem artichokes) starchy

TABLE 3
Potential Alcohol Yield from Commercial Crops

Crop	Carbohydrate (tons/ha)	Alcohol (liters/ha)
Sugarcane	6—12	3600—7200
Sugarbeet	5—7	3000—4200
Sorghum	2—7	1500—4400
Potatoes	3—7	1800—4200
Corn	2—4	1200—2400
Cassava	3—12	1800—7200

From Venkatasubramanian, K. and Keim, C. R., *Starch Conversion Technology,* Van Beynum, G. M. A. and Roels, J. A., Eds., Marcel Dekker, New York, 1985, 143. With permission.

commodities (potato, cassava, sweet potato, corn, grain sorghum and millets, wheat, etc.), cellulosics (wood, wheat and rice straws, cotton stalks, cane trash, bagasse, etc.), and municipal solid waste. However, the economic viability of bioethanol would be largely governed by the availability and cost of the feedstocks. In Table 3 the potential alcohol yield from some of the important carbohydrate energy crops is presented.[55]

As potatoes are a rich source of starch, small or surplus tubers, cull, greened, damaged, or discarded potatoes, and potato waste can be readily converted to alcohol. In the U.S., the need for utilization of potatoes for fermentative alcohol existed during and immediately after World War II as an emergency measure. In 1946, the American alcohol distilleries used 29 million bushels of potatoes for alcohol.[56] Today, the important feedstock for alcohol production in the U.S. is corn because it offers many advantages over the other starchy commodities. In Germany, the conversion of starch-containing crops, especially potatoes and grains, to bioethanol can be traced back at least to the middle of the last century.[57] Dellweg and Luca[57] cite examples of agricultural distilleries in Germany which produced 360,000 m^3 of ethanol, the main part of which was used as fuel and for other nonfood uses. The concept of bioethanol in developed countries is, however, limited to highly specific situations of small volume potential. Also, the exact market potential for ethanol in the long run would depend upon a number of competing factors. On the other hand, developing countries, largely dependent upon agro-based industries, cannot afford to relegate the commercial exploitation of common carbohydrate crops to the distant future. Starchy commodities such as potato, sweet potato, and cassava have the potential for making industrial fermentative alcohol in the foreseeable future.

B. PROCESSES

Although corn represents an important biomass resource for alcohol production, large quantities of starch-rich agro-industrial residues such as potato processing wastes represent another important resource which could be fermented to yield ethanol.[58] For example, an estimated 1.3 million tons of potato processing wastes which is produced annually in the U.S. is potentially convertible by fermentation to a large quantity of alcohol.[58] The main processes for bioconversion of all starch-containing raw materials to ethanol involve starch gelatinization with steam, enzymatic hydrolysis of starch, yeast fermentation of resulting fermentable sugars, and recovery of alcohol. The plants designed to carry out the above operations are generally capable of handling potatoes, other tubers, or grains for alcohol production.

The starch granules of raw material absorb water during gelatinization by steam heating and then burst, making the starch molecules susceptible to subsequent hydrolysis carried out by means of amylolytic enzymes, α-amylases, and glucoamylase. In recent years the use of concentrated, purified microbial enzymes has been gaining importance because of technical and economic advantages. The fermentable sugars are transferred to a fermentor where yeast is introduced to continue fermentation to alcohol in the batch mode. The typical manufacturing of alcohol from starch-containing raw materials is exemplified by German batch (GB) process, American batch (AB) process, and continuous process (CP). These processes as depicted by Novo Industry A/S[59] are presented in Figure 2. The GB process is characterized by gelatinization of raw material without disintegration in a Henze cooker and also without enzyme and stirring. The cooked mash is passed through a strainer valve into the mash tub where liquefaction is achieved by either high- or low-temperature (HT or LT) mode. For HT liquefaction the blowdown is carried out rapidly. Then the mash is cooled in the mash tub to 80°C, at which point liquefaction is catalyzed by an enzyme (0.3 to 0.6 l/t starch) for 20 min. In the LT mode, the lower part of the cooling coil of the mash tub is covered with cold water, enzyme (0.1 to 0.2 l/t starch) is added with stirring, and then the mash is allowed to enter the mash tub under agitation and cooling to maintain a temperature between 55 and 60°C. Saccharification is carried out at 60°C with 1.5 to 2.0 l of enzyme per ton of starch. Yeast fermentation is continued at 30°C. The AB process deals with disintegrated raw material. Cooking and liquefaction of starch are carried out under agitation in the cooker. Preliquefaction of starch with enzyme (0.2 to 0.4 l/t starch) controls its viscosity during gelatinization. After cooling the mash to 80°C, postliquefaction for 10 to 20 min is carried out with the same enzyme (0.4 to 0.8 l/t starch). Postliquefaction may be carried out at 60°C if the bacterial enzyme is replaced by fungal enzyme. The liquefied mash is then treated with saccharifying enzyme (1.5 to 2 l/t starch) along with yeast in the fermentor section. In the CP, a continuously stirred tank (CST) is employed for preliquefaction at 80 to 90°C of the disintegrated raw material-water slurry. The mixture is brought to cooking temperature (105 to 106°C) by injecting steam at the cooker inlet. The slurry is then flash cooled to 80 to 90°C and subjected to postliquefaction with a fresh dose of enzyme in the tube reactor. The liquefied mash is pumped to the fermentor through the mash cooler. The liquefaction enzyme dose is nearly the same as that used for the AB process. The saccharification and fermentation steps are carried out in batch fermentors and their parameters are similar to those observed in the GB and AB processes.

In an attempt to reduce energy consumption in the alcohol manufacturing process, Lutzen[60] eliminated the starch cooking step and introduced the concept of simultaneous cold saccharification/liquefaction and yeast fermentation in a continuous mode (Figure 3). The continuous fermentation was proceeded by separation of yeast and unreacted starch, their recycling to the fermentor, and a continuous distillation of alcohol from the liquid stream followed by recycling of the stillage water to the fermentor. This process is an example of a single CST reactor (CSTR) with cell and material recycle. The other scheme uses multiple series-connected stirred tank reactors known as the cascade system. However, it is readily applicable to fermentable sugar-based feedstocks. Another choice is the use of immobilized yeast cells for fermentation. In this system, yeast cells immobilized in calcium alginate beads are employed in a fluidized bed tower fermentor for continuous fermentation of hydrolyzed starch slurry to ethanol. Dense cell populations and high productivity are two distinct advantages that immobilized cell systems have over CSTR systems.[61] Fukushima and Yamade[62] designed two types of immobilized bioreactors for continuous alcohol production from liquefied starch materials. For saccharifying and fermenting the liquefied starch, a bioreactor was designed by connecting a tubular unit packed with immobilized glucoamylase pellets with a rhomboid unit packed with 1.5 mm Al alginate-yeast beads. The other system was a three-stage rhomboid bioreactor packed with biocatalysts entrapping the mixture of glucoamylase and cells. It is generally believed that the continuous fermentation process offers many technical and economical advantages over the conventional batch process. Tegtmeier[63] also suggested a process design in order to minimize the energy consumption for the production of alcohol from starch-containing raw materials.

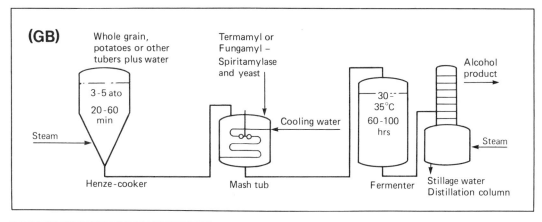

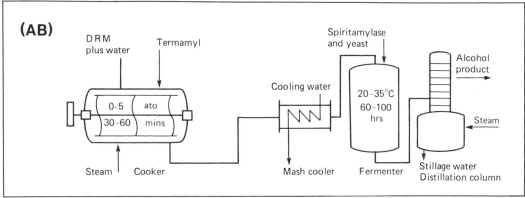

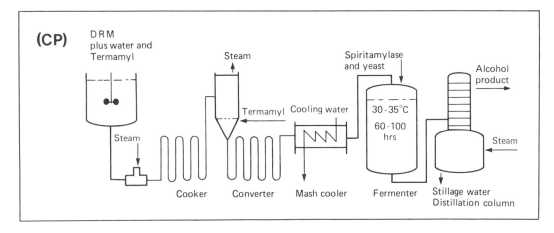

FIGURE 2. Schemes for production of alcohol from starch containing raw materials by German batch (GB) process, American batch (AB) process, and continuous process (CP). DRM = Disintegrated raw material. (From Novo Industry A/S, Technical Information Brochure, Bagsvaerd, Denmark, 1981.[59] With permission.)

Recently, Dellweg and Luca[57] described some developments in the process technology of ethanol fermentation starting from starch materials. A continuously operating starch hydrolyzing plant was designed (Figure 4) which was capable of completely suppressing starch viscosity under the simultaneous influence of shearing forces in the colloid mill. Also, the use of temperature-stable α-amylase from *B. licheniformis* was made. The operation is called the "SUPRAMYL-Process".[64,65] As indicated in Table 4, the SUPRAMYL-Process

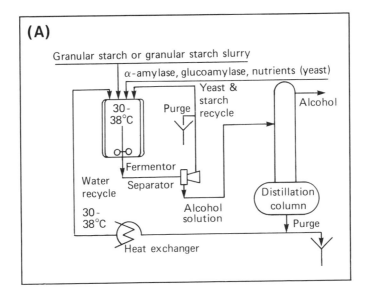

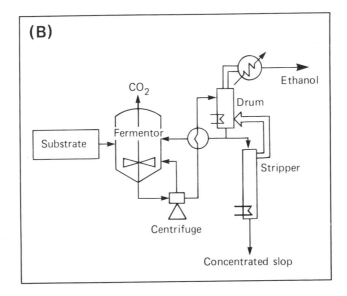

FIGURE 3.   (A) Flowsheet for continuous operation of cold saccharification/ liquefaction and for (B) 100-l pilot fermentation equipment used for continuous cold process. (From Lutzen, N. W., *Proc. Int. Ferment. Symp. on Adv. Biotechnol.*, 6th, 2, 161, 1980. With permission.[60])

has the lowest energy consumption of all known comparable processes.[66] Thus, the optimized starch hydrolysis process, energy savings, and improved performance of the yeast strain 0762 have led Dellweg and Luca[57] to consider the possibility of a two-stage continuous fermentation process with a recycling system which at present is under investigation.

C. POTATO RESEARCH

A number of publications on alcohol from potatoes, potato starch, or potato processing wastes have appeared during the last few years. These reports discuss various aspects of alcohol production which are briefly summarized in the following section.

Yamamoto et al.[67] suggested that potatoes and sweet potatoes require maceration with

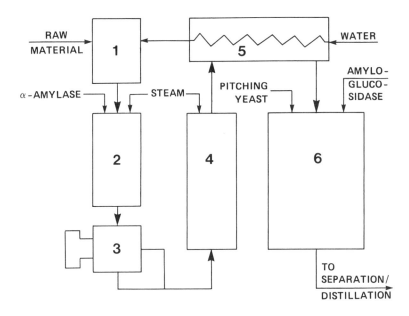

FIGURE 4. Outline of the SUPRAMYL process. 1. Crushing mill; 2. mash tub; 3. homogenizer, 90°C (e.g., Supraton mill); 4. holding reactor for dextrination at 90 to 105°C; 5. heat exchanger; 6. saccharification and fermentation at 40°C. (From Dellweg, H. and Luca, S. F., *Process Biochem.*, 23, 100, 1988. With permission.[57])

TABLE 4
Energy Consumption for Manufacture of Ethanol from Starch Materials
(MJ per Liter of Dry Ethanol)

Process	Optimized process	Traditional distillery
Starch degradation (SUPRAMYL Process)	0.9	
Azeotropic distillation at increased temperature/pressure	1.9	20
Vapor compressor for distillation column (0.3 kWh)	1.2	
Drying of slops after recycling	4	11
Total	8	31

Energy gain expected from biogas: 5 to 8 MH/liter of ethanol

From Misselhorn, K., Paper presented to *Oktoberagung der Wersuchs-und Lehranstalt Fur Brauerei,* Berlin, October 14, 1987. With permission.

pectic enzyme followed by dextrinization of starch with α-amylase at 88°C, before gluco-amylase and yeast could be added to obtain 13 to 14% ethanol as it is done in the case of cassava. Recent studies[58] showed that a simultaneous single-step system for the enhanced fermentation of potato starch to ethanol by using symbiotic cocultures of *A. niger,* which hydrolyzes starch to glucose, and *S. cerevisiae,* which is nonamylolytic, but efficiently ferments glucose to ethanol, is feasible. The synergistic effects of these microorganisms eliminate the enzymatic starch hydrolysis step, thereby improving the economy of alcohol production from starchy biomass. The efficiency of alcohol production from starchy tubers of potatoes, tapioca, and sweet potatoes was 68, 81, and 75%, respectively.[68] The tubers were gelatinized in boiling acidified solution, saccharified with *R. niveus* and fermented with *S. ellipsoideus* in the presence of the saccharifying fungus. Cheese whey can be fermented in conjunction with potatoes, grain, or other raw materials.[69]

In Norway, an alcohol plant was constructed for utilization of wastes from a potato starch factory, a potato flake plant, and a potato chip line.[70] The conceptual ethanol plant was designed[71] to process potatoes, sugar beets, and wheat for a portion of the year using geothermal resources in the Raft River area of Idaho, but could operate year-round on any of them. The processing facility had a capability of 75.7 million l ethanol per year. The data on small-scale production of fuel alcohol from potato culls[72] indicated that net economic and energy gains can be realized through the use of an appropriate plant size and processing technology. Hammaker et al.[73] studied the feasibility of converting a sugar beet plant to fuel ethanol production from potatoes, corn, and hybrid beets. A plant for ethanol fermentation of potato and wheat starch-processing waste with special attention to potato pulp is described by Marihart.[74] The pulp was subjected to acid hydrolysis and the other 50% to treatment with cellulose. After 24 to 48 h, amyloglucosidase was added and the mixture was combined with a nonviscous, noncellulosic substance for ethanol fermentation. It was stated that conventional fuel ethanol production from potatoes, corn, and candidate crops available in Suffolk county, New York is not strongly attractive because of storage problems, seasonality, and high cost.[75] Huang[76] studied the feasibility of producing ethanol from potato processing waste. The alcohol production facility tested by Kuby et al.[77] produced nearly 3.785×10^6 l ethanol per year from cull potatoes as a feedstock.

VI. BIOMETHANE GENERATION

Conversion of biomass to methane gas by microbial degradation has been recognized for centuries. The biological process is known as anaerobic digestion in which the organic matter is converted to methane and carbon dioxide in the absence of air (oxygen). In nature, an anaerobic conversion occurs at places such as marshes, lake sediments, etc., where favorable conditions exist. The mechanism of anaerobic decomposition is represented in Figure 5, which indicates both acetogenic dehydrogenation and hydrogenation.[78] Most of the methane gas is formed from decarboxylation of acetate, while the remainder is being formed by the reduction of carbon dioxide using hydrogen. The effective conversion of complex organic matter to methane depends upon the combined activity of a miscellaneous microbial population consisting of diverse genera of obligate and facultative anaerobic bacteria. The production of methane from waste material has a twofold advantage: pollution control and energy recovery. Various types of wastes have been studied for the production of methane. These include mainly wastes from the paper mill, dairy, distillery, sewage, animal slurries, fruit and vegetable solids, slaughter house, and food industry. The nature and the volume of biomass, reactor design, and process parameters are important factors in methane gas generation.

Production of biogas from potato waste both on laboratory and pilot plant scale have been reported. In such an attempt, the addition of potato waste or sawdust in small amounts increased methane production from poultry manure.[79] The anaerobic contact process at 35°C was capable of producing 2 m^3 methane per cubic meter per day from food processing plant wastes with volatile solids loading at a rate of 8 kg/m^3/d. The wastes consisted of bean blanching, pear and potato peeling, waste ground peeling, and rum stillage.[80] The efficiency of conversion of organic solids to methane was about 80%. The average methane production in the pilot plant, based on an upflow anaerobic sludge blanket (UASB) process, was 0.35 m^3/kg chemical oxygen demand (COD) converted when operated on potato and corn product wastewater.[81] The water purification efficiency was nearly 90% with a loading of about 10 kg COD removed per cubic meter per day. Jackson[82] presented the economics of large-scale production of methane from solid and soluble wastes from a potato processing plant together with flowsheets of potential processes and some design data for plants. The Anoxal process, an upflow anaerobic filter process, with spall ring packing was used to

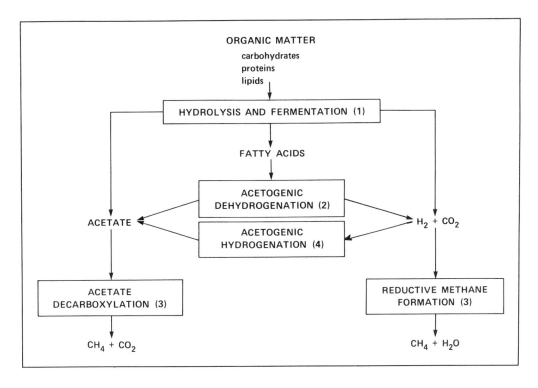

FIGURE 5. Microbial pathways in anaerobic digestion. From McInerney, M. J., Bryant, M. P., and Stafford, D. A., in *Anaerobic Digestion,* Applied Science Publishers, London, 1980. With permission.[78])

treat liquid industrial effluents containing potato-bleaching wastewater.[83] In this treatment, a daily average biogas production (containing 65% methane) was 5800 m^3 at a COD loading of 8 $kg/m^3/d$ and hydraulic residence time (HRT) of 24 h.

VII. OTHER USES

An organic cleaner from potato processing plant waste was found to be biodegradable.[84] It contained very low phosphorus and nitrogen, had good washing and cleaning properties, and apparently could be prepared in granular or bar soap forms to meet the various use requirements. The potato waste when mixed with wheat starch milk can give an adhesive with improved properties.[85] Ady[86] described a simple procedure for preparing a synthetic detergent from potato scraps and waste oil from potato processing. It is possible to obtain pectin in a high yield from potato waste from starch manufacturing.[87] Pectin was obtained after removal of residual starch from the waste by heating in water and subsequent treatment with glucoamylase. The starch product, obtained by drying the debris recovered from peeling potatoes, is recommended for its use in drilling muds.[88] The by-products of potato processing were considered as a source of extraction of L-ascorbic acid which could be used in an agar gel.[89]

VIII. EFFLUENT TREATMENTS

A. BACKGROUND

Various industries such as chemical, food, pharmaceutical, leather, and paper, etc., are responsible for the degradation of the environment through pollution. In the food industry, the main source of pollution is liquid effluents released by this industry. Liquid effluents may affect the environment in a number of ways. The aquatic flora and fauna usually survive at the pH of 6 to 8. Small variations outside these limits caused by effluents endanger their life. Suspended solids in effluents can be either inorganic or organic. The former are heavy and create silting problems in rivers and streams. The latter undergo biodegradation and cause depletion in dissolved oxygen, thus again endangering aquatic life. Bacteria, present in water streams, proliferate on organic materials and use up dissolved oxygen. In more severe cases, anaerobic conditions develop with the liberation of H_2S and methane gas, both of which can be toxic. Any toxic chemicals if present in the effluents released in the stream can cause severe hazard to life. The presence of coloring matter may lead to aesthetic problems and interfere with aquatic life.

B. PLANT MANAGEMENT

In view of the greater social awareness about environmental pollution and legal sanctions imposed by government agencies, it is essential for the industry to reduce the volume of effluents to be discharged into the water resources. In this connection the following should act as guidelines for achieving the desired objective of a clean environment:

1.   Efficient inplant control;
2.   Efficient recovery of solids and suspended material from the effluents;
3.   Maximum recycling of water;
4.   Continual research and development efforts to reduce waste and to reduce the cost of treatment.

It has been emphasized that conservation, reclamation, and reuse of solids and water are essential in the potato processing industry.[99] The authors[90] pointed out that up to about 11 m^3 of water were used to process 1 ton of raw potatoes. In frozen French fry plants, waste treatment is a continuing problem. Moreover, the situation is complicated by the use of a 10 to 20% caustic peeling solution; the unnecessary fat contamination of the alkali waste by improper fryer design; the failure to segregate wastes from washing, peeling, and blanching from waste containing mainly suspended starch; and the failure to reduce water consumption by starch recovery and water reuse.

Another view[91] points out that it is possible to meet production requirements while remaining compatible with the environment by concentrating on potato plant reclamation and waste product reuse. In general, it is felt that certain guidelines can be developed to determine the actual state of wastewater discharge and reduce the organic load. The latter goal can be achieved by the modification of the washing process. It has been suggested that the inhibitory effect of potato juice on the rate of particle settling should be accounted for in designing settling tanks for treating wastewater from a potato starch plant.[92]

Hindin[93] proposed a scheme for a more complete reuse (limited recycling) of the wastewater in processing based on three French fried potato plants in the state of Washington. As per the scheme, the fluming water from inspection tables, sizers, or graders was collected and reused for up to 7 h after the settling of solids and the chlorination of the water. The cutter deck wastewater was collected, allowed to settle, and the supernatant was reused. However, water must be replenished to compensate for the quantity carried away by the potato cuts. The blancher water, which contains high amounts of organic and inorganic matter, can be used in the barrel washer for the removal of peeling and NaOH. Also, wastewater from cooling the blanched potatoes can be used as the rinse water for NaOH removal from the peeled potatoes in the barrel washer. The wastewater from the peeling and NaOH rinse sections may be used as makeup water for the fluming process. Cooling water from the freezer condensers is useful anywhere in the process line after chlorination and is ideal as boiler feed water. After every shift, intermittent flows of water are discharged from the trim table, the cutter deck, the freeze tunnel, and plant and equipment cleanup. A continuous supply of fresh water is required for cooling blanched potatoes and as cooling water for the freezing equipment.

C. PRIMARY TREATMENT

After removing large solids by a conventional screening method, wastewater is allowed to flow into the primary treatment system, which involves the removal of floatable and settleable solids. The treatment system is essentially a rectangular or circular clarifier equipped with a scraper mechanism to remove solids that settle to the bottom or float to the top. The clarifier is generally provided with an overflow outlet.

D. SUBSEQUENT TREATMENTS
1. Aerated Stabilization

It is an aerated pond in which oxygen is mechanically supplied by compressed-air diffusion or surface aerators for BOD removal by dispersed bacteria. A special device such as the aero-hydraulic gun increases the oxygenation capacity in the treatment of potato waste.[94] The wastewater from a potato processing plant with a biological oxygen demand (BOD) of 6500 mg/l can be treated to give odorless water with a BOD of 6 mg/l by means of a device[95] consisting of four spherical tanks with activated sludge and aeration pipes in series for intermittent 10-min aeration and 20-min settling. Richter et al.[96] demonstrated that full-scale aerobic digestion can be used in conjunction with basket centrifugation or vibratory screening with the addition of a small amount of cationic polymer to obtain an increased concentration of solids from potato wastewater. The aerobic secondary treatment employed by Richter et al.[97] showed a removal of 90% BOD and of 96% coliform bacteria in the activated sludge. High pH values of influent process wastewaters were buffered in the aeration basins. The clarifier thickener removed nearly 100% of the suspended solids when operated on water alone. The jet aeration system[98] provided the required dissolved oxygen and the results showed no fecal coliforms in treated effluents during two operating seasons.

2. Activated Sludge Process

In this system, soluble organic material is brought to a low level in a short time as a result of a high concentration of flocculant biological solids maintained in an oxygenated tank in which the waste and bacteria are in contact. The organic matter becomes part of the bacterial mass known as activated sludge, which is removed by sedimentation, thus providing a highly treated effluent. The mixed fluid is allowed to flow from the aeration tank to the clarifier and the settled solids are returned to the aeration basin to maintain a high concentration of biological solids. This concept can be implemented in a variety of ways, such as stage or plug flow, completely mixed, and contact stabilization types. Nutrients (i.e., nitrogen and phosphorus) are added to increase the activity of microbes.

The efficiency of this method in reducing BOD by at least 90% is revealed by several reports. The protein water from potato starch manufacturing containing suspended solids 820, total nitrogen 830, COD 3300, and BOD 6000 ppm resulted in 96.2% COD and 98.8% BOD removal when treated with 5% activated sludge at 25°C and 1.3 kg/m^3 · d of COD loading.[99] In spite of a high organic load, the purification of combined wastewater from the potato processing industry and municipal waste in an activated sludge plant was successful.[100] In a bench-scale completely mixed continuous activated sludge reactor,[101] 90% removal of COD with improved ammonia removal from the potato processing effluent was observed at 5 d. However, additional treatment was suggested to produce effluents suitable for inplant reuse. Another laboratory-scale activated sludge process[102] at an HRT of 1.43 d, at a sludge age of 5 d, and with additions of powdered activated carbon (4000 ppm) reduced the COD values of the potato processing effluent by 94 to 96%, making the process more effective than aerated lagoon treatment. Yeh and Hung[103] observed a slight improvement in the performance of the activated sludge process for the treatment of wastewater containing potato juice when it was augmented with bacterial culture consisting of seven strains of bacteria commonly found in soil.

3. Anaerobic Process

Anaerobic treatment of effluents in closed reactors or in the bottom of open lagoons is the simplest procedure as it requires no aeration and the least maintenance. It involves the reduction of sulfates, nitrates, and organic molecules. Oxygen from the sulfate and nitrate ions is utilized for organic oxidation. Moreover, organic molecules which are difficult to degrade aerobically are easily degraded anaerobically and methane gas can be utilized. However, high temperatures and high biomass concentrations are required for reasonable rates of biological degradation. A long generation time for methane bacteria leads to a high solids retention and acclimatization. The concerned bacteria are sensitive to shock loads, toxic material, and environmental conditions. Effluents with low BOD (50 ppm) with good characteristics are difficult to produce and an obnoxious odor is unavoidable. The application of anaerobic process for wastewater treatment is attractive in the event of large volumes of effluents which are forced through the system in a relatively short period of time. The anaerobic digestion process is extensively used for the stabilization of sludge from sewage works and for energy production from a wide variety of biomass. The loading potentials of anaerobic wastewater treatment processes are largely governed by the sludge retention in the anaerobic reactor. In the past, various sludge retention techniques have been evaluated including the anaerobic contact process and the anaerobic filter. More recently, a very promising technique has been developed in the Netherlands, viz., the UASB process.[104] The system is shown schematically in Figure 6. Today, the industrial-scale UASB systems are utilized in treating soluble carbohydrate and protein-type wastewaters from the potato, potato starch, wheat starch, beet sugar, and liquid sugar and from candy-making, brewery industries, yeast production, molasses fermentation and canary industries. Chemical industry wastewaters can also be treated in the UASB process while it has potential application to treat effluents from corn wet milling. The anaerobic digestion processes as applied to wastewaters from potato processing are summarized in Table 5.

4. Anaerobic-Aerobic Process

Occasionally, it becomes necessary to combine the anaerobic and aerobic processes in order to combat their disadvantages while retaining their best features for significant results. In such a combination, anaerobic treatment is followed by aerobic treatment. It is evident that potato processing wastewaters can be treated effectively by an anaerobic-aerobic system.

Dostal[122] conducted pilot plant experiments on secondary treatment of potato processing effluents and found costwise superiority of the combined treatment over the separate anaerobic or aerobic treatment. The treatment afforded 90% BOD reduction. Mixing was essential for proper aeration while covering[122,123] of the lagoon prevented a drop in temperature and bad odor development. The addition of an anaerobic lagoon and a mechanically aerated basin before the stabilization pond had average removal efficiencies of 99% BOD, 96.2% COD, and 98.5% suspended solids (SS).[123] Isik and Slack[124,125] developed a novel effluent treatment based on the combined principles of the anaerobic-aerobic process which was capable of producing large quantities of biogas from food processing effluents and decreasing BOD to the river discharge permissible levels. Even at a temperature as low as 4°C, the combination of anaerobic and aerobic reactors provided very stable and satisfactory effluents from potato processing wastewaters.[126]

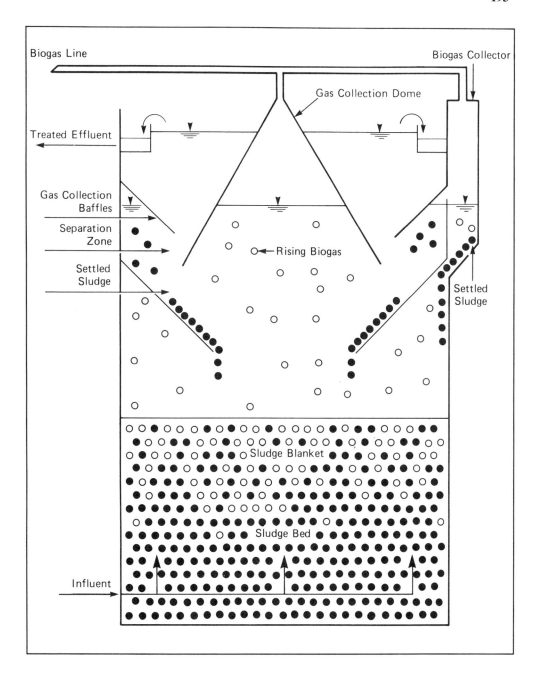

FIGURE 6. UASB reactor schematic. (From Hack, P., Personal communication with SJJ., 1987. With permission.)

E. OTHER TREATMENTS

1. Chemical Treatment

The use of chemical methods in effluent treatment, in general, is limited and is intended for such specific chemical reactions as neutralization, precipitation, oxidation, reduction, decomposition, or ion exchange to meet particular requirements. However, pH control or neutralization is considered a critical factor in certain treatments.

In a simultaneous treatment of community and industrial wastewater, including that from potato processing, control of pH was an important factor for consideration.[127] Purification

TABLE 5
Anaerobic Treatment of Potato Processing Wastewater

Treatment/system	Remarks	Ref.
1. Anaerobic contact process	Effect of variable HRT (1 to 20 d), added nutrients, temperature, overloading, and feed break	105
2. UASB process	Adopted for treatment with high loading rate 25 kg/m^3 · d at 30°C and detention times of 3 h	106
	Design description: load capacity 16 kg COD/m^3 · d; highly suitable for industrial waste; design and operation details: significant COD reduction; methane production for process energy	107 108
3. Anaerobic filters	Require liming (200 ppm) of effluents for effectiveness	109
4. Anaerobic lagoon/anaerobic filter system	Equilibrium between deposition and degradation of VSS at loading rate of 0.2 kg VSS/m^3 · d	110
	Efficient and consistent performance during low temperature over a range of loading rates (0.14 to 0.719 BOD/m^3 · d, 8 to 20°C)	
5. Laboratory scale horizontal anaerobic filter	Loading 1.35 kg COD/m^3 · d and HRT 0.6 d	111
	Reduction of 91% BOD, 91% COD, and 85% SS	
	Earthen construction	
6. Bench scale anaerobic	Between clarification and first-stage biotowers to upgrade treatment plant in French fry and instant potato flake plant	112
	COD removal 67.6 to 86.8% and higher removal with more loading rate; SS removal 16.6 to 70.8%	
7. Bulk volume fermentor	Capable of treating effluents with high SS and producing methane at 20 to 25°C	113
	Solids disposal and energy recovery	
8. Bulk volume fermentor with a floating membrane, aeration, nitrification, and final clarification	Overall removal of 98% COD, 99.5% BOD, 98% SS, and 80% total N	114
9. Anaerobic fermentation filtration	Reduction of COD 73.3 to 86.8% on clarifier waste at 1 kg COD/m^3 · d loading	115
10. Anaerobic digestion and liming	Removal of P; details on full-scale plant design, performance, costs, utilization of biogas, and land disposal of P sludge	116
11. Two-stage unified anaerobic fermentor-filter lab. model with horizontal flow for 114 d at 21°C	Removal of COD, BOD, and SS 96% at COD load 1.16 kg/m^3 · d and HRT 9.5 d	117
12. Anaerobic fixed bed containing corrugated plastic plates	Removal of 85.8% COD and biogas production 5.4 m^3/m^3 · d at 7.44 kg VSS/m^3 · d loading and 0.4 d RT	118
13. Pilot scale anaerobic digester	Removal of 68% volatile solids during 213 d	119
	Results useful in designing full-scale treatment plant with biogas and process heat recovery	
14. Anaerobic reactor with reactive support media	Removal of 70% COD at 3 kg COD/m^3 · d loading and 20 h HRT	120
15. Anaerobic fixed-bed reactors	Removal of 90—95% COD	121
	Effluents from treated wastewater and solid waste can be economically treated aerobically because of biogas recovery and energy saving	

Note: HRT = hydraulic retention time; COD = chemical oxygen demand; VSS = volatile suspended solids; BOD = biological oxygen demand; SS = suspended solids; N = nitrogen; P = phosphorus; RT = retention time.

of starch-containing wastewater required a pH of 7 before the addition of metallic salts.[128] Bentonite is known as a settling aid and has been used for the treatment of sludge in wastewater from potato washing.[129] Fussum and Cooley[130] collected basic data on the coagulation of lye peel potato wastewater using gels made from modification of humic acids

TABLE 6
Effects of Coagulating and Flocculating Agents on Potato Processing Water[131]

Chemical	Effluent	Remarks
1. FeCl$_3$ · 6H$_2$O (150 mg/l) + Purifloc® anionic polymer no. 23 (20 mg/l)	Abrasive peeled	Enhanced coagulation and flocculation including removal of 90% COD, TSS, and turbidity
2. CaCl$_2$ (300 mg/l) + Purifloc®-23 (25 mg/l)	Lye peeled	Removal of 69% COD and 76% TSS
3. CaCl$_2$ (350 mg/l) + NaClO 7122 (mg/l)	Steam peeled	Removal of 60% COD and 90% TSS

alone and with other coagulants and suggested water pollution abatement by improved coagulation. Karim and Sistrunk[131] conducted a systematic study to evaluate the effectiveness of coagulating and polymeric flocculating agents for the reduction of the wastewater strength in abrasive-peeled, lye-peeled, and steam-peeled potato processing wastewater (Table 6). These authors[131] concluded that the strength of waste effluents from the processing of potatoes can be substantially reduced by treatment with coagulating and flocculating agents before discharging into the municipal or other treatment system.

2. Activated Carbon Treatment

Activated carbon effectively adsorbs the contaminants in the wastewater mainly because of its large interfacial area and porous nature. It has the capability of removing color, BOD, COD, SS, odor, foam, and bacterial buildup from food process waters.[132,133] Unfortunately, the treatment suffers a high cost and its inability to adequately adsorb small polar compounds. However, it has attracted much attention because of the facts that effluent quality criteria have become increasingly stringent and removal of organic residues by this process is the only method beyond biological treatment. The process has been applied to potato processing effluents as a tertiary treatment after the activated sludge treatment.[134,135]

3. Filtration and Membrane Separation

Filtration devices used for the treatment of effluents include: contact bed, sand filters, and trickling filters. In the case of trickling filters, the effluent is allowed to pass through certain media where the biological growth attached to the media consumes the waste organics. The excessive loading rates on trickling filters can lead to poor performance of the potato processing effluent treatment plant. This was evidenced by the poor BOD removal.[136] Hence reduction of the loading rate on the trickling filters to 5 kg BOD per cubic meter per day was suggested. Larken et al.[137] were able to eliminate 70 to 90% BOD from corn and potato chip processing wastewater by using a fixed-film biofiltration system at a loading rate of 41.65 kg/m^3 · d. In the opinion of these authors,[137] a full-scale biofiltration system was better than a pilot-scale system in lessening the shock load and temperature variation effects.

On the other hand, membrane separation by reverse osmosis, ultrafiltration, electrodialysis, and dialysis is very interesting as it requires no evaporation. Nonetheless, membrane fouling can adversely affect flux rates and lead to a serious economic problem. Potato processing wastewater can be treated by ultrafiltration and reverse osmosis with prior degassing to avoid foaming of albumins.[138] However, the objective in this treatment is to remove the solvent and recover proteins.

4. Irrigation

A practical way of utilizing dilute industrial wastewaters is its disposal on land. It can be achieved by solar evaporation in ponds and various kinds of irrigation techniques. Biological solids, filter cakes, or slurries are usually landfilled. Irrigation is compatible with waste streams of a low flow rate and readily biodegradable components. It can be done by means of sprinkler, flood-flow, and infiltration methods. The rate of application needs to be controlled in order to provide water to the vegetation with the idea of proper root zone percolation and leaf evapotranspiration. However, texture, pH levels, and sodium ion content of soil need to be controlled.

IX. CONCLUSIONS

It has been pointed out that the most serious pollution in potato processing industries is from starch mill effluents, although lye peeling also produced a highly polluted water. The quantity of water for washing and rinsing of potatoes amounts to about 4.5 m^3/ton which increases to 5 to 9 m^3/ton for starch production with an additional 4 to 10 m^3/ton as wash water. The high cost of effluent purification makes its use as an irrigation and land disposal water important.[139-143] The use of potato processing water increased the biomass and N, P, K, Ca, Na, Mg, Fe, Mn, Zn, and Cu contents in certain forest plants[144] and saved[142] water and nutrients. However, long-term irrigation with potato starch mill effluents should be planned with caution.[145] On the other hand, effective soil reclamation was noticed over a period of 5 years.[146] Smith et al.[147] studied redox potential in a field irrigated with potato processing wastewater and noted that such irrigation from April to July promoted denitrification, thereby decreasing the potential for groundwater pollution. The identification of offensive odor compounds of the potato processing effluents showed the formation of skatole, geosmin, and short-chain organic acids, mostly in the irrigation fields.[148-150]

REFERENCES

1. **Knorr, D.,** Protein quality of the potato and potato protein concentrations, *Lebensm. Wiss. Technol.,* 11, 109, 1978.
2. **Knorr, D.,** Protein recovery from waste effluents of potato processing plants, *J. Food Technol.,* 12, 563, 1977.
3. **Krochta, J. M., Rumsey, T. R., and Farkas, D. F.,** Defining food R & D needs as a guide for the future, *Food Technol.,* 29, 74, 1975.
4. **De Noord, K. G.,** *Voedingsmiddelen Technol.,* 8, 23, 1975.
5. **Meister, E. and Thompson, N. R.,** Physico-chemical methods for the recovery of protein from waste effluent of potato chip processing, *J. Agric. Food Chem.,* 24, 919, 1976.
6. **Finley, J. W. and Hautala, E.,** Recovery of soluble proteins from waste streams, *Food Prod. Dev.,* 10(4), 92, 1976.
7. **Vlasblom, M. and Peters, H.,** Recovery of Proteins from Potato Wash Water, P5540g, 20616h, Dutch Patent 87 150, 15.01, 1958.
8. **Strolle, E. O., Cording, J., and Aceto, N. C.,** Recovering potato proteins coagulated by steam injection heating, *J. Agric. Food Chem.,* 21, 974, 1973.
9. **Lindner, P., Keren, R., and Ben-Gera, I.,** Precipitation of proteins from potato juice with bentonite, *J. Sci. Food Agric.,* 32, 1177, 1981.
10. **Stabile, R. L., Turkot, V. A., and Aceto, N. C.,** Economic analysis of alternative methods for processing potato starch plant effluents, *Second National Symp. on Food Processing Wastes,* Denver, 1971.
11. **Stringer, D. A.,** Acceptance of single cell protein for animal feeds, in *Comprehensive Biotechnology,* Vol. 4, Robinson, C. W. and Howell, J. A., Eds., Pergamon Press, New York, 1985, 685.
12. **Vecher, A. S. and Prokazov, G. F.,** Accumulation of food-yeast biomass on media from potato-starch production wastes, *Biokhimiya,* 1, 191, 1973.
13. **Skripnichenko, V. S., Navrodskaya, L. I., Ovrutskaya, I. Y., and Baitina, N. M.,** Biosynthesis of feed proteins on wastes from potato processing, *Mikroorganizmy-Produtsenty Biologika Aktiv. Veshchesty,* Samtsevich, S. A., Ed., Hauka: Tekhnika:Minsk, U.S.S.R., 1973, 203.
14. **Tomlinson, E. J.,** The production of single-cell protein from strong organic waste waters from the food and drink processing industries. 1. Laboratory cultures, *Water Res.,* 10, 367, 1976.
15. **Deschamps, F. and Meyer, F.,** Single-Cell Protein from Starchy Substrates such as Potatoes, E. German Patents 2,285,080 and 18,09,74 and 16.04.76, 1974.
16. **Lemmel, S. A., Heimsch, R. C., and Edwards, L. L.,** Optimizing the continuous production of *Candida utilis* and *Saccharomycopsis fibuligera* on potato processing waste water, *Appl. Environ. Microbiol.,* 37, 227, 1979.

17. **Stakheev, I. V., Kolomiets, E. I., and Romanovskaya, T. V.,** Use of starch and alcohol production residues for the production of fodder yeast, *Vestsi Akad. Navuk B. SSR, Ser. Biyal. Navuk,* 6, 60, 1984.
18. **Calleja, G. B., Yaguchi, M., Levy-Rick, S., Sequin, J. R. H., Roy, C., and Lusena, C. V.,** Single cell protein production from potato starch by yeast, *Schwanniomyces alluvius, J. Ferment. Technol.,* 64, 71, 1986.
19. **Lobanok, A. G., Babitskaya, V. G., and Kostina, A. M.,** Enzymic activity of mycelial fungi during their cultivation on wastes of the potato processing industry, *Prikl. Biokhim. Mikrobiol.,* 13, 475, 1977.
20. **Stakheev, I. V. and Babitskaya, V. G.,** Fungi of the *Penicillium* genus as producers of protein on media containing agricultural waste, *Mikol. Fitopatol.,* 12, 490, 1978.
21. **Stakheev, I. V., Kolomiets, E. I., and Lobankk, A. G.,** Biological value of crude fat of the fungus *Penicillium digitatum* 24 p grown on potato processing wastes, *Vestsi Akad. Navuk B. SSR, Ser. Sel'skagaspad. Navuk,* 2, 82, 1977.
22. **Staron, T. J.,** Food Proteins from Fungi or Multicellular Microorganisms, E. German Patents 2,393,535 and 07,06,1977 and 05,01, 1979.
23. **Senez, J. C.,** Solid state fermentation of starchy substrates, *Food Nutr. Bull.,* Suppl. 2, 127, 1979.
24. **Senez, J. C.,** Protein enrichment of starchy materials by solid state fermentation, *EEC Rep. Prod. Feed Single Cell Protein,* 1983, 101.
25. **Mori, K., Yanagimoto, M., Okada, N., and Yanai, S.,** Protein enrichment of potato waste by solid-state fermentation, *Shokuhin Sogo Kenkyusho Kenkyu Hokoku,* 48, 15, 1986.
26. **El-shimi, N. M., Mohsin, S. M., and El-Magied, M. M. Abd.,** Bioconversion of some agricultural wastes by *Aspergillus foetidus, Egypt. J. Food Sci.,* 15, 121, 1987.
27. **Stevens, C. A. and Gregory, K. F.,** Production of microbial biomass protein from potato processing wastes by *Cephalosporium eichorniae, Appl. Environ. Microbiol.,* 53, 284, 1987.
28. **Bergkvist, R.,** Symba process for making yeast protein, *Elementa,* 57, 98, 1974.
29. **Skogman, H.,** Production of Symba-yeast from potato wastes, in *Food Waste,* Birch, G. G., Parker, K. J., and Worgan, J. T., Eds., Applied Science Publishers, Barking, England, 1976, 167.
30. **Lines, K. K.,** Single-cell protein from potato processing waste, U.S. Patents 4,144,132 and 11,08, 1977 and 13,03, 1979.
31. **Kling, L. J., Bushway, R. J., Cleale, R. M., and Bushway, A. A.,** Nutrient characteristics and gly-coalkaloid content of potato distiller by-products, *J. Agric. Food Chem.,* 34, 54, 1986.
32. **Stakheev, I. V. and Vil'dflush, R. I.,** Selection of conditions for cultivating fodder yeast on potato starch production waste, *Izv. Vyssh. Uchebn. Zaved. Pishch. Tekhnol.,* 2, 86, 1971.
33. **Kramarz, M.,** Fodder yeast strains for the production of biomass in deproteinated potato juice, *Pr. Nauk. Akad. Ekon. im. Oskara Langego Wroclawiu,* 199, 69, 1982.
34. **Vecher, A. S., Levitskaya, M. V., Dvadtsatova, E. A., and Komarova, G. I.,** Changes in the carbo-hydrate complex of potato pulp under the action of enzymes of a culture liquid from *Rhizopus oligosporus* and *Endomycopsis fibuligera* cultures, *Dokl. Akad. Nauk. B. SSR,* 19, 746, 1975.
35. **Forney, L. J. and Reddy, C. A.,** Fermentative conversion of potato-processing wastes into a crude protein feed supplement by *Lactobacilli, Dev. Ind. Microbiol.,* 18, 135, 1977.
36. **Skafar, S., Perdih, A., and Cimerman, A.,** Possibilities of wastewater treatment with molds and use of their mycelium as animal feed, *Hem. Ind.,* 31, 3 & 120, 1977.
37. **Stakheev, I. V., Orlova, L. A., and Grinevich, T. A.,** Biomass and protein synthesis in two stage continuous culture of *Penicillium terlikowskii* BIM G-123 on a potato juice and pulp medium, *Vestsi Akad. Navuk B. SSR, Ser. Biyal Navuk,* 5, 67, 1985.
38. **Zacek, M.,** Potato Flakes for Animal Feeds, Czechoslovakia Patent 128,988 and 20.03, 1965 and 15.09, 1968.
39. **Zacek, M., Vojta, L., and Novak, K.,** Cattle Feed, Czechoslovakia Patent 142,012 and 23.10, 1969 and 15.07, 1971.
40. **Florea, O., Blaga, L., Noraru, M., and Florea, V.,** Feed from Food Industry Waste, Rome, Italy, Patent RO 87,265 and 04.08, 1982 and 30.08, 1985.
41. **Tornow, K. D., Krumnow, E., Schmidt, R., Golea, C., Niestroj, B., Trapp, H., and Meilcke, R.,** Protein-Containing and Mineral-Rich Feed from By-products and Wastes of the Potato Starch Industry, E. Germany Patents 215,935 and 02,05, 1983 and 28,11, 1984.
42. **Green, J. H., Paskell, S. L., Goldmintz, D., and Schisler, L. C.,** Utilization of fish solubles, a fishery by-product, as a means of pollution abatement, in *Food Processing and Waste Management, Proc. Cornell Agric. Waste Manage. Conf., 5th 1973,* NY State College, Agriculture and Life Science, Cornell University, Ithaca, N.Y., 51.
43. **Gee, M., Huxsoll, C. C., and Graham, R. P.,** Acidification of dry caustic peeling waste by lactic acid fermentation, *Am. Potato J.,* 51, 126, 1974.
44. **Gerry, R. W.,** Dehydrated soluble potato solids in broiler rations, *Poult. Sci.,* 56, 1947, 1977.
45. **Fritze, H.,** Use of drum driers for processing various industrial wastes into high grade animal feeding stuffs, *Escher Wyss News,* 49, 36, 1976.

46. **Rawlings, R. M. and Procter, D.,** Separating vegetable waste solids from aqueous slurries of vegetable waste materials and feeding the separated solids to ruminant animals, U.S. Patents 4,144,355 and 22.12, 1975 and 13.03, 1979.

47. **Stanhope, D. L., Hinman, D. D., Eversion, D. O., and Bull, R. C.,** Digestibility of potato processing residue in beef cattle finishing diets, *J. Anim. Sci.,* 51, 202, 1980.

48. **Aritsuka, T.,** Feed preparation from starch waste, Japan Patent 8,023,905 and 07.08, 1978 and 20.02, 1980.

49. **Hulan, H. W., Proudfoot, F. G., and Zarkadas, C. G.,** Potato waste meal. II. The nutritive value and quality for broiler chicken, *Can. J. Anim. Sci.,* 62, 1171, 1982.

50. **Hulan, H. W., Proudfoot, F. G., and Zarkadas, C. G.,** Potato waste meal. I. Compositional analyses, *Can. J. Anim. Sci.,* 62, 1161, 1982.

51. **Schneider, P. L., Stokes, M. R., Bull, L. S., and Walker, C. K.,** Evaluation of potato meal as a feedstuff for lactating dairy cows, *J. Dairy Sci.,* 68, 1738, 1985.

52. **Bushway, A. A., Bureau, J. L., Bergeron, D., Stickney, M. R., and Bushway, R. J.,** The nutrient and glycoalkaloid content of a new potato meal, *Am. Potato J.,* 62, 301, 1985.

53. U.S. Environmental Protection Agency, Diquat: tolerances for pesticides in animal feeds, *Fed. Regist.,* 46, 30339, 1981.

54. U.S. Environmental Protection Agency, Metalaxyl: tolerances for pesticides in animal feeds administered by the environmental protection agency, *Fed. Regist.,* 48, 3587, 1983.

55. **Venkatasubramanian, K. and Keim, C. R.,** Starch and energy: technology and economics of fuel alcohol production, in *Starch Conversion Technology,* Van Beynum, G. M. A. and Roels, J. A., Eds., Marcel Dekker, NY, 1985, 143.

56. **Treadway, R. H. and Cordon, T. C.,** The chemicals we get from potatoes, in *Yearbook of Agriculture,* U.S. Department of Agriculture, 1950—51, 190.

57. **Dellweg, H. and Luca, S. F.,** Ethanol fermentation: suggestions for process improvements, *Process Biochem.,* 23, 100, 1988.

58. **Abouzied, M. M. and Reddy, C. A.,** Direct fermentation of potato starch to ethanol by cocultures of *Aspergillus niger* and *Saccharomyces cerevisiae, Appl. Environ. Microbiol.,* 52, 1055, 1986.

59. Novo Industries A/S, Novo Enzymes for the Alcohol Industry, Brochure Enzymes Division, DK-2880, Bagasvaerd, Denmark, 1981.

60. **Lutzen, N. W.,** Enzyme technology in the production of ethanol-recent process development, *Proc. Int. Ferment. Symp. on Adv. Biotechnol.,* (6th), 2, 161, 1980.

61. **Luong, J. H. T. and Tseng, M. C.,** Process and technoeconomics of ethanol production by immobilized cells, *Appl. Microbiol. Biotechnol.,* 19, 207, 1984.

62 **Fukushima, S. and Yamade, K.,** Continuous rapid alcohol fermentation from carbohydrates in a novel immobilized bioprocess, *Proc. Pac. Chem. Eng. Conf.,* (3rd), 3, 425, 1983.

63. **Tegtmeier, U.,** Process design for energy saving ethanol production, *Biotechnol. Lett.,* 7, 129, 1985.

64. **Misselhorn, K.,** *Chem. Ing. Technik.,* 53, 47, 1981.

65. **Misselhorn, K.,** Technological aspects of ethanol production, *Starch,* 34, 325, 1982.

66. **Misselhorn, K.,** Paper presented at the *Oktobertagung der Versuchs-und Lehranstalt fur Breauerei,* Berlin, October 14th, 1987.

67. **Yamamoto, T., Matsumura, Y., and Uenakai, K.,** Alcohol fermentation of uncooked starch tubers and roots, *EEC Rep. Energy Biomass.,* 979, 1983.

68. **Sreekantiah, K. R. and Rao, B. A.,** Production of ethyl alcohol from tubers, *J. Food Sci. Technol.,* 17, 194, 1980.

69. **Friend, B. A. and Shahani, K. M.,** Fuel alcohol production from waste materials, *Fuels Biomass Wastes* (Pap. Symp.), 343, 1981.

70. **Borud, O. J.,** Utilization of waste products in the potato processing industry, *Die Staerke,* 23, 172, 1971.

71. **Stenzel, R. A., Yu, J., Lindemceth, T. E., Soo-Hoo, R., May, S. C., Yim, Y. J., and Houle, E. H.,** Ethanol production for automotive fuel usage, Report 1980 DOE/ID/12050-3, *NTIS from Energy Res. Abstr.,* (1), Abstr. No. 667, 1981.

72. **Al-Taweel, A. M., Woodside, R. G., MacKay, G. D. M., and Gates, W.,** Technoeconomics of small-scale fuel alcohol production from agricultural wastes, *Proc. Inst. Ferment. Symp. on Adv. Biotechnol. 6th,* 2, 153, 1981.

73. **Hammaker, G. S., Fost, H. B., David, M. L., and Marino, M. L.,** Feasibility of converting a sugarbeet plant to fuel ethanol production, Report 1981 DOE/CS/83010-T1, *NTIS from Energy Res. Abstr.,* 6(14), Abstr. No. 19959, 1981.

74. **Marihart, J.,** Production of ethanol from starch industry by-products, especially potato pulp, *Starch,* 34, 290, 1982.

75. **Kalter, R. J., Boisvert, R. N., Gabler, E. C., and Walker, L. P.,** Ethanol production in Suffolk County, New York: technical and economical feasibility, Report 1981 NYSERDA-81-13, *NTIS from Govt. Rep. Announce. Index* (U.S.) 82(10), 2028, 1982.

76. **Huang, P. J. D.,** The feasibility of producing ethanol from potato processing waste, Univ. Microfilms Int. Order No. DA 8303806, from *Diss. Abstr. Int. B,* 43(9), 2845, 1983.
77. **Kuby, W., Nackord, S., and Wyss, W.,** Testing and evaluation of alcohol production facility utilizing potatoes as a feed stock, *NTIS from Govt. Rep. Announce. Index* (U.S.) 84(15), 214, 1984.
78. **McInerney, M. J., Bryant, M. P., and Stafford, D. A.,** Metabolic stages and energetics of microbial anaerobic digestion, in *Anaerobic Digestion,* Applied Science Publishers, London, 1980.
79. **Hassan, H. M., Belyea, D. A., and Hassan, A. E.,** Characterization of methane production from poultry manure, Managing Livestock Wastes, *Proc. Int. Symp. 3rd, Am. Soc. Agri. Eng.,* St. Joseph, Mich., 244, 1975.
80. **Van den Berg, L. and Lentz, C. P.,** Methane production during treatment of food plant wastes by anaerobic digestion, *Food Fert. Agric. Residues,* Proc. Cornell Agric. Waste Manage. Conf. 9th, Ithaca, NY, 381, 1977.
81. **Sax, R. I., Holtz, M., and Roberts, J. E.,** Pilot study of BIOTHANE upflow anaerobic sludge blanket process for methane production, *Energy Biomass Wastes,* 5, 413, 1981.
82. **Jackson, M. L.,** Methane production and recovery using potato waste solubles and solids, *Proc. Ind. Waste Conf.,* 36, 241, 1982.
83. **Cutayar, J. M. and Mouliney, M.,** The Anoxal process: anaerobic treatment of liquid industrial effluents, *EEC Rep, Energy Biomass,* 527, 1985.
84. **Bhagat, S. K., Ongerth, J. E., Ettling, B. V., and Ady, E. W.,** Organic cleaner prepared from potato processing plant waste, *Eng. Bull., Purdue Univ., Eng. Ext. Ser.,* 145, (Pt. 1) 319, 1974.
85. **Sochor, V. and Smid, J.,** Adhesive from Potato Waste, Czechoslovakia Patents 157,980 and 26.07, 1972 and 15.05, 1975.
86. **Ady, E. W.,** Cleaning Composition Using Plant Starch, U.S. Patents GO 2,612,501 and 04,04, 1975 and 07,10, 1976.
87. **Morita, S., Tamaya, H., Hakamata, S., Suzuki, T., and Sato, K.,** Pectin Separation from Potato Starch Waste, Japan Patents 61,37,098 and 31,07, 1984 and 21,02, 1986.
88. **Soepenberg, J., Soepenberg, G., and Laaraman, A.,** Starch Product Suitable for Use in Drilling Muds and a Drilling and Containing Said Product, Europe Patents 21515 and 14,06, 1979 and 07,01, 1981.
89. **Nankowski, C., Poulin, G., and Dean, P. R.,** Extraction of L-Ascorbic Acid from Plant Tissues, California Patent 1,132,586, 1982.
90. **Cansfield, P. E. and Gallop, R. A.,** Conservation, reclamation, and reuse of solids and water in potato processing, *Proc. Ont. Ind. Waste Conf.,* 17, 42, 1970.
91. **Alsager, M. D.,** Approach to total management of potato processing wastes, *Eng. Bull., Purdue Univ., Eng. Ext. Ser.,* 141 (Pt. 2), 747, 1972.
92. **Krivtsun, L. V., Makushchenko, N. I., and Mozal 'Kova, D. A.,** The effect of the addition of coagulants on the sedimentation capacity of suspended matter in suspensions of different soils, *Sakh. Promst.,* 6, 48, 1986.
93. **Hindin, E.,** Waste water utilization in the potato processing industry, *Washington State University, College of Engineering Circ.,* Pullman, No. 34, 1970.
94. **Dutton, C. S. and Fisher, C. P.,** The use of aero-hydraulic guns in the biological treatment of organic wastes, *Eng. Bull., Purdue Univ., Eng. Ext. Ser.,* 121, 403, 1966.
95. **Schneider, N. and Zink, J.,** Device for Biological Waste Water Treatment, Patent Ger. Offen. 2,125,983 and 26,05, 1971 and 14,12, 1972.
96. **Richter, G. A., Sirrine, K. L., and Tollefson, C. I.,** Conditioning and disposal of solids from potato waste water treatment, *J. Food Sci.,* 38, 218, 1973.
97. **Richter, G. A., Pailthorp, R. E., and Sirrine, K. L.,** Aerobic Secondary treatment of potato processing waste, *Eng. Bull., Purdue Univ., Eng. Ext. Ser.,* 140 (Pt. 2), 684, 1973.
98. **Singh, T. and Porterfield, C. E.,** Treating vegetable cannery waste water in Hallwood, Virginia, Eng. Bull., *Purdue Univ., Eng., Ext. Ser.,* 145 (Pt. 2), 744, 1974.
99. **Michio, D., Makoto, O., and Hideo, O.,** Treatment of industrial wastes by activated sludge. X. Treatment of sweet and white potato starch wastes, *Kogyo Gijutsuin, Hakko Kenkyusho Kenkyu Hokoku,* 29, 61, 1966.
100. **Jarvensiva, H.,** Treatment of waste water from the potato processing industry, *Vatten,* 35, 208, 1979.
101. **Hung, Y. T. and Wang, L. I.,** Secondary treatment of effluent streams from potato processing plant, *Proc. N.D. Acad. Sci.,* 34, 8, 1980.
102. **Hung, Y. T.,** Feasibility of activated sludge treatment of potato processing waste waters, *Agric. Wastes,* 3, 215, 1981.
103. **Yeh, R. Y. and Hung, Y. T.,** Bio-augmented activated sludge treatment of potato waste water, *Acta Hydrochim. Hydrobiol.,* 16, 213, 1988.
104. **Hack, P.,** Paper for distilleries-India, I., 1987, (personal communication).
105. **Van den Berg, L. and Lentz, C. P.,** Food processing waste treatment by anaerobic digestion, *Proc. Ind. Waste Conf.,* 32, 252, 1978.

106. **Lettinga, G., Van Velsen, A. F. M., De Zeeuw, W., and Hobma, S. W.,** Feasibility of the upflow anaerobic sludge blanket (UASB) process, *Natl. Conf. Environ. Eng.,* (Proc.), 35, 1979.

107. **Pette, K. C. and Versprille, A. I.,** Application of U.A.S.B. concept for wastewater treatment, *Anaerobic Dig., Proc. Int. Symp. 2nd,* 121, 1982.

108. **Christensen, D. R., Gerick, J. A., and Eblen, J. E.,** Design and operation of an upflow anaerobic sludge blanket reactor, *J. Water Pollut. Control Fed.,* 56, 1059, 1984.

109. **Brown, G. J., Lin, K. C., Landine, R. C., Cocci, A. A., and Viraraghavan, T.,** Lime use in anaerobic filters, *J. Environ. Eng. Div.,* 106(EE4), 837, 1980.

110. **Brown, G. J., Cocci, A. A., Lindine, R. C., Viraraghavan, T., and Lin, K. C.,** Sludge accumulation in an anaerobic lagoon-anaerobic filter system treating potato processing wastewater, *Proc. Ind. Waste Conf.,* 35, 610, 1981.

111. **Landine, R. C., Brown, G. J., Cocci, A. A., and Viraraghavan, T.,** Potato processing wastewater treatment using horizontal anaerobic filters, *Can. Inst. Food Sci. Technol. J.,* 14, 144, 1981.

112. **Landine, R. C., Cocci, A. A., Viraraghavan, T., and Brown, G. J.,** Anaerobic pretreatment of potato processing wastewater—a case history, *Proc. Ind. Waste Conf.,* 36, 233, 1982.

113. **Landine, R. C., Cocci, A. A., Viraraghavan, T., and Brown, G. J.,** Anaerobic treatment key to pollution control and solids disposal with energy recovery for a food processor, *Proc. Ind. Waste Conf.,* 37, 47, 1983.

114. **Landine, R. C., Degarie, C. J., Cocci, A. A., Steeves, A. L., Brown, G. J., and Viraraghavan, T.,** Anaerobic pretreatment facility also provides sludge disposal capability and source of renewable energy for food processor, *Proc. Ind. Waste Conf.,* 38, 805, 1984.

115. **Landine, R. C., Viraraghavan, T., Cocci, A. A., Brown, G. J., and Lin, K. C.,** Anaerobic fermentation-filtration of potato processing wastewater, *J. Water Pollut. Control. Fed.,* 54, 103, 1982.

116. **Parker, J. G., Lyons, B. J., and Parker, C. D.,** An integrated low cost system for treatment of potato processing waste water incorporating anaerobic fermentation and phosphorus removal, *Water Sci. Technol.,* 14, 675, 1982.

117. **Landine, R. C., Brown, G. J., Cocci, A. A., and Viraraghavan, T.,** Anaerobic treatment of high strength, high solids potato waste, *Agric. Waste,* 7, 111, 1983.

118. **Braun, R.,** Anaerobic fixed bed wastewater treatment in a potato processing factory, *Conserv. Recycl.,* 8, 221, 1985.

119. **Landine, R. C., Pyke, S. R., Brown, G. J., and Cocci, A. A.,** Low-rate anaerobic treatment of a potato processing plant effluent, *Proc. Ind. Waste Conf.,* 41, 511, 1987.

120. **Goronszy, M. C. and Eckenfelder, W. W.,** Anaerobic pretreatment of high-strength wastewaters using a reactive growth support media, *Proc. Ind. Waste Conf.,* 42, 757, 1988.

121. **Dorfler, J. and Brau, R.,** Waste treatment and utilization in potato processing, in *Proc. 4th European Congress on Biotechnology,* Vol. 1, June 14 to 19, 1987 (Amsterdam), Neijssel, O. M., Van der Meer, R. R., and Luyben, K. Ch. A. M., Eds., Elsevier, Amsterdam, 1987, 261.

122. **Dostal, K. A.,** Pilot plant studies on secondary treatment of potato processing wastes, *Potato Waste Treat. Proc. Symp.,* 27, 1968.

123. **Dornbush, J. N., Rollag, D. A., and Trygstad, W. J.,** Anaerobic-aerobic lagoon system for potato processing wastes, *J. Water Pollut. Control. Fed.,* 50 (3, Pt. 1), 524, 1978.

124. **Isik, H. and Slack, J. C.,** Novel potato processing effluent treatment, *Effluent Water Treat. J.,* 24, 154, 1984.

125. **Isik, H. and Slack, J. C.,** Novel potato processing effluent treatment, *Effluent Water Treat. J.,* 24, 115, 1984.

126. **Lin, K. C.,** Aeration of anaerobically treated potato processing wastewater, *Agric. Waste,* 12, 1, 1985.

127. **Nordstrom, B.,** Some experiences with the simultaneous treatment of community and industrial waste water, *Vatten,* 32, 96, 1976.

128. **Sugiyama, N. and Ando, T.,** Purification of Starch-Containing Waste Water, Japan Patents 73,104,347 and 17,02, 1972 and 27,12, 1973.

129. **Kyoritsu, Yuki Co. Ltd.,** Treatment of Sludge in Wastewater from Starch Processing as Food, Japan Patents 80,94,697 and 11,01, 1979 and 18,07, 1980.

130. **Fussum, G. O. and Cooley, A. M.,** Water pollution abatement by improved coagulation on effluents from lye-peel potato processing plants, *U.S. Gov. Res. Dev. Rep.,* 70, 125, 1970.

131. **Karim, M. I. A. and Sistrunk, W. A.,** Treatment of potato processing wastewater with coagulating and polymeric flocculating agents, *J. Food Sci.,* 50, 1657, 1985.

132. **Hydamaka, A., Stephan, P., Gallop, R. A., and Carvalho, L.,** Control of color problems during recycling of food process waters, *Proc. Natl. Symp. Food Process Wastes, 7th,* 698, 237, 1976.

133. **Viraraghavan, T., Cocci, A. A., and Landine, R. C.,** Study indicates activated carbon can treat potato wastewater, *Ind. Wastes,* 24, 30, 1978.

134. **Hung, Y. T.,** Tertiary treatment of potato processing waste by a biological activated carbon process, *Am. Potato J.,* 60, 543, 1983.

135. **Hung, Y. T.,** Treatment of potato processing wastewaters by activated carbon adsorption process, *Am. Potato J.,* 61, 9, 1984.
136. **Viraraghavan, T., Landine, R. C., and Cocci, A. A.,** Performance of a potato processing wastewater treatment plant. A case history, *Proc. Ind. Waste Conf.,* 38, 789, 1984.
137. **Larkin, J. J., Sherman, R. J., and McIntire, M. G.,** Pilot vs full-scale performance of a fixed film biological filtration system, *Proc. Int. Conf. Fixed-film Biol. Processes,* 2, 1693, 1985.
138. **Wavin, B. V.,** Apparatus and method for the removal of solvent by membrane filtration, The Netherlands Patent 73,06,558 and 10,05, 1973 and 12,11, 1974.
139. **Smith, J. H. and Oates, J. H.,** Closed cycle industrial wastes control, *Ind. Wastes,* 28, 30, 1982.
140. **Peters, H.,** Measures taken against water pollution in starch and potato processing industries, *Pure Appl. Chem.,* 29, 129, 1972.
141. **De Haan, F. A. M., Hoogeveen, G. J., and Riem, V. F.,** Agricultural use of potato starch waste water, *Neth. J. Agric. Sci.,* 21, 85, 1973.
142. **Smith, J. H.,** Treatment of potato processing waste water on agricultural land, *J. Environ. Qual.,* 5, 113, 1976.
143. **Smith, J. H., Robbins, C. W., Bondurant, J. A., and Hayden, C. W.,** Treatment of potato processing wastewater on agricultural land: water and organic loading, and the fate of applied plant nutrients, land waste management alternative, *Proc. Cornell Agric. Waste Manage. Conf.,* 8, 769, 1977.
144. **Bialkiewicz, F.,** Lysimetric studies of purification and use of potato processing waste waters for forest cultures, *Rocz. Nauk Roln. Ser. F.,* 79, 123, 1976.
145. **Diez, T. and Sommer, G.,** Effect of long term irrigation with potato starch manufacturing wastewater on the soil and plant growth, *Bayer. Landwirtsch. Jahrb.,* 53, 643, 1976.
146. **Stehlik, K. and Musil, J.,** Agricultural and reclamation effectiveness of irrigation with starch-plant wastewater, *Sci. Agric. Bohemoslov,* 11, 145, 1979.
147. **Smith, J. H., Gilbert, R. G., and Miller, J. B.,** Redox potentials in a cropped potato processing wastewater disposal field with a deep water table, *J. Environ. Qual.,* 7, 571, 1978.
148. **Buttery, R. G., Guadagni, D. G., and Garibaldi, J. A.,** Identification of offensive odor compounds from potato processing plant waste effluent irrigation fields, *J. Agric. Food Chem.,* 27, 646, 1979.
149. **Buttery, R. G. and Garibaldi, J. A.,** Odorous compounds from potato processing waste effluent irrigation fields: volatile acids, *J. Agric. Food Chem.,* 28, 158, 1980.
150. **Gerick, J. A.,** Land application odor study, *J. Water Pollut. Control Fed.,* 56 (3, Pt. 1), 287, 1984.
151. **Peters, H.,** Measures taken against water pollution in starch and potato processing industries, *Pure Appl. Chem.,* 29, 1 and 129, 1972.

Chapter 8

GLYCOALKALOIDS

S. J. Jadhav, A. Kumar, and J. K. Chavan

TABLE OF CONTENTS

I. INTRODUCTION

Plants belonging to the family *Solanaceae* contain a variety of glycoalkaloids which have been characterized for their potential toxic effects. Much of the available information on such alkaloids pertains to α-solanine and α-chaconine, the major toxic constituents in potatoes. Although present in rather small quantities in various parts of the tuber and several other parts of the plant, these alkaloids possess mild to moderate animal toxicity. However, episodes of poisoning have been reported due to an increased alkaloid content in certain conditions. There is considerable evidence concerning potato-related poisoning in man and farm animals attributed to the ingestion of large amounts of glycoalkaloids in the green tubers or sprouts. The potato cultivar Lenape was rapidly withdrawn because it contained high levels of steroidal glycoalkaloids,[1] a fact that focused attention on the need to test all new potato varieties. Börner and Mattis[2] stated that potato tubers with more than 20 mg of solanine (glycoalkaloids) per 100 g of fresh weight exceeded the upper safety limit for food purposes. It is interesting to note that the glycoalkaloids of potatoes are not destroyed during boiling, baking, frying, or drying at high temperatures.[1,3] A controversial hypothesis[4] concerning the relationship between imperfect potatoes and birth defects has renewed interest in the pharmacological and toxicological aspects of potato glycoalkaloids.

II. OCCURRENCE AND COMPOSITION OF GLYCOALKALOIDS

Until 1954, the cultivated form of potato was thought to contain only one glycoalkaloid, solanine, discovered nearly 170 years ago. Kühn and Löw[5] discovered another glycoalkaloid, α-chaconine, in the leaves and shoots of cultivated potato and in the leaves of the wild potato, *Solanum chacoense,* from which it was named. The glycoalkaloid solanine was a mixture of two classes of glycosides, the solanines and the chaconines. α-Solanine and α-chaconine have the same aglycone alkaloid solanidine, but differ with respect to the composition of the sugar chain (Figure 10, Chapter 2). Besides these main glycoalkaloids, which represent as much as 95% of the total alkaloids in leaves of *S. tuberosum* L. and *S. chacoense,* forms of solanine and chaconine with a shortened chain were found. The occurrence of leptinines and leptines have been reported only in wild potatoes, *S. chacoense.*[6,7] Zitnak[8] detected free solanidine in concentrations up to 33% of the total glycoalkaloid level in bitter Netted Gem (Russet Burbank, Idaho Russet) potatoes. Additionally, other alkaloids, such as α-solamarine, β-solamarine, demissidine, and 5β-solanidan-3 α-ol have been identified in *S. tuberosum* L. Sinden and Sanford[9] studied tuber tissues of 123 common cultivars for their ability to synthesize solamarine glycoalkaloids. They found that 11 cultivars, including Kennebec and White Rose, synthesized major concentrations of solamarine when the tuber slices were exposed to light during wound healing. The several alkaloids in potatoes may have emerged when species such as *S. chacoense* and *S. demissum* were hybridized with a cultivated potato plant.[10] Although α- and β-solanines were initially reported only in incubated slices of the Kennebec variety,[10] Osman et al.[11] also found them in tuber tissue from cultivated species of *S. curtilobum* and *S. juzepczukii.* The incorporation of selected plant introduction lines of *S. chacoense* and *S. commersonii,* which contain two major glycoalkaloids, demissine and a new compound, commersonine,[12] in a breeding program may introduce compounds other than α-solanine and α-chaconine. The occurrence and broad spectrum of these alkaloids are summarized in Table 1.[11]

III. DISTRIBUTION IN PLANTS

A. ALKALOID DISTRIBUTION IN TISSUES

In the potato plant, most of the tissues, including leaves, shoots, stems, blossoms, tubers, tuber eyes, peels, and sprouts, contain the major glycoalkaloids. Wolf and Duggar[14]

TABLE 1
Occurrence and Composition of Glycoalkaloids

Plant species	Glycoalkaloid	Aglycone alkaloid	Sugar moiety
Solanum tuberosum L.	α-Solanine	Solanidine	—D-Galactose〈 D-glucose / L-rhamnose
	β-Solanine	Solanidine	—D-Galactose-D-glucose
	γ-Solanine	Solanidine	—D-Galactose
	α-Chaconine	Solanidine	—D-Glucose〈 L-rhamnose / L-rhamnose
	β-Chaconine	Solanidine	—D-Glucose-L-rhamnose
	γ-Chaconine	Solanidine	—D-Glucose
	α-Solarmarine	Tomatidenol	—D-Galactose〈 D-glucose / L-rhamnose
	β-Solamarine	Tomatidenol	—D-Glucose〈 L-rhamnose / L-rhamnose
	—	Tomatidenol	—
	—	Solanidine	—
	—	5β-Solanidan-3α-ol	—
	—	Demissidine	—
S. chacoense Bitt	Leptine I	*O*(23)-Acetylleptinidine	—D-Glucose〈 L-rhamnose / L-rhamnose
	Leptine II	*O*(23)-Acetylleptinidine	—D-Galactose〈 D-glucose / L-rhamnose
	Leptine III[a]	*O*(23)-Acetylleptinidine	Unknown
	Leptine IV[a]	*O*(23)-Acetylleptinidine	Unknown
	Leptinine I	Leptinidine	—D-Glucose〈 L-rhamnose / L-rhamnose
	Leptinine II	Leptinidine	—D-Galactose〈 D-glucose / L-rhamnose
	Leptinine III[a]	Leptinidine	Unknown
	Leptinine IV[a]	Leptinidine	Unknown
	α,β,γ-Solanines	Solanidine	As above
	α,β,γ-Chaconines	Solanidine	As above
	Commersonine[b]	Demissidine	—D-Galactose D-glucose〈 D-glucose / D-glucose
	Demissine[b]	Demissidine	—D-Galactose D-glucose〈 D-glucose / D-xylose

[a] Tentative identification by chromatographic methods.

[b] Detected as only major glycoalkaloids in selected plant introduction lines of *S. chacoense* and *S. commersonii.*

From Jadhav, S. J., Sharma, R. P., and Salunkhe, D. K., *CRC Crit. Rev. Toxicol.,* 9, 21, 1981. With permission.

noted that glycoalkaloid content is high in the meristematic regions, such as leaf buds and young leaves down to about the eighth node, and decreases markedly beyond that point. It is known that sprouts are an excellent source of alkaloids. Morgenstern[15] showed that at flowering the concentration of the glycoalkaloids was low in the tuber, stem, and petiole, but much higher in the roots, runners, and leaves, the highest in the flowers. As the haulms

TABLE 2
Distribution of Total Glycoalkaloids in the Potato
Plant and Various Tuber Tissues

Potato part	Total glycoalkaloids (mg/100 g fresh weight)
Plant	
Sprouts	200—400
Flowers	300—500
Stems	3
Leaves	40—100
Normal tuber tissue	
Skin, 2 to 3% of tuber	30—60
Peel, 10 to 15% of tuber	15—30
Peel and eye, $1/_8$ in. (3 mm) disk	30—50
Peels from bitter tubers	150—200
Flesh	1.2—5
Whole tuber	7.5
Bitter tubers	25—80

From Wood, F. A. and Young, D. A., *Agriculture Canada*, Publication 1533, 1974. With permission.

began to die down, the glycoalkaloid content of runners, roots, stems, and leaves decreased, but the content of the flowers changed only slightly. Lampitt et al.[16] also studied the distribution of the glycoalkaloids in Arran Signet potato plants; the concentration increased from 8 to 10 weeks in the flowers (280 mg % at 8 weeks as compared to 416 mg % at 11 weeks), leaves (66 mg % as compared to 61 mg %), and new tubers (2.5 mg % as compared to 9 mg %). Thus, the glycoalkaloid content of *S. tuberosum* is the highest in the flowers;[15,16] in the tubers, glycoalkaloids are concentrated in the peels and sprouts and around the eyes.[16] In sprouting tubers, the highest concentration is in the shoot tips.[14] Paseshnichenko[17] found that a decrease in glycoalkaloid content in the leaves and tubers affected both α-solanine and α-chaconine during the vegetation process. However, a drop in the glycoalkaloid content in the seedlings was at the expense of α-chaconine. Distribution of total glycoalkaloids in the potato plant and their normal levels in various tuber tissues[18] are shown in Table 2.

Guseva et al.[19] reported that α-solanine represents about 40% of the total glycoalkaloids of sprouts, and α-chaconine represents about 60%. However, the solanine-chaconine ratio in the tubers, as determined by gas chromatography of the hydrolyzed sugars,[10] differed markedly in the cultivars Bintje, Clivia (solanine-chaconine, 1:2) and Amigo, Hansa I, and Hansa II (1:7). Several reports on variations in the glycoalkaloid composition of potatoes are now available.[12,20-22] Glycoalkaloids accumulated continuously in the tubers of all cultivars studied.[23] The glycoalkaloids are formed in the parenchyma cells of the periderm and cortex of the tubers and in areas of high metabolic activity, such as the eye regions.[24,25] The concentration decreases towards the center of the tuber. Little or none is found in the pith and only small amounts are present in the intermediate regions.[16] Zitnak[8] studied the distribution and concentration of solanine (glycoalkaloid) and its aglycone, solanidine, in peels and peeled potatoes (Netted Gem) of high glycoalkaloid content and found that potato peels contained solanidine in amounts equal to those in the peeled tuber, although the peels represented only one seventh the weight of the tuber. Zitnak[8] also detected β-chaconine, up to 30% of the total glycoalkaloids, in several lots of Sebago and Kennebec tuber peels. According to Zitnak and Johnston,[26] the glycoalkaloids diffuse through the entire tuber after reaching a high concentration.

Roddick[27] examined the glycoalkaloid distribution in vegetative cells of potatoes by

TABLE 3
Distribution of Glycoalkaloids in Cell Fractions from Potato

Plant part		Tissue residue	500 × g pellet	2,500 × g pellet	16,000 × g pellet	105,000 × g pellet	Supernatant
Potato leaf	Alkaloid (mg)	10.3	N.D.[b]	N.D.	0.7	0.9	19.3
(60 g)[a]	Protein (g)	10.57	0.12	0.16	0.35	0.22	1.42
	Alkaloid conc (mg/g protein)	1.0	—	—	2.0	4.1	13.6
Potato root	Alkaloid (mg)	18.3	N.D.	N.D.	N.D.	N.D.	1.3
(83 g)[a]	Protein (g)	9.98	0.14	0.02	0.14	0.18	0.55
	Alkaloid conc (mg/g protein)	1.8	—	—	—	—	2.4
Potato tuber	Alkaloid (mg)	0.9	N.D.	N.D.	0.05	0.09	4.7
(202 g)[a]	Protein (g)	9.75	0.11	0.11	0.14	0.1	1.6
	Alkaloid conc (mg/g protein)	0.1	—	—	0.4	0.9	2.9

[a] Fresh weight of tissue extracted.

[b] N.D., not detectable by assay or TLC.

From Roddick, J. G., *Phytochemistry*, 16, 805, 1977. With permission.

homogenization in a 0.1 M phosphate buffer solution and differential centrifugation (Table 3). Glycoalkaloid accumulated principally in the soluble fraction. Smaller amounts were present in the microsomal fraction. The glycoalkaloids as detected by thin-layer chromatography (TLC), appeared sporadically in the mitochondrial fraction. The anatomical distribution of glycoalkaloids[28] is shown in Figure 1. Microscopic examination of a potato tuber (cv. Russet Burbank) indicated alkaloid concentrations in the peridermal and cortical regions of the tuber.[29] The glycoalkaloids were distributed mainly in and just below the compact phellem cells. Secondary deposits were found in the cortical parenchyma cells just below the skin layers of the tuber.

Han[30] and Han et al.[31] used electron microscopy and cytochemistry to localize the glycoalkaloids in sprout tips, unsprouted meristemic tips, and epidermal layers of the potato tubers. Han et al.[31] observed a solanidine-digitonin complex as darkly stained needles or spicules (Figure 2). These spicules were observed mainly in the vacuoles of sprouted tips with a few also noted in the cytoplasm and peridermal tissue. The minute quantities of solanidine observed may reflect its rapid conversion to glycoalkaloids or insensitivity of the method.

B. DISTRIBUTION IN DIFFERENT SPECIES OR VARIETIES

Limited information is available on the relative distribution of various glycoalkaloids in different species of Solanaceae or varieties of potatoes. Samples from different species or even clones of the same species exhibit a wide variation in the alkaloid contents. Osman et al.[32] reported heterogeneity among species or different clones, both in the total glycoalkaloid content and also in the relative distribution of different glycoalkaloids. The results reported by these workers are illustrated in Table 4. The same workers reported variability of total glycoalkaloid content in various varieties. Wauseon and Houma varieties contained lesser amounts of glycoalkaloids (3.2 and 3.8 mg/100 g in fresh slices) than found in the Kennebec variety (9.4 mg/100 g). The glycoalkaloid content increased dramatically both immediately after slicing or if the slices were stored for 4 d in controlled storage (44°F, 85% relative humidity). Levels as high as 163 mg/100 g of potato slice were observed in the Kennebec variety after 24 weeks of storage and aging of slices. The total glycoalkaloid content did

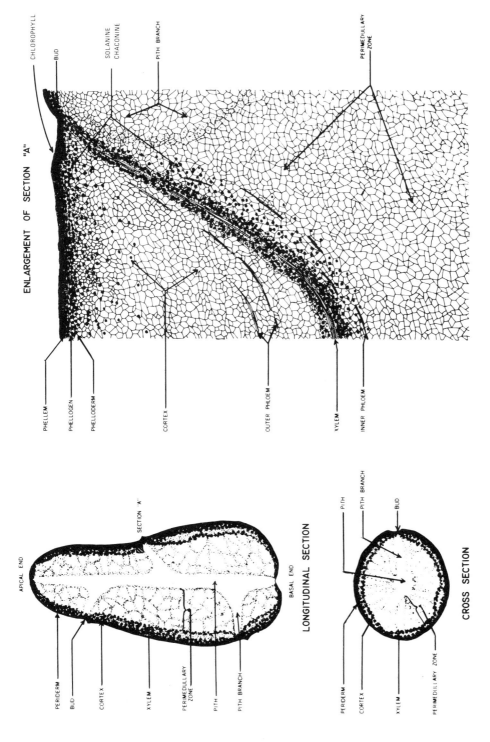

FIGURE 1.   Anatomical distribution of solanum glycoalkaloids in the periderm, upper cortex, and vascular region in a potato tuber. (From Jeppesen, R. B., Salunkhe, D. K., and Jadhav, S. J., *33rd Annu. Inst. Food Technol. Meet.*, Miami, FL, 1973. With permission.)

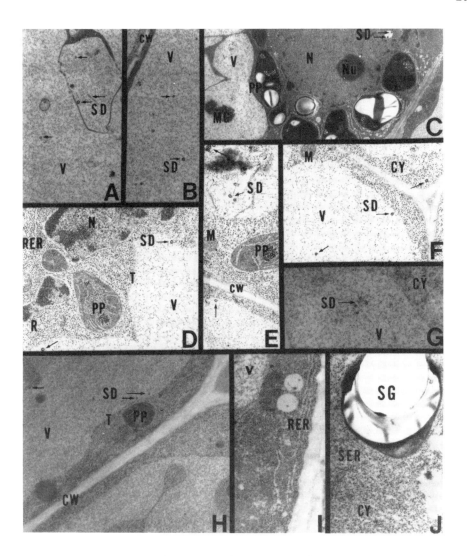

FIGURE 2.   Glutaraldehyde-osmium tetroxide-digitonin fixation of light-treated potato sprouts. Spicules (arrows) are seen within the cytoplasm, which are sites of solanidine-digitonide complex. (X of A, B, D, E, F, I, J = 14,550; C = 5,580; G, H = 12,610). V, vacuole; SD, solanindine-digitonide complex; CW, cell walls; N, nucleus; Nu, nucleolus; MG, melinin granule; PP, proplastid; R, ribosome; RER, rough endoplasmic reticulum; T, tonoplast; M, mytochondrion; CY, cytoplasm; SER, smooth endoplasmic reticulum; and SG, starch granule. (From Han, S. R., Campbell, W. F., and Salunkhe, D. K., *J. Food Biochem.*, 13, 377, 1989. With permission.)

not increase after 34 weeks of whole-tuber storage. Gregory et al.[33] examined glycoalkaloids of foliage from 16 wild, tuber-bearing Solanum species. Although most species contained only α-chaconine and α-solanine, unusual glycoalkaloids, probably solamarine and solasonine, were found in *S. berthaultii*. The data collected by these authors indicated the need for careful evaluation of glycoalkaloids as a potato quality factor in breeding programs, which exploit wild solanum species. Also, Gregory[34] has reviewed the diverse array of total glycoalkaloid levels and individual glycoalkaloid corpus of solanum species with emphasis on potato breeding.

Bushway et al.[35] reported the alkaloid contents in different varieties of fresh, baked, fried, or baked-fried peels. There was apparently no effect by any of the cooking procedures. Raw peels contained 1.3 to 56.7 mg/100 g peel (fresh weight) α-chaconine and 0.5 to 50.2

TABLE 4
Glycoalkaloid Contents of Selected *Solanum* Species

Species[a]	Glycoalkaloid						
	β-Chaconine[b]	α-Chaconine	α-Solanine	Solamarines[c]	Demissine	Tomatine	TGA (mg/100 g FW[d])
Solanum ajanhuiri	3.5	39.0	57.3				—
S. curtilobum		34.8	46.4	5.3	13.4		<0.1
S. stenotonum	5.5	69.8	24.7				<0.1
S. juzepczukii		14.0	37.8	7.7	40.4		0.7—1.3
S. acaule 1[e]					95.5		1.7
S. acaule 2					62.1	30.9	1.4—5.0
S. acaule 3					88.2	11.6	1.4—5.0
S. acaule 4					64	34	1.4—5.0

a   All species are cultivated except for *S. acaule.*
b   Values represent percent of total glycoalkaloids.
c   Combined value for α- and β-solamarine.
d   Fresh weight.
e   Four clones of species *S. acaule* were analyzed.

From Osman, S. F., Zacharius, R. M., Kalan, E. B., Fitzpatrick, T. J., and Krulick, S., *J. Food Prot.*, 42, 502, 1979. With permission.

TABLE 5

α-Chaconine and α-Solanine Levels in Fresh Flesh or Peels of Potato Tubers of Different Commercial Varieties

Product	Variety[a]	Range (mg glycoalkaloid/100 g of product)		
		α-Chaconine	α-Solanine	α-Chaconine + α-solanine
Flesh	Russet Burbank	0.98—2.32	0.58—2.18	1.58—4.50
	Kennebec	0.75—2.13	0.55—2.10	1.30—4.23
	Katahdin	0.15—0.35	0.15—0.23	0.30—0.58
	Superior	0.07—0.23	0.05—0.10	0.12—0.33
	Allagash Russet	0.19—0.45	0.25—0.48	0.44—0.93
	Round White	0.18—0.60	0.03—0.38	0.21—0.98
	Green Mountain	0.05—0.30	0.05—0.18	0.10—0.48
	Round White	0.02—0.06	0.01—0.11	0.03—0.17
	Bel Rus	0.08—0.38	0.03—0.15	0.11—0.53
	Russet	0.03—0.10	0.01—0.10	0.04—0.20
	Bel Rus	0.08—0.38	0.03—0.15	0.11—0.53
	Russet	0.05—0.30	0.05—0.18	0.10—0.48
Peels	Russet Burbank	24.33—42.83	13.50—27.00	37.83—69.83
	Kennebec	46.67—56.67	36.17—50.16	82.84—106.83
	Katahdin	21.33—46.83	9.33—25.83	30.66—72.66
	Superior	14.83—17.90	7.00—7.90	21.33—25.80
	Allagash Russet	14.83—14.90	9.50—10.17	24.33—25.07
	Round White	12.90—13.60	3.90—5.10	16.80—18.70
	Green Mountain	14.40—17.50	6.70—8.00	21.70—25.50
	Round White	8.60—12.20	3.40—5.80	12.00—18.00
	Bel Rus	3.67—6.80	2.00—4.10	5.67—10.90
	Russet	2.40—4.20	1.70—2.40	4.10—6.60
	Bel Rus	4.20—5.50	2.40—3.10	6.60—8.60
	Russet	1.30—4.90	0.50—3.60	1.80—8.50

[a] All varieties analyzed three times in duplicate.

From Bushway, R. J., Bureau, J. L., and McGann, D. F., *J. Food Sci.*, 48, 84, 1983. With permission.

mg/100 g peel (fresh weight) α-solanine. Raw flesh from the same potatoes contained 0.02 to 2.32 mg/100 g and 0.01 to 2.18 mg/100 g flesh (fresh weight) α-chaconine and α-solanine, respectively. The ranges of glycoalkaloids in fresh peel or flesh of different potato varieties are indicated in Table 5.

In a report by Maga,[36] the total glycoalkaloid content of slices from stored potatoes ranged between 11.8 and 29.1 mg/100 g (fresh weight). The total alkaloid concentration did not differ markedly if the slices were incubated at 5 or 25°C. However, the presence of water (soaking of slices) decreased the development of glycoalkaloids during storage. Storage of slices for 7 h nearly doubled the alkaloid contents (26 to 30 mg/100 g fresh weight). The initial α-chaconine/α-solanine ratio of 1.42 changed with the time and temperature of storage to 2.4 to 2.8 in unsoaked slices and 1.5 to 1.6 in soaked slices.

The glycoalkaloid content in freeze-dried potato meal was reported by Bushway et al.[37] The alkaloid content in meal prepared by low-temperature dehydration was 15.53 mg α-chaconine and 4.75 mg α-solanine per 100 g meal compared to 15.79 mg α-chaconine and 7.83 mg α-solanine in meal produced by high-temperature dehydration. The results suggested that low-temperature processing may have slowed the development of α-solanine, but not of α-chaconine.

Parfitt et al.[38] analyzed the pressed potato vine silage suggested as a feed for ruminants. Again a large variation was observed depending upon the source of variety of vine. The

PSDT 5 clone contained 88 mg/100 g (dry vine) of total glycoalkaloids, whereas the amount of alkaloids in PSDT 41 was 295 mg/100 g dry vine. In view of the possible occurrence of potato fruits as toxic contaminants in frozen peas and beans, Coxon[39] reported the glycoalkaloid content of potato berries from six of the main U.K. compound varieties.

IV. BIOSYNTHESIS AND METABOLISM

Biosynthetically all steroidal compounds, such as sterols, certain sapogenins, terpenes, hormones, and alkaloids, are interrelated and pathways leading to the synthesis of a structurally similar compound could be postulated on the basis of known ones. Thus, the regular pathway starting from acetate via mevalonate, isopentyl pyrophosphate, farnesyl pyrophosphate, squalene, and cholesterol is applicable to steroidal alkaloids. Several authors have reviewed the biochemistry and possible biogenetic relationships of steroids and steroidal alkaloids of the *Solanum* group.[40-45]

The first tracer work on the biogenesis of potato alkaloids was initiated by Guseva and Paseshnichenko,[46] who demonstrated the uptake and utilization of radioactive acetate by potato sprouts. The glycoalkaloids isolated from such sprouts grown under normal illumination had the labeled carbon chiefly in the aglycone, while the labeled carbon of sprouts grown in the dark was in both aglycone and sugar portions of the glycoalkaloids. Maximum radioactivity in the glycoalkaloids occurred when labeled acetate was fed for 2 d. In a later experiment, Guseva et al.[19] found that α-chaconine had nearly twice as much specific activity as α-solanine. Mevalonate was more effectively utilized in the biosynthesis of glycoalkaloids by potato seedlings than acetate.[47] Wu and Salunkhe[48] observed that light-exposed tubers incorporated a higher percentage of label from mevalonate into α-chaconine than mechanically injured tubers. The tuber surface had been exposed to light (2152 lx) at 13°C, and the precursor was applied on a 9-cm² skin surface area confined by a thin line of Vaseline® and then exposed to light for 2 d. The mechanical injury treatment involved half of a normal tuber similarly coated with the precursor and incubated in the dark at 13°C and 80% relative humidity for 2 d.

The biosynthesis of cholesterol has been investigated by the isolation of radioactive cholesterol from *S. tuberosum* fed with mevalonic acid-2-^{14}C.[49] In plants, the biosynthetic pathway from squalene is thought to proceed via 2,3-oxidosqualene, cycloartenol, lanosterol, lophenol to cholesterol, or a closely related phytosterol. The widespread distribution of cycloartenol in the plant kingdom has indicated that cycloartenol may be the first product of cyclization in higher plants. Cycloartenol and lophenol, as well as cholesterol, have been isolated from the potato.[50-52] Ripperger et al.[53] showed that cycloartenol and lanosterol are the precursors of *Solanum* alkaloids. Tschesche and Hulpke[54] reported that cholesterol, when applied to leaf surfaces of potato plants, was metabolized to solanidine. During the formation of solanidine from cholesterol, 16 β-H is lost from the latter.[55]

The hypothesis that labeled carbon atoms are distributed in all the steroidal rings of solasodine, synthesized from radioactive acetate, or mevalonate by *S. aviculare*, is consistent with the known biosynthetic and cyclization scheme of squalene[56] and may be applied to α-solanine and α-chaconine because of their structural similarity to solasodine. The origin of the nitrogen atom in potato alkaloids is unknown. However, Heftmann[42] hypothesized that cholesterol may undergo cyclization in the side chain subsequent to the formation of 26-hydroxycholesterol; the hydroxyl group is then replaced by an amino function. The presence of 26-hydroxycholesterol in growing potato plants has been reported by Heftmann and Weaver.[57] Tschesche et al.[58] found that the C-26 or C-27 hydroxyl group is directly replaced by an amino group during the biosynthesis of C-27 alkaloids of *S. lycopersicum* and *S. laciniatum* administered with 25 (RS)-25, 26 ^{3}H$_2$-4-^{14}C-cholesterol, a finding consistent with the hypothesis of Heftmann. Kaneko et al.[59] suggested that L-arginine is the

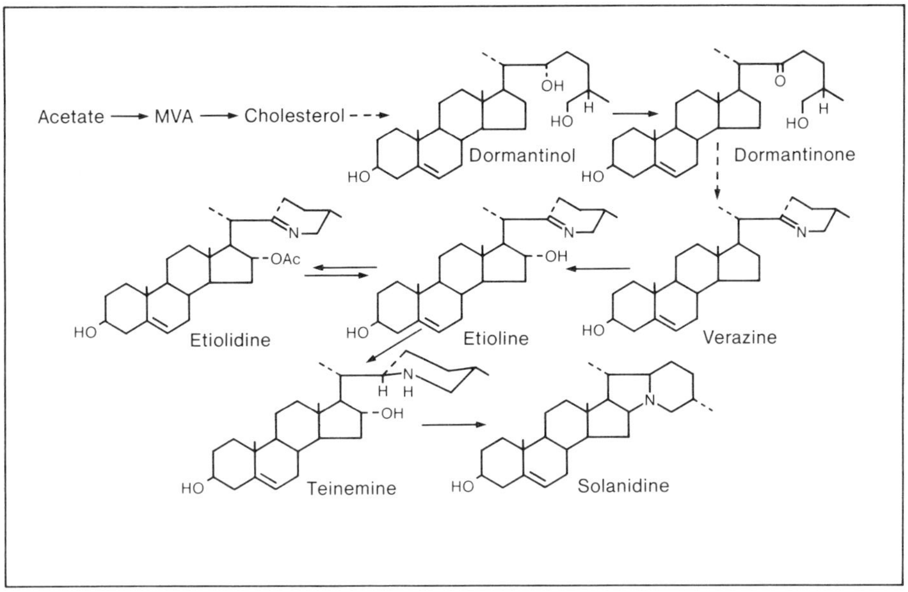

FIGURE 3. Hypothetical formation of solanidine from sterols and alkaloids lacking ring E. (From Kaneko, K., Tanaka, M. W., and Mitsuhasi, H., *Phytochemistry*, 15, 1391, 1976. With permission.)

most likely source of nitrogen for solanidine biosynthesis in *Veratrum grandiflorum* and postulated a biogenetic pathway for solanidine involving sterol intermediates and steroidal alkaloids lacking ring E (Figure 3).

Jadhav et al.[60] studied the incorporation of labeled carbon from β-hydroxy-β-methyl-glutaric acid, L-leucine, L-alanine, and D-glucose into the glycosidic steroidal alkaloids of potato sprouts. The higher amount of radioactivity in the glycoside moiety than in the aglycone part of the glycoalkaloids indicated predominant glycosylation when labeled D-glucose was administered to the potato sprouts. Subsequently, glycosylation of solanidine was apparently catalyzed by the crude enzyme preparation from sprouts of potato tubers.[61] This enzyme system is known to glucosylate a sterically unhindered 3-β-hydroxy group of a steroid if it belongs to the 5α-H or Δ^5-series. These results and *in vitro* synthesis of γ, β, and α forms of solanine and chaconine (one, two, and three sugars in the glycosidic part, respectively) by potato tissue[62] supported the hypothesis that α-solanine and α-chaconine are synthesized in a stepwise manner from solanidine.

Potatoes contain an enzymatic system capable of hydrolyzing the glycoalkaloids. Their role in plant metabolism is not clearly understood. Guseva and Paseshnichenko[63] observed that enzyme preparations from the juice of potato sprouts hydrolyze α-solanine to β- and γ-solanines by a stepwise mechanism. However, the same enzyme system removes the rhamnose substituent at the two position of the glucose residue in chaconine, giving rise to β_2-chaconine, which is further hydrolyzed to solanidine without forming γ-chaconine as an intermediate product. Swain et al.[64] confirmed the presence of rhamnosidase, glucosidase, and galactosidase activities along with the nonstepwise hydrolysis of α-chaconine by the enzyme mixture from sprouts. They also were the first to report that the enzyme preparation from dormant tubers is capable of producing β_1-chaconine (2-rhamnosylglucoside of solanidine), β_2-chaconine, γ-chaconine, and solanidine from α-chaconine in a stepwise manner, while producing β-solanine and solanidine from α-solanine by a concerted (anomalous) mechanism. A large accumulation of solanidine accompanying cellular destruction of tuber tissue has been attributed to the hydrolytic enzyme activity on the glycoalkaloids.[65] However, Holland and Taylor[66] found that the blight fungus *Phytophthora infestans* itself can convert

TABLE 6
Effects of pH and Incubation Time on the Pattern of Glycoalkaloids in
Potato Blossom Homogenates

Buffer pH	Incubation (37°C) hours	α-Solanine	α-Chaconine	β-Chaconine	Solanidine
4.00	0	High	High	0	0
	24	High	0	High	0
	72	High	0	High	0
5.00	24	High	0	Medium	0
	48	High	High	0	0
	72	Medium	0	Trace	0
6.00	72	Low	0	0	Low
7.00	24	High	Low	High	0
	72	Medium	0	0	Low

Note: Detected by resolution of ammonia precipitates by ascending paper chromatography (Whatman 1 buffered with 5% KH_2PO_4, pH 4.3 to 4.5; butanol-acetic acid, water system at 70:7:23). Alkaloid levels: 2 to 5 µg/5 to 10 g = traces; >50 µg = high.

From Zitnak, A., *Proc. Can. Soc. Hortic. Sci.*, 3, 81, 1964. With permission.

α-solanine to solanidine. Rumen microorganisms can initially hydrolyze glycoalkaloids to solanidine[67] and then reduce the double bond at position 5.

Zitnak[68] made several observations concerning the heat sensitivity, pH dependency, water extractability, and light sensitivity of the hydrolytic enzyme system associated with glycoalkaloids in potato tissue homogenate. The potato blossom rhamnosidase initiated a selective hydrolysis of α-chaconine to β-chaconine within 2 to 3 h of incubation without affecting α-solanine and completed the reaction after 24 h. The rate of hydrolysis increased at pH 5, leading to a complete degradation of the steroidal system, apparently by a separate enzyme, after 72 h of incubation. At pH 6 and 7, the degradation of the steroid ring occurred more slowly. α-Solanine was affected by a pH between 5 and 7 (Table 6). A mixture of sprout-blossom homogenate increased the rate of terminal degradation at a faster rate than that in blossom tissue alone. Netted Gem (cv. Russet Burbank) had rhamnosidase activity only in photostimulated eye tissues and produced solanidine as an end product. Cultivar differences in the enzyme systems, hydrolytic activity in protein precipitates of acetone and ammonium sulfate, and a reaction on α-solanine were noted.

V. FACTORS INFLUENCING GLYCOALKALOID DISTRIBUTION AND CONTENT

A. PHOTOINDUCTION

Potatoes exposed to light in the field or after harvest develop a green pigmentation at the surface. This condition, known as "greening", indicates the formation of chlorophyll, which is harmless and tasteless. Green potatoes are usually associated with an increased level of glycoalkaloids, although chlorophyll and glycoalkaloid formation are independent processes. The relationship between toxic glycoalkaloids and potato greening is discussed in detail.[69,70] The probability that tubers will be exposed to a certain quantity and duration of light, including daylight, sunshine, ultraviolet, fluorescent, or incandescent light, varies with environmental factors and marketing conditions. Light intensity as low as 53.8 lx produces greening; greening increases with increased light intensity, but the increase is not directly proportional to light intensity.

Lilijemark and Widoff[71] studied the effect of light intensity (25 to 3600 lx) on greening and found that the glycoalkaloid content of Majestic potatoes almost paralleled chlorophyll development, despite high total glycoalkaloid (TGA) levels at the onset of illumination. Patil et al.[72] hypothesized that the insignificant differences in the high glycoalkaloid content of Kennebec tubers after exposure to four light intensities (538, 1076, 1614, and 2152 lx) resulted from long storage at low temperatures. Zitnak[73] exposed the tubers of Netted Gem (2.8 mg % TGA) and Katahdin (2.5 mg % TGA) to a Mazda lamp (100 W, 76-cm distance) for 8 d and found excessive amounts of TGA. Netted Gem developed 17 mg % at 6.7 to 8.9°C and 11.8 mg % at 12.2 to 16.6°C.

Freshly harvested tubers may remain exposed to solar radiation prior to storage. Zitnak[74] exposed tubers of Netted Gem (1.69 mg % original TGA) and Katahdin (1.46 mg % original TGA) to sunlight during clear sunny days (October 14 to 22, 1952) for 6, 12, and 36 h and found exceptionally rapid synthesis of glycoalkaloids (12 and 4.95 mg % TGA; 13.18 and 7.10 mg % TGA; and 20.38 and 16.28 mg % TGA). These treatments were conducted for 6 h/d followed by storage (10 to 15°C) in the dark until the next suitable day. Traces of greening after a 12-h exposure and intense greening after a 36-h exposure were observed in the tubers of both the cultivars.

Curing of Sebago potatoes for 10 d at 25°C prior to storage at 5°C reduced responsiveness of the tubers to photoinduced glycoalkaloid synthesis.[75]

Conner[76] exposed tubers to different wavelengths of light. The blue end of the spectrum encouraged glycoalkaloid formation the most, while the yellow-red end of the spectrum was most efficient for chlorophyll synthesis, but did not increase glycoalkaloids. Zitnak[73] found that the glycoalkaloid content of dormant tubers of Netted Gem cultivar increased rapidly from 5.72 mg % to 11.00, 18.88, 22.42, 21.54, and 23.17 mg % when exposed to an efficient (13,000 Å) infrared light source (18 to 22°C) for 4, 6, 8, 10, and 16 d, respectively. Glycoalkaloid concentrations further increased after 2 months of storage at 4 to 8°C. Greening was noted in all samples, and an intense color developed after only 4 d of exposure. Glycoalkaloid content of tubers stored for 3.5 months and then irradiated with ultraviolet light (14 to 18°C) for 4, 6, 8, or 10 d increased gradually and was higher than control values by 47, 141, 190, and 222%, respectively. At a temperature of 7 to 10°C, however, glycoalkaloid content decreased when tubers were stored and irradiated for more than 4 d. Wavelengths other than at 3654 Å emitted by the ultraviolet source in the visible spectrum of light may have increased glycoalkaloid content. No greening was observed in the tubers irradiated with ultraviolet light.

B. CULTIVAR

Several cultivars have been known for their glycoalkaloid contents. Wintgen[77] reported that the glycoalkaloid content of 11 German cultivars ranged from 2 to 10 mg/100 g of fresh material. Morgenstern[15] found the range of 39 cultivars to be from 4.6 to 35 mg/100 g of fresh tissue. The range of glycoalkaloid content was wider in nine cultivars tested by Bömer and Mattis.[78] Wolf and Duggar[23] found that the glycoalkaloid content of 32 cultivars ranged from 1.8 to 13 mg/100 g of fresh tuber. Most of the results discussed above were based on the gravimetric method of determination of glycoalkaloid contents. Gull and Isenberg[79] determined the glycoalkaloid content of peel colorimetrically and noted that Katahdin, although greening the least, contained the most glycoalkaloids (88 mg/100 g of fresh peel), while the glycoalkaloid content of Cherokee, the cultivar that greened most rapidly, increased only moderately (52 mg/100 g of fresh peel). The highest and lowest mean glycoalkaloid concentration of 15 cultivars studied by Zitnak[73] in Canada occurred in East Dewey (7.6 mg/100 g of fresh tuber) and in Cultivar 177 (1.9 mg/100 g of fresh tuber), respectively. Sanford and Sinden,[80] Sinden and Webb,[81] and Lepper[82] found that cultivars with high mean glycoalkaloid contents are more likely to produce excessive glycoalkaloid contents than are cultivars with low mean glycoalkaloid contents when subjected to less than ideal environ-

TABLE 7
Glycoalkaloid Content of 11
Cultivars Exposed to White
Fluorescent Light (1076 lx) for a
Period of 5 d

Cultivar[a]	Glycoalkaloids[b] (mg/100 g fresh peel)
LaChipper	44.81
Platte	55.86
Cascade	65.69
LaRouge	73.06
Sioux	58.93
Norchip	79.20
Red LaSoda	69.99
Shurchip	44.50
Russet Burbank[c]	69.07
Kennebec	96.40
Bounty	70.30

[a] Initial glycoalkaloid contents were not reported.
[b] Determined by the method of Gull and Isenberg.[79]
[c] 24 mg of initial glycoalkaloid content per 100 g of fresh peel as determined by Wu and Salunkhe.[229]

From Patil, B. C., Salunkhe, D. K., and Singh, B., *J. Food Sci.*, 36, 474, 1971. With permission.

mental conditions or when handled improperly. Average total glycoalkaloid contents of five cultivars grown in 1970 and in 39 locations in the U.S. were as follows: Kennebec, 9.7 mg/100 g fresh weight of tubers, Russet Burbank, 7.9 mg/100 g, Katahdin, 7.9 mg/100 g, Irish Cobbler, 6.2 mg/100 g, and Red Pontiac, 4.3 mg/100 g. B-5141-6 (Lenape) had the highest glycoalkaloid content (about 29 mg/100 g of fresh tuber) of all cultivars studied by Zitnak and Johnston[26] and Sinden and Webb.[81] Patil et al.[72] reported that various cultivars differ significantly in glycoalkaloid formation (Table 7). It appears, therefore, that glycoalkaloid content is a genetically controlled characteristic that varies with the cultivar.

Sinden and Deahl[83] reported that a taste panel described several clones with glycoalkaloid contents in excess of 14 mg/100 g as having a bitter taste. The clones produced a mild to severe burning sensation in the mouth and throat when the glycoalkaloid levels exceeded 22 mg/100 g of edible tissue (Table 8). While the wild species of *Solanum acaule* is known to contain high levels of glycoalkaloids, 35 to 126 mg/100 g fresh tuber, the glycoalkaloid content of cultivated tuber-bearing species of *S. tuberosum* subsp. *Andigena*, *S. ajanhuiri*, *S. curtilobum*, *S. juzepczukii*, and *S. stenotonum* do not appear to pose any hazard to human health.[11] The high correlation between levels of glycoalkaloids in foliage and tubers and increasing interest in use of wild species in potato breeding programs as sources of desirable characteristics increase the importance of knowing the foliar levels of glycoalkaloids of wild potato species.[84]

C. LOCATION, CLIMATE, AND ENVIRONMENT
Sinden and Webb[81] reported significant differences in tuber glycoalkaloid contents of five commercial cultivars and of Lenape grown at 39 locations in the U.S. The effects of

TABLE 8
Effects of Glycoalkaloid and Phenolic Contents on Potato Flavor

Clone[a]/ cultivars	Bitterness rating[b] (0 to 4 scale)	Burning rating[b] (0 to 4 scale)	Glycoalkaloid content (mg/100 g)	Phenolic content (mg/100 g)
1	2.4	3.4	58.0	29
2	2.2	3.2	51.0	43
3	1.8	2.0	25.0	17
4	0.9	1.7	23.0	23
5	1.3	1.7	22.0	41
6	1.9	1.7	22.0	33
7	0.8	0.6	14.0	59
8	0.0	0.1	7.3	27
9	0.2	0.2	5.9	30
10	0.1	0.0	4.4	31
11	0.1	0.1	2.0	29
12	0.1	0.0	0.9	24
13	0.0	0.0	0.7	21
LSD[c] (0.05)	0.64	0.71	5.7	6.7

[a] Clones 1 through 8 are breeding lines; 9 through 13 are cultivars.
[b] Means of 18 evaluations; 0 = no bitterness or burning, 4 = very strong bitterness or burning.
[c] Least significant difference.

From Sinden, S. L. and Deahl, K. L., *J. Food Sci.*, 41, 520, 1976. With permission.

location were also significant. Glycoalkaloid contents rarely increased to excessive levels (above 20 mg/100 g) in the five commercial cultivars at certain locations, and the increase usually could be explained by abnormal growing conditions (environment) or improper handling. In contrast, Zitnak[73] observed that location had an insignificant influence, and reported that slightly higher levels of glycoalkaloids in potatoes were a result of conditions incurred during transportation of the tubers. The variability in glycoalkaloid concentrations in certain cultivars was attributed to natural (environmental variations rather than to differences in soil conditions.

Hutchinson and Hilton[85] showed that very bitter tuber samples from several locations in Alberta (Canada) contained toxic levels of glycoalkaloids, but showed no greening. It was suggested that one or more unfavorable climatic conditions had probably prevailed such as nutritional imbalance and frost, or hail damage to the tops of the plants before tubers matured. An unusually cool growing season accompanied by an abnormally high number of overcast days can also result in excessive glycoalkaloid content of a potato crop.[81] In spite of the obvious relationships between planting date, tuber size, and tuber maturity, Braun[86] reported no correlation between glycoalkaloid content and planting date. Arutyunyan[87] claimed that potatoes grown in mountainous sections always contain less glycoalkaloids than those grown in hot climates.

There are conflicting reports of the effects of the major elements, moisture, and organic content of the soil on the glycoalkaloid content of potatoes.[86-91] Magnesium fertilizers[92,93] increased the TGA, while foliar application of indoleacetic acid[94] caused a decrease in TGA contents of potatoes. The nature of physiological stress at different locations, climates, and environments may lead to variations in the TGA contents.

D. MATURITY AND SPECIFIC GRAVITY

Environmental factors that tend to retard the maturation process are associated with higher than normal levels of glycoalkaloids.[81] According to Bömer and Mattis,[2] the glycoalkaloid content of immature and small potatoes increases more when exposed to light

TABLE 9
Effects of Storage Temperature on Tuber Bitterness in Netted
Gem Potatoes after Cooking

Storage temp [a] (°C)	Taste score		Mean
	Tubers originally not bitter	Tubers originally bitter	
0—5	2.3 (9)	2.9 (9)	2.60
5—10	3.0 (18)	2.5 (18)	2.75
10—15	2.1 (18)	2.2 (18)	2.15
15—20	1.9 (9)	1.7 (9)	1.80

Note: Abnormally high glycoalkaloid content, evaluated by organoleptic tests based on taste score, such as: 0 = not discernible; 1 = trace; 2 = slight; 3 = moderate; 4 = marked; 5 = very marked. Values in parentheses are number of readings.

[a]   Stored for 5 months at the indicated temperatures.

From Hilton, R. J., *Sci. Agric.*, 31, 61, 1951. With permission.

than do old and large tubers. Wolf and Duggar[23] noticed an inverse relationship between tuber size (31.3, 135, 210, and 270 g average weight of White Rural) and glycoalkaloid concentration (18, 14.8, 8.9, and 7.3 mg/100 g fresh weight).

Patil et al.[72] reported the formation of chlorophyll and glycoalkaloids in Kennebec potatoes of the following three different specific gravities: 1.06 to 1.08, 1.08 to 1.10, and 1.10 and above. Chlorophyll development was inversely related to specific gravity, but glycoalkaloid synthesis (98, 98, and 85 mg/100 g of fresh peel) was independent of specific gravity.

E. STORAGE AND TEMPERATURE

Gull and Isenberg[79] observed that potato tubers stored for 8 months at 4.4°C contained more glycoalkaloid than those stored for 3 months. In contrast, storing of tubers at 4 to 5°C for 6 to 7 months after harvest did not appreciably affect the glycoalkaloid content of the cultivars tested by Wolf and Duggar.[23] Hilton,[24] on the basis of his results (Table 9), generalized that low-temperature storage maintained or caused more bitterness of the tubers than did storage temperatures above 10°C. Subsequently, Zitnak[74] found that high glycoalkaloid concentrations could develop at low temperatures (4 to 8°C) during postharvest storage of Netted Gem tubers whether continuously illuminated or stored in darkness (Table 10).

The effects of storage on glycoalkaloid content of tubers were reported by Wilson et al.[95] The potatoes were purchased from roadside stands and stored for 1 to 3 months at 12.2°C. There was no apparent influence of storage on α-chaconine, α-solanine, or TGA contents. The authors suggested an interaction between location and storage period and concluded that the level of glycoalkaloids in tubers is controlled by a complex mechanism. In no case, however, did the glycoalkaloid content even approach the presumed toxic level, i.e., 20 mg/100 g fresh weight of tubers. In a similar study, Bushway et al.[96] reported the effects of MH-30-treated potato tubers stored for 9 months at 3.3, 7.7, and 20°C. The glycoalkaloid content of all potato varieties increased during the first few months of storage; but by the ninth month the concentrations of both α-chaconine and α-solanine had decreased to near their original values. Again, in all cases the maximum glycoalkaloid contents were well below the maximum safe level for human consumption.

F. PHYSIOLOGICAL STRESS

Many microbial and/or nonmicrobial stresses contribute to the formation and accumulation of abnormal metabolites in plant tissues.[97] Stressed potatoes may contain certain

TABLE 10
Effects of Various Storage Conditions on Glycoalkaloid
Content of Netted Gem Potatoes

Storage time in weeks	Cold, humid storage at 4 to 8°C		Dry, warm storage at 12 to 15°C	
	Dark	Light[a]	Dark	Light[a]
1	7.93	19.83	7.50	7.20
2	5.63	19.01	7.96	5.04
3	11.30	18.51	3.52	3.20
4	13.75	17.38	5.78	7.26
5	11.09	16.34	4.01	3.65
6	15.42	23.50	8.72	6.98
Average	10.86	19.09	6.26	5.86

Note: Glycoalkaloid content determined colorimetrically using sulfuric acid-formaldehyde reagent and expressed in mg% of untreated tissue; 5.9 mg% average value of normal 'Netted Gem' tubers.[74]

[a] A weak Mazda light (15 W) at a distance of 30 in. from tubers.

From Zitnak, A., M.S. thesis, University of Alberta, Edmonton, 1953. With permission.

compounds that are either normal constituents or ones not normally found in the tissue.[98] McKee[99] reported that the glycoalkaloids were found around sites of wounded potatoes. Although the glycoalkaloids are localized in the peel of whole tubers, mechanical injury caused by slicing could increase their synthesis and accumulation in peeled tubers.[100] Locci and Kuć[101] regarded this phenomenon in response to physiological stress, which is observed at the surface of cut tissue.[102,103] Tubers injured by either bruising or mechanical grading after harvest synthesized glycoalkaloids.[104] Kuć[105] reported that the concentration of glycoalkaloids in fresh potato slices increased from an undetectable amount to 20 mg/100 g after storage at room temperature (22°C) in the dark for 3 d. Similarly, Salunkhe et al.[106] indicated that potato slices (0.3 mg of glycoalkaloids per 100 g) held in the dark at relatively high temperatures (15 or 24°C) for 2 d synthesized glycoalkaloids (1.3 and 2.05 mg/100 g). The rate of glycoalkaloid formation increased (4.94 and 7.4 mg/100 g) when the slices were stored under high-intensity light (2152 lx). Similar reports have appeared concerning various factors affecting the accumulation of glycoalkaloids in potato slices.[107-110] In many potato processing plants, slices, cubes, mash, string, strips, and shreds are often stored at relatively high light intensity and temperature before cooking or processing. This may cause synthesis and subsequent accumulation of glycoalkaloids—an effect that Salunkhe et al.[106] attributed to a physiological defense mechanism. In comparison with light and mechanical injury, hollow heart and blackheart disorders seem to be much less potent factors in stimulating glycoalkaloid synthesis.[107]

Wu and Salunkhe[108] studied the relationship between glycoalkaloid content and the type of mechanical injury sustained by potato tubers. Mechanical injuries such as bruising, cutting, dropping, puncturing, and hammering significantly stimulated glycoalkaloid synthesis in both peel and flesh of the tubers (Table 11). The extent of glycoalkaloid formation depended on the cultivar, the type of mechanical injury, the storage temperature, and the duration of storage. High-temperature storage stimulated more glycoalkaloid formation than low-temperature storage. Most of the injury-stimulated glycoalkaloid formation occurred within 15 d after treatments. Mechanical injury caused by cutting of tubers resulted in the highest contents of glycoalkaloids in both flesh and peel. Fitzpatrick et al.[22] noted a similar

TABLE 11
Effects of Mechanical Injuries of Russet Burbank Tubers on Glycoalkaloid Content of Flesh and Peel During Storage at 4 and 21°C

Mechanical injury	Storage temp. (°C)	Flesh, days storage					Peel, days storage				
		0	15	30	60	90	0	15	30	60	90
		Total glycoalkaloids (mg/100 g dry weight)									
Control	4	3.0	4.0	4.2	5.9	6.2	123.9	126.8	129.0	130.1	134.3
(nontreated)	21	3.0	4.0	4.1	7.8	9.7	123.9	129.1	132.0	133.8	136.7
Brushing	4	3.0	29.2	35.5	38.0	43.1	123.9	212.6	235.1	237.2	239.5
	21	3.0	43.6	45.7	45.0	46.0	123.9	257.5	279.7	282.1	286.4
Hammering	4	3.0	35.0	39.5	46.1	46.4	123.9	203.3	210.0	215.3	224.0
	21	3.0	60.5	60.0	63.2	64.1	123.9	208.0	240.2	241.0	245.2
Dropping	4	3.0	74.0	93.4	98.9	103.1	123.9	169.8	172.0	177.1	180.0
	21	3.0	85.9	103.5	106.0	110.3	123.9	174.3	196.6	212.5	215.8
Puncturing	4	3.0	91.9	105.2	127.5	137.5	123.9	176.0	183.8	185.0	194.1
	21	3.0	135.8	152.1	152.5	156.6	123.9	192.2	208.3	227.0	232.4
Cutting	4	3.0	163.9	191.7	195.3	201.0	123.9	255.8	268.0	275.7	282.1
	21	3.0	212.6	220.0	225.2	236.4	123.9	292.1	315.7	317.0	321.9

Note: LSD at 1% level: 10.3, 9.2 (flesh 4°C, 21°C) and 14.5 18.4 (peel 4°C, 21°C).

From Wu, M. T. and Salunkhe, D. K., *J. Am. Soc. Hortic. Sci.,* 101, 329, 1976. With permission.

phenomenon in 4-d-old slices made from fresh as well as stored potatoes. The buildup of glycoalkaloids on aging peaked early in storage and then gradually declined over the storage period. α-Solanine consistently increased more than α-chaconine in these slices. In addition, α- and β-solamarine appeared only in the aged slices of the Kennebec cultivar, and levels gradually decreased over the storage period. Despite excessive synthesis of glycoalkaloids in the above model systems, damaged commercial potato samples did not contain deleterious levels of glycoalkaloids.[112] The high correlation between the initial glycoalkaloid level and the increased level after damage showed that selection of low-glycoalkaloid potato clones is important.[111] Accumulation of stress metabolites is discussed elsewhere.[113,114]

Accumulation of glycoalkaloids in potato tuber slices can be suppressed when the cut surface is inoculated with *P. infestans*.[115] In this instance, the additional stress induced by the microorganism alters the biogenetic pathway of isoprene compounds, thus leading to accumulation of terpenoids.[116] In a chemically defined and complex media, however, *P. infestans* synthesized glycoalkaloids.[117] Cheema and Haard[118] have proposed a mechanism of terpene and glycoalkaloid induction in sliced tubers based on repressor proteins. The buildup of free solanidine in potato tuber tissue has been regarded as a hydrolytic product of the related glycoalkaloids under certain stresses; cellular disruption or tissue liquefaction was caused by *Erwinia atroseptica* as well as by *P. infestans*.[65]

G. INFLUENCE OF COOKING METHODS ON GLYCOALKALOID CONTENT

A variety of cooking methods have been investigated for their effects on individual or TGA contents. It appears that most cooking modes have little or no effect on the levels of glycoalkaloids, suggesting a moderate temperature stability of glycoalkaloids. Bushway and Ponnampalam[119] reported on the stability of glycoalkaloids during four cooking procedures—frying, baking, microwaving, and boiling. Only a slight loss of alkaloids by frying was reported. In a similar study, Bushway et al.[35] reported the effects of frying, baking, or bake-frying on alkaloid levels in potato peels. As indicated earlier, there were no effects by any of the cooking procedures. On the other hand, Ponnampalam and Mondy[120] analyzed the

TGA contents of cortex and pith of three potato varieties after baking and frying. A significant decrease in the cortical tissue was noted by both methods of cooking. The decrease in the pith was not significant. The significant decrease in the cortex (although not remarkable) could be due to its high glycoalkaloid content, which in some varieties exceeded the level considered safe even after cooking. Except at added moisture of 38% and 130 or 160°C, extrusion of potato flakes at 59 or 48% moisture and 70 to 160°C did not result in a significant reduction of their TGA content.[121] The changes in glycoalkaloids of potatoes during processing for chips, puree, and dried products are mainly attributed to peeling.[122,123] Since several processed potato products are being marketed, studies relating to the TGA contents of these products have been attracting the attention of many researchers.[124-130]

VI. PHYSIOLOGICAL FUNCTIONS

Physiological functions of glycoalkaloids are particularly important in potato-breeding programs and the development of new cultivars since these compounds are involved in disease and insect resistance.[99] Many pathologists and breeders prefer to develop multigenic resistance (field resistance) to late blight. When attacked by the pathogen, potato plants with this characteristic remain in production several weeks longer than susceptible plants. Although the nature of the late blight resistance is unknown, nonspecific defense mechanisms such as physical barriers[131] and toxic plant constituents[132] in multigenic resistance have been suggested as possible factors. The lack of correlation between glycoalkaloid contents of blight-infected and healthy plants[133,134] suggests that it might be possible to select multigenic late blight resistance without increasing glycoalkaloid contents of clones. It is interesting to note that leptines of *S. chacoense* lose their ability to protect from the potato beetle larvae if they occur in the deacetylated form (leptinines). Potato cultivars differ markedly in their rates of glycoalkaloid formation in response to wounding, which may be a significant part of disease resistance. Older leaves of the potato plant contain a lower concentration of glycoalkaloids, so the increased susceptibility of the older leaves to lesion development suggests that the potato glycoalkaloids may be involved in temporary resistance of leaves to lesion development in the early blight disease. The possible and suspected role in flowering, species compatibility, male-sterility phenomenon, seed formation, sprouting, and postharvest tuber response to environmental factors or cold tolerance are some of the interesting functional properties related to the glycoalkaloids in potatoes. Studies on reciprocal grafts of potato and tomato suggested that alkaloid transport between root and shoot does not occur in these species.[135]

VII. PHARMACOLOGY

The first attempt to establish the biological effects of glycoalkaloids by using experimental animals was made by Meyer in 1895.[136] He could not isolate either α-solanine or solanidine from the urine of a dog that had been fed α-solanine. Willimott[137] unsuccessfully studied the effects of feeding potato plants to adult rats. Golubeva[138] examined 35 patients with food allergy, mainly referring to nightshade. He treated these patients with α-solanine and observed curative effects. Satoh[139] observed that the pretreatment of rats with adrenergic blockers or reserpine prevented the glycemic effects of solanine. Nishie et al.[140] reported that α-solanine prolonged sleeping times induced by pentobarbital, reduced spontaneous motor activity in mice, and caused contraction of guinea pig ileum strips with certain minor exceptions. The metabolic rate and distribution of tritiated α-solanine in rats recorded by these authors indicated: (1) partial as well as complete gastrointestinal hydrolysis, (2) poor absorption from the gastrointestinal tract; (3) rapid urinary and fecal excretion of metabolites; and (4) buildup of high concentrations in various tissues such as spleen, kidney, liver, fat, heart, brain, and blood (in descending order).

Less radioactivity from labeled α-solanine administered parenterally (5 to 15 mg/kg) was excreted in feces and urine than when it was administered orally. As the dose increased to 15 mg/kg, elimination was slightly impaired. A sudden decrease in the urinary and fecal excretion, together with a corresponding increase in radioactivity, appeared in liver, spleen, kidneys, and intestines at a dose of 25 mg/kg. The major metabolite of α-solanine in the excreted feces after *per os* administration was solanidine. After intraperitoneal injection, urine and feces contained two metabolites from a partial hydrolysis of α-solanine in addition to the aglycone solanidine.

The inhibitory effect of α-solanine on cholinesterase was established by Orgell et al.,[141] Pokrovskii,[142] Harris and Whittaker,[143,144] and Orgell.[145] Orgell et al.[141] demonstrated the presence of cholinesterase inhibitor in the aqueous extracts of potato tissue (tuber, sprouts, leaves, flowers, and stem). The tuber peel contained 10 to 40 times more inhibitor than the innermost flesh, while extracts of tuber sprouts were as active as the extracts of tuber peel. Subsequently, Harris and Whittaker[143] showed that this naturally occurring inhibitor differentially inhibited the three serum cholinesterase phenotypes in much the same way as dibucaine. At that time, however, it was not realized that the glycoalkaloid α-solanine (known to be present in appreciably amounts in potatoes and other Solanaceae) inhibited serum cholinesterase. On the basis of the possible presence of α-solanine (or solanidine in its numerous forms) in potato plant extract and because inhibition of cholinesterase coincided with that of glycoalkaloid distribution, Zitnak[146] interpreted potato alkaloid as the enzyme inhibitor in the experiments described by Orgell et al.[141] Zitnak[146] further predicted the presence of free alkaloid solanidine in the extract as a result of enzymatic hydrolysis of glycoalkaloids. Pokrovski[142] proved that the inhibitor described by Orgell et al.[141] and indicated by Harris and Whittaker[143] was α-solanine. It was further proposed that the antiesterase property of α-solanine could be used to develop methods to identify biologically harmful concentrations of glycoalkaloids in potatoes.[142]

The anticholinesterase action of α-solanine was further supported by the experiments of Harris and Whittaker,[144] who showed that α-solanine and solanidine differentially inhibited the serum cholinesterase of individual persons of "usual", "intermediate", and "atypical" phenotypes. This effect was similar to that obtained with dilute aqueous extracts of potatoes. They concluded that if the inhibition of serum cholinesterase is attributed to the toxic effects of α-solanine, then presumably persons with the "atypical" enzyme will be less susceptible than persons with the "usual" enzyme. Patil et al.[147] studied the pattern of plasma and erythrocytic cholinesterase inhibition of α-solanine in rabbits. The reduced inhibition of erythrocytic rather than plasma cholinesterase was considered to be caused by (1) the different distribution of α-solanine at these two sites; (2) differences in the mechanism of inhibition of the enzymes as it was produced by plasma and erythrocytes; and (3) dilution of red blood cells during assay, if the inhibition was reversible. Small multiple doses of α-solanine quickly inhibited serum cholinesterase, but levels rapidly recovered, and red cell cholinesterase was not inhibited. They speculated that small doses of α-solanine may cause discomfort on ingestion; repeated doses will have little noticeable effect due to the cholinesterase inhibition.

Several animals have been tested for sensitivity to a total potato alkaloid extract as well as to α-solanine. The data are presented in Table 12. Apparently, the high activity of α-solanine injected into the bloodstream may be due to cholinesterase inhibition, while the lesser effects of orally administered α-solanine could reflect its poor absorption from the gastrointestinal tract.[140,147] Nishie et al.[140] found α-solanine to be very toxic to mice, rats, and rabbits by parenteral administration and to chick embryo by yolk-sac injection. The aglycone solanidine was considerably less toxic than the glycoside α-solanine. Swinyard and Chaube[148] found that the TGA (via an intraperitoneal dose of 10 mg/kg/d) was seven times more toxic to fetuses of rats (5 to 12 d of gestation) than α-solanine and produced similar fetal abnormalities. The post-mortem examination in the experiments of Patil et al.[147]

TABLE 12
Toxicity of α-Solanine in Various Species of Laboratory Animals after Different Routes of Exposure

Experiment	Dose of α-solanine[a]		Effects
	Administration	**Amount**	
Man[b]	Oral	≥2.8 mg/kg[c]	Toxic[d]
	Oral	20—25 mg[c]	Toxic
Sheep	Oral	225 mg/kg	Toxic
	Oral	500 mg/kg	Lethal
	Intravenous	17 mg/kg	Toxic
	Intravenous	50 mg/kg	Lethal
Pregnant rat	Oral	10% of sprout diet	Death of all pups before weaning age
Rat	Gastric intubation	590 mg/kg	50% Death within 24 h
	Intraperitoneal	75 mg/kg	50% Death in a few hours
5—12 or 5—17 of gestation	Intraperitoneal	5—10 mg/kg/d	Nontoxic to material and fetal rats
5—12 d of gestation	Intraperitoneal	20 mg TGA/kg/d	100% Fetal mortality
5—6 d of gestation	Intraperitoneal	40 mg TGA/kg/d	100% Maternal mortality 2 h after 2nd injection
6—15 d of gestation	Gavage dose	0.3—3 mg/kg	5—12% Fetal resorption
8—11 d of gestation	Gavage dose	2—25 mg/kg	2—7% Fetal resorption
	Intraperitoneal	67 mg/kg	50% Death
	Intraperitoneal	60 mg TGA/kg	50% Death
	Intraperitoneal	40 mg/kg/d	1/6 Death rate after 3 d of 2nd injection
	Intraperitoneal	40 mg TGA/kg/d	3/8 Death rate after 2nd injection
5—12 d of gestation	Intraperitoneal	10—20 mg/kg/d	18—86% Fetal resorption
	Oral	1000 mg/kg	Nontoxic
Mice	Oral	1000 mg/kg	Nontoxic
	Intraperitoneal	42 ± 1.8 mg/kg	50% Death in 7 d
	Intraperitoneal	10 mg/kg	Toxic
	Intraperitoneal	32.3 mg/kg	50% Death
	Intraperitoneal	≥50 mg/kg	Lethal
	Intraperitoneal	30.0 ± 1.95 mg/kg	50% Death
	Intraperitoneal	30 mg/kg	50% Death
Chick embryo 96 h incubation	Injection into yolk sac	18.8 ± 1 mg/kg	50% Mortality in 18 d
20—22 h incubation	Injection into yolk sac	5—20 mg/kg	63—90% Mortality within 72 h
0 h incubation	Injection into yolk sac	0.5—1.5 mg/kg	No effect on hatchability
96 h incubation	Injection into yolk sac	19 mg/kg	Reduction in hatchability
0 h incubation	Injection into yolk sac	1 mg/kg[c]	No effect on hatchability
	Injection into subgerminal yolk	0.015—1.5 mg/egg	0—70% Mortality after 72 h of incubation
	Injection into subgerminal yolk	0.26 TGA/egg[f]	26% Mortality after 72 h of incubation
3rd day of embryogenesis	Injected into embryo	0.15 μg	3/6 Death or malformation rate
3rd day of embryogenesis	Injected into embryo	0.20 μg	8/10 Death or malformation rate
4th day of embryogenesis	Injected into embryo	0.15 μg	4/10 Death or malformation rate
4th day of embryogenesis	Injected into embryo	0.20 μg	8/10 Death or malformation rate

TABLE 12 (continued)
Toxicity of α-Solanine in Various Species of Laboratory Animals after Different
Routes of Exposure

Experiment	Dose of α-solanine[a]		Effects
	Administration	**Amount**	
Rabbit	Interperitoneal	20 mg/kg	Overnight death; death in 2.5—24 h; recovery, if survived for at least 24 h
	Intraperitoneal	30 mg/kg	Death in 6.25 h; death in 50 min
	Intravenous	10 mg/kg	Death in 2 min
	Intraperitoneal	30, 40, 50 mg/kg	0/1, 1/1, 1/1 Death rate within 8—24 h
0—8 d of gestation	Intraperitoneal	5 mg/kg/d	Abortion in 1/3 animals; 23% resorptions
Rhesus monkey	Intraperitoneal	20 mg/kg/d	Death 2 h after 2nd injection
	Intraperitoneal	40 mg TGA/kg	Death 48 h after treatment
	Ad libitum for 25 d during 0—42 d after maturing	Diced potatoes containing 260 mg α-solanine/ kg tuber	No maternal toxicity

[a]   Source of α-solanine: 1. Aldrich Chemicals, Milwaukee, WI;
                                  2. K & K Laboratories, Inc. Plainview, NY
                                  3. Sigma Chemical Co., St. Louis, MO

[b]   In a case of potato poisoning (total alkaloid).
[c]   Determined from potatoes consumed (total alkaloid).
[d]   General symptoms of food poisoning.
[e]   Solanidine.
[f]   Isolated from extract of *P. infestans*-infected potatoes.

From Jadhav, S. J., Sharma, R. P., and Salunkhe, D. K., *CRC Crit. Rev. Toxicol.*, 9, 21, 1981. With permission.

and also reported by Gull et al.[149] did not reveal any well-defined symptoms directly attributable to the toxic effects of α-solanine. However, an autopsy of rhesus monkeys that received toxic i.p. doses of α-solanine or TGA showed nasal and periorbital hemorrhage, accumulation of serosanguineous pleural and peritoneal fluids, and mild hepatic and spleen congestion.[148] Histology of the lung, liver, and spleen indicated areas of hemorrhagic congestion. Except for the anticholinesterase activity of leptine I, pharmacological and toxicological properties of leptines and leptinines are unknown.

α-Solanine has been shown to possess local anaesthetic properties. It induces hemolysis, diminishes blood catalase, acts as a mitotic poison, and has been shown to inhibit oxygen uptake of mouse ascites tumor cells.

To counteract α-solanine toxicity, Patil et al.[147] explored the possible therapeutic uses of atropine sulfate (2 mg/kg), pargyline hydrochloride (5 mg/kg), and amphetamine sulfate (5 mg/kg). A prior dose of atropine sulfate lowered the mortality associated with i.p. injection into mice of 40 mg of α-solanine per kilogram from nine of ten to five of ten. Similar applications of pargyline hydrochloride and amphetamine sulfate resulted in eight of nine and ten of ten mortality. The mice injected with amphetamine sulfate were very active and stimulated before the administration of α-solanine. Unlike the effects of other drugs, this period of stimulation persisted for some time after the α-solanine injection. Thus, amphetamine and pargyline had no important effects, while the atropine influence appeared antagonistic to and able to counteract the α-solanine toxicity.

Although α-chaconine is known to occur in potatoes nearly twice as much as α-solanine, information on pharmacological and toxicological action of α-chaconine in test animals was not available before 1975.[150-153] The parasympathetic effect on guinea pig ileum, the positive inotropic action on the isolated electrically stimulated frog ventricle, and the recordings of EEG, ECG, respiration, and blood pressure in the rabbit produced by α-chaconine were similar to those results which were caused by α-solanine.[154] The cardiotonic activity of glycoalkaloids, including α-chaconine on the heart, depended much upon the nature of the aglycone and the number of sugars instead of the kinds of sugars or their stereochemical configuration.[155] The similar LD_{50} values of α-chaconine and α-solanine following parenteral administration to the mouse and rabbit indicated that other compounds may have potential toxic effects in men and animals.[154] However, Sharma et al.[156] found the toxicity of α-chaconine slightly greater than α-solanine; the LD_{50} for the former was nearly two thirds that of α-solanine. The lethal effect was accompanied by severe congestion in kidney and liver of the mouse.[156] Similar, but less pronounced effects of α-solanine were noted in kidney and liver along with occasional leukocytic infiltration of liver.

Norred et al.[153] investigated the metabolic fate and distribution of tritiated α-chaconine after oral or parenteral administration of rats. A low oral toxicity of α-chaconine (5 mg/kg, 3×10^8 dpm/kg), like α-solanine,[140] was the result of poor gastrointestinal absorption and rapid (80% within 48 h) fecal elimination. The level of radioactivity was highest in liver; intermediate in kidney, spleen, and lung; and low in blood, brain, fat, heart, muscle, and the glands. When administered intraperitoneally (3×10^8 dpm/kg, 5 to 25 mg/kg), tritium was eliminated chiefly in urine. At doses higher than 10 mg/kg, fecal loss of tritium became negligible, and the activity increased disproportionately in the gastrointestinal tract. Liver, spleen, kidney, pancreas, fat, lung, and thymus were major tissues that showed a tendency to accumulate α-chaconine or its metabolites. At the acutely toxic doses of α-chaconine (15 and 25 mg/kg), higher levels of tritium in various tissues were associated with impaired fecal and/or urinary excretion. Both α-chaconine and α-solanine provided a higher tritium content in various tissues as shown by intraperitoneal/*per os* ratio of greater than 1; biliary excretion was observed at lower intraperitoneal doses, and metabolism to solanidine plus at least two other metabolites of polarity between solanidine and the glycoside was reported. The toxicity data on α-chaconine is presented in Table 13.

In hamsters, α-chaconine-³H was well absorbed from the gastrointestinal tract and nearly 25% of the label was excreted in 7 d via urine and feces after oral administration.[150] More was excreted in urine (24%) than in feces (1%). Tissue concentration of radioactivity peaked 12 h following administration, with the highest concentrations being in lungs, liver, spleen, skeletal muscle, kidney, and pancreas; heart and brain contained moderate amounts (Figure 4). The concentrations of radioactivity in tissues following intraperitoneal administration were significantly higher than following an oral treatment. Excretion of chloroform-soluble metabolites in the feces was 100 times higher than that of chloroform-insoluble metabolites after oral or intraperitoneal administration. In urine, the activity was predominantly in the chloroform-insoluble form, and relatively little activity was in the chloroform-soluble metabolites (0.29, 0.85, and 2.45% vs. 0.005, 0.14, and 0.19% for 12, 24, and 72 h, respectively). After 7 d, the chloroform-soluble metabolites in urine increased to 20% of the excreted radioactivity, while the amount of chloroform-insoluble metabolites was less than 1%. Subcellular distribution of the labeled compound indicated the highest concentration of radioactivity occurred in the nuclear and microsomal fractions of brain, liver, and heart tissues. A small amount of radioactivity, shown by a minor peak, was also observed in the fraction between the mitochondrial and microsomal fractions on a sucrose gradient. Binding of radioactivity was observed in brain, testes, kidney, lung, liver, and heart. All of the

TABLE 13
Evaluation of α-Chaconine Toxicity

Experiment	Administration	Amount	Effects
Mice	Intraperitoneal	27.5 ± 1.74 mg/kg[a]	50% Death
	Intraperitoneal	19.2 mg/kg	50% Death
Chick embryo			
(0 h incubation)	Injection into yolk sack	<0.5 mg/kg	50% Death
(96 h incubation)	Injection into yolk sack	15.5 ± 3.98 mg/kg	50% Death
(0 h incubation)	Injection into yolk sack	1—1.5 mg/kg	Reduction in hatchability
(96 h incubation)	Injection into yolk sack	12,20,30 mg/kg	Reduction in hatchability
Rabbit	Intraperitoneal	30,40 mg/kg	0/2 Death rate within 8—24 h
		50 mg/kg	1/3 Death rate within 8—24 h
		60 mg/kg	1/1 Death rate within 8—24 h
Rat	Intraperitoneal	10,30,60,90 mg/kg[b]	0/3, 1/3, 2/3, 3/3 Death in 4 h
	Intraperitoneal	84 mg/kg[c]	50% Death
	Intraperitoneal	20 mg/kg/d	3/7 Death rate after 3 d of 8th injection
	Intraperitoneal	40 mg/kg/d	2/5 Death rate after 3 d of 2nd injection
(5—12 d of gestation)	Intraperitoneal	5—20 mg/kg	0—43% Maternal death; 16—97% of fetal resorption
(5 and 6 d of gestation)	Intraperitoneal	40 mg/kg	67% Maternal death; 100% fetal resorption

Note: α-Chaconine was isolated from leaves of *S. chacoense*,[a] potato sprouts,[b] and blossoms[c] by potato starch-silica gel column chromatography[a,c] and preparative TLC[b].

labels in brain appeared to be in a bound form. The results indicated that α-chaconine is slowly absorbed from the gastrointestinal tract after oral administration and persists in various tissues, much of it in bound (nonextractable) form in the microsomal fraction.

The excretion of α-chaconine-^{3}H and its metabolites was also investigated after oral and intraperitoneal administration in hamsters.[152] The separation of the glycoalkaloid and its metabolites in feces and urine by thin-layer chromatography and strip-counting techniques showed that their concentrations increased with time. In urine, over 50% of the eliminated radioactivity during the initial 24 h was due to unaltered α-chaconine and solanidine. The fraction of the total oral dose which was excreted represented only 22% (21% in urine and less than 1% in feces) during the 7-d test period. Contrary to the general belief that potato glycoalkaloid absorption is poor following oral administration, only 4% or less was excreted in feces 72 h later, and this was also supported by the binding of radioactivity in tissues.

The inhibition of purified bovine erythrocyte acetylcholinesterase (AChE) and horse serum cholinesterase (ChE) by α-chaconine was found to be a mixed type, with inhibition constants (K_i) for both enzymes of $8.3 \times 10^{-6} M$ and $4.0 \times 10^{-4} M$, respectively.[151] Subcellular acetylcholinesterase activity in nuclear-free rat brain homogenate showed that the highest activity was equally distributed between the mitochondrial and microsomal fractions; the least activity was in the nuclear fraction. Inhibition of the subcellular enzyme activity *in vitro* by $0.016 M$ α-chaconine were as follows: whole homogenate, 43%; nuclear fraction, 55%; mitochondria, 35%; and microsomes, 33%. *In vivo* administration of α-chaconine (10, 30, and 60 mg/kg) to adult male rats reduced brain AChE activity to 79, 55, and 18%, respectively, of the control activity (Figure 5). Heart and plasma AChE activity

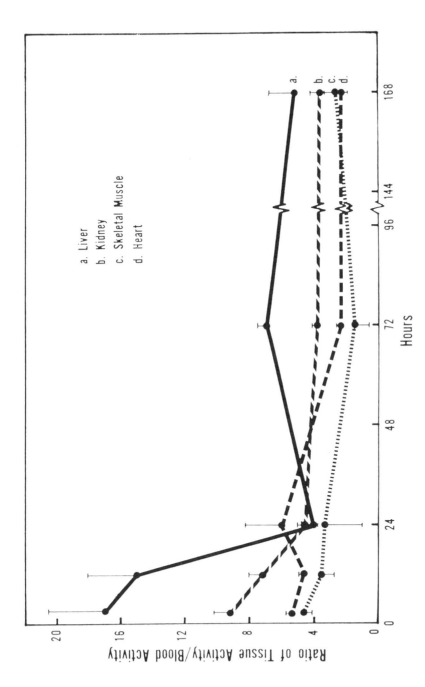

FIGURE 4.   Tissue concentration/blood concentration ratio of radioactivity in the male hamster after p.o. administration of 10 mg/kg of the labeled glycoalkaloid (α-chaconine-³H). (From Alozie, S. O., Sharma, R. P., and Salunkhe, D. K., *J. Food Safety*, 1, 257, 1978. With permission.)

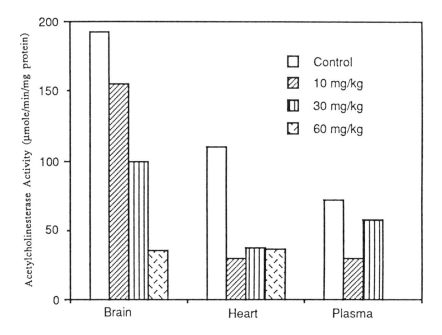

FIGURE 5.   Comparison of acetylcholinesterase activity in brain plasma and heart homogenates of adult male rats injected intraperitoneally with several doses of α-chaconine (0, 10, 30, and 60 mg/kg). (From Alozie, S. O., Sharma, R. P., and Salunkhe, D. K., *J. Food Biochem.*, 2, 259, 1978. With permission.)

ɔf treated animals did not show a dose-related response as observed in the brain. Inhibition of heart AChE was 61%, while that of plasma was 51% for rats administered 10 mg/kg; there was no further inhibition in rats given 30 mg/kg α-chaconine. Isoenzymes of rat brain cholinesterases, separated by acrylamide gel electrophoresis, were inhibited by 30 and 60 mg/kg when α-chaconine was administered intraperitoneally. An electrophoretic separation of plasma from the treated animals showed five anodally migrating zones that hydrolyzed acetylthiocholine and α-naphthylacetate. Inhibition of enzyme activity of isoenzyme bands 1 and 2 was 40 and 77% for animals given 10 mg/kg α-chaconine, while inhibition in rats given 30 mg/kg was 100 and 75% for bands 1 and 2, respectively. Isoenzyme bands 3 and 4 were completely inhibited in the plasma of alkaloid-treated animals. Nonspecific cholinesterase isoenzyme inhibition by α-chaconine was also demonstrated. *In vitro* inhibition of rat plasma, erythrocyte, and brain esterase isoenzymes was estimated by incubating gels with 10^{-4} M α-chaconine after electrophoretic separations. The slower-moving isoenzyme bands were inhibited completely, while the faster-migrating isoenzyme bands in plasma and erythrocytes were least affected by the concentration of α-chaconine (Table 14).

In another series of experiments,[157] preliminary tests with six rats showed no significant differences in norepinephrine, dopamine, and serotonin levels following administration of as much as 20 mg/kg of α-chaconine. Acetylcholine levels in α-chaconine-treated rats may be elevated, if the esterase isoenzymes in the brain are affected by the low levels of toxin reaching the brain. Casual inspection of EEG records did not reveal substantial differences in response to high α-chaconine doses, but power spectrum analysis showed a measurable shift to slightly lower frequencies at medium to high doses of the alkaloids.

Daróczy and Hernadi[158] investigated the effects of glycoalkaloids and α- and β-solamarine on the inducible tryptophan pyrrolase synthesis in the liver of adrenalectomized and normal rats, but did not find any direct glucocorticoid effect. However, β-solamarine showed antitumor activity against Sarcoma 180 in mice.[159] Intraperitoneal injections of 10 mg/kg

TABLE 14
In Vitro Inhibition of Total Esterase Isoenzyme of Various Tissues in the Rat Following Incubation of Electrophoretic Gels with α-Chaconine

Isoenzymes	Inhibition of total esterase activity (%)		
	Plasma	Erythrocytes	Brain
1	0	0	22.96
2	0	41.97	27.10
3	78.49	41.77	15.60
4	81.73	100.00	19.50
5	90.00	100.00	67.31
6	100.00	100.00	100.00

Note: Concentration of α-chaconine, 10^{-4} M.

From Alozie, S. O., Sharma, R. P., and Salunkhe, D. K., *J. Food Biochem.*, 2, 259, 1978. With permission.

α-solanine in Sprague-Dawley rats increased the pentobarbital sleeping time by 275%. This treatment caused a slight decrease in the *in vitro* activities of benzphetamine *N*-demethylase and aniline hydroxylase and the cytochrome P-450 contents of liver microsomes.[160] The administration of α-solanine, however, did not influence the phenobarbital-induced increase in either of the above parameters.

VIII. TOXICOLOGY

A. HUMANS

Accidental consumption of potatoes containing high amounts of glycoalkaloids has caused severe illness and occasionally death. Damon[161] recounted cases of potato poisoning reported by several investigators. Most cases occurred in Europe, mostly among soldiers. Although poisoning was attributed to potatoes, glycoalkaloid contents were not determined. Harris and Cockburn[162] investigated an epidemic among civilians of Glasgow, Scotland. They examined 61 cases, one of which was fatal (a boy 5 years of age). Potatoes contained 410 mg of glycoalkaloids per kilogram. Rothe[163] reported an outbreak in Leipzig, Germany that affected 41 persons. The circumstances clearly implicated potatoes, which contained as much as 430 mg/kg of glycoalkaloids. Since then, several cases of glycoalkaloid poisoning from potatoes have been reported.[161,164,165] In Canada, there were several complaints from potato growers and consumers concerning unpalatable and bitter-tasting potatoes linked to high glycoalkaloid contents.[8] Losses of livestock and poultry caused by ingestion of potato vines, sprouted potatoes, cull potatoes, and potato peels containing glycoalkaloids have been reported.[165] These products had been exposed to light when discarded by the processing plant or left in the field by farmers. Wilson[166] summarized the literature on outbreaks of poisoning due to glycoalkaloids in potatoes, and also reported an occurrence in Great Britain. He reported that such outbreaks are extremely uncommon and apparently occur only when potatoes containing unusually high levels of glycoalkaloids were ingested. The symptoms are similar to an acute gastrointestinal upset—with abdominal pain, vomiting, and diarrhea.

McMillan and Thompson[167] recently reported an outbreak of potato poisoning in school children in Southeast London involving 78 boys 7 to 19 h after lunch at the beginning of the autumn term. Of these, 17 required hospitalization, including 3 who were seriously ill. Sickness began with headache and vomiting along with diarrhea and abdominal pain. Neurological symptoms reported in most patients included apathy, restlessness, drowsiness, mental confusion, rambling, incoherence, stupor, hallucinations, dizziness, trembling, and visual disturbances. Although most of the boys recovered in a few days, some required hospitalization for nearly 1 week. The problem was traced to the potatoes served to affected children which contained 33.3 mg/100 g of glycoalkaloids. Potatoes left over from the meal had significant anticholinesterase activity, approximately equivalent to their glycoalkaloid contents. The episode served as a reminder that poisoning caused by potatoes high in alkaloid content was possible in spite of precautions.

Renwick[4] hypothesized that two severe birth defects, anencephaly and spina bifida cystica (ASB), could result from ingestion of imperfect potatoes. He based the hypothesis on the similarity between the incidence of ASB and potato blight caused by *P. infestans,* and speculated that the incidence of ASB may be prevented by preventing pregnant women from ingesting the teratogen. It was speculated that the alleged teratogen could be fungal-synthesized coumarins, tuber-synthesized pyhotoalexins, or steroidal alkaloids.

B. TERATOLOGY IN ANIMALS

Poswillo et al.[168] established a causal relationship between blighted potatoes and anencephaly. In spite of gross abnormalities in 4 of 11 fetuses in 6 pregnant monkeys and severe teratogenic defects in those conceived after a prolonged period of consumption of the "blighted" potato concentrate, no cases of ASB (anencephaly or spina bifida) were reported. Moreover, the experimental procedure did not show that *P. infestans* was responsible for an increase in the glycoalkaloid content of potatoes. In a later report, Poswillo et al.[169] did not find the evidence of ASB in marmosets fed a mixture of potato concentrate rejected by industry and tubers infected with *E. carotovora* during early pregnancy.

The bacterial pathogen is responsible for blackleg in potatoes and induces the synthesis of antifungal terpenoids, rishitin, and phytuberin. Although no gross abnormalities were found, behavioral deficits were observed in some offspring of marmosets that had consumed the industry-reject concentrate.[169] The feeding trials in Wistar rats and marmosets with blighted potatoes under similar conditions revealed no abnormalities except cranial osseous defects in 4 of 11 fetal marmosets. These authors, however, found a compound common to both blighted Kerr's Pink and poor-quality potato concentrate. The significance and identity of this compound has not yet been established. Swinyard and Chaube[148] reported that gavaging of pregnant rats with blighted potatoes on days 5 to 11 or daily i.p. injections of α-solanine or glycoalkaloids (5 to 10 mg/kg extracted from B-5141-6 potatoes given on days 5 to 12 or 7 to 17 of gestation) resulted in minor skeletal and renal tract abnormalities, but failed to produce neural-tube defects. Wu and Salunkhe[170] tested nine races of *P. infestans* and ten isolates of *Alternaria solani,* grown on potato dextrose broth and sterilized potato slices, on chicken eggs to assay the toxicity of culture filtrates, chloroform, and water extracts. None of the extracts and culture filtrates from the races of *P. infestans* was toxic to chick embryos. Chloroform extracts of 4 out of 10 isolates of *A. solani* grown on potato dextrose broth were somewhat toxic to chick embryos. Production of glycoalkaloids by *P. infestans in vitro* has recently been reported.[117] These compounds, however, are linked with malformations in the chick embryo.[171,172]

Chaube et al.[173] showed no teratogenicity to rats fed blighted potatoes. Also, Ruddick et al.[174] reported no teratogenic or embryolethal effects when rats were fed freeze-dried blighted potatoes. When α-solanine, α-chaconine, and other potato constituents were incubated orally during organogenesis there was no evidence of teratogenicity. Keeler et al.[175] tested freeze-dried preparations from Russet Burbank potatoes vigorously infected with *P. infestans* or *A. solani* on rats, mice, hamsters, and rabbits; no ASB was produced in the test animals, thus eliminating the induction of phytoalexin-teratogen by *A. solani*-infected tubers. Similarly, Sharma et al.[176] reported no teratogenic effects from diets containing blight-infested potatoes or their extracts in hamsters. Lesions of neurulation and anencephaly were noticed in rabbits and swine, respectively.

Massive doses of the early and late blight-infected tubers or an extract of the former containing terpenoid phytoalexins did not cause congenital deformities in rats.[177] There was an absence of the neural-tube defects in rhesus monkeys and marmosets.[178] Occurrence of some congenital deformities when one strain of hamsters was fed potato sprouts is not convincing enough to link these problems to humans.[179] Renwick et al.,[180] however, reported similar teratogenesis in hamsters by feeding potato sprouts or their glycoalkaloid fractions. The doses of α-chaconine and α-solanine that produced consistent teratologic effects were obviously toxic to the dams. A preliminary report indicating the toxic effects of α-chaconine on developing nervous system of mice was presented by Pierro et al.[181] Serum concentrations of solanidine and glycoalkaloids in pregnant women with a fetus, subsequently shown to be affected by a neural tube defect (NTD), failed to demonstrate the theory that potato alkaloids are responsible for NTD.[182]

Keeler et al.[183] eliminated the possibility of solasodine, a spirosolane, as a teratogen since neither solasodine nor its glycosides have been isolated from potatoes. Solasodine, tomatidine, or their glycoalkaloids, as well as potato glycoalkaloids, do not satisfy the structural requirement of fused furan and piperidine rings similar to veratrum teratogen cyclopamine.[184] The position of nitrogen or saturation at Δ^5 as in α-tomatine caused no teratogenicity.[185] Although the spirosolane alkaloid, solasodine, was nonteratogenic in rats,[184] it has produced such deformities as spina bifida, anencephaly, and cranial bleb in hamsters when considerably higher doses than those given to rats were administered.[185]

There is generally a relationship between a biological activity and the structural requirement of an organic compound. Keeler et al.[185] proposed that a basic nitrogen of an alkaloid should be accessible to the steroid α face for activity. This hypothesis was tested by using solanidan epimers in relation to steroid teratogenicity.[186] There was no direct evidence that a compound of the solanidine structure might cause birth defects. The structure-activity of steroid teratogens is justified in two varatrum alkaloids, jervine and cyclopamine, whose fused furanopiperidine rings, E and F, are at right angles to the plane of the steroidal ring system because of spiro attachment of C-17.[187-188] Kuć[189] reviewed the literature and found no relationship between certain potato constituents, including glycoalkaloids and terpenoids, to spina bifida cystica.

C. MICROORGANISMS AND INSECTS

In an attempt to find a quick, simple, and inexpensive method to determine α-solanine toxicity, Patil et al.[147] noted a depression in the growth of fungus *Trichoderma viride.* The daily rate of growth was the same on the control medium, potato dextrose agar (PDA), without α-solanine and on the medium containing 10 mg α-solanine per 100 ml. Increasing the concentration of α-solanine in the medium progressively inhibited growth. The lethal concentration (LC_{50}) was 102.2 mg/100 ml of the medium. Many workers have attributed the disease-resistance character of potato cultivars to the presence of glycoalkaloids in the tuber and plant. Allen and Kuć[100] reported that α-solanine and α-chaconine were highly fungitoxic to *Helminthosporium carbonum,* and that these compounds accounted for at least 90% of the fungitoxicity of potato peel extracts to this fungus at pH 5.6. Inhibition of the growth of *A. solani* on potato dextrose agar by potato alkaloids (α-solanine, α-chaconine, and solanidine) has been reported by Sinden et al.[190] Solanidine was most inhibitory, followed by α-chaconine and α-solanine. At a concentration of 250 ppm, α-solanine inhibited growth by 33% after 96 h of incubation at 24°C.

McKee[99] assessed α-solanine toxicity to *Fusarium caeruleum* (Lib.) Sacc. on the basis of its concentration, period of exposure, H$^+$ ion concentration, spore density, Na/Ca ions ratio, and the previous history of the spores. LC_{50} values ranged from 20 to 36 mg α-solanine per liter with a mean value of 33 mg/l at pH 7.6. Toxicity of α-solanine to macrospores of *F. avenaceum* and *F. culmorum* and sporangia of *P. infestans* was similar to that of *F. caeruleum* spores. Zoospores of *P. infestans* were very sensitive to α-solanine, with an LC$_{50}$ less than 5 mg/l. The evaluation of toxicity of α-chaconine and solanidine to *F. caeruleum* indicated LC$_{50}$, 11, and 100 mg/l, respectively. The spores of *Streptomyces scabies* were not affected by several hours of exposure to 2000 mg α-solanine per liter. Also, bacteria from young cultures of *Bacillus subtilis*, *Micrococcus lueteus*, *Erwina* spp., and *Pseudomonas* spp. were insensitive to this concentration of α-solanine.[191] The fungitoxic nature of α-solanine on *Cladosporium fulvum* has also been reported.

Although Paquin and Lachance[192] reported that α-solanine inhibited the growth of *Corynebacterium sepedonicum* (Spieck and Kott) Skapt. and Burkh in culture, it does not appear to protect potatoes from ring rot.[193] Leptines are known to repel the Colorado potato beetle and its larvae. The presence of commersonine or dehydrocommersonine rather than α-solanine and α-chaconine as the major glycoalkaloid in foliage of *Solanum chacoense* seems to be as important as the level of glycoalkaloids in limiting damage and numbers of these insects.[194] α-Solanine, solanidine, and demissidine reduced the rate of initial imbibition of the potato leaf hopper, *Epoasca fabae* (Harris), but not its survival time.[195] However, leptine I markedly reduced both. In general, potato glycoalkaloids act as resistance factors against two insect pests of potato, the Colorado potato beetle and the leaf hopper.[84,194]

Bentley et al.[196] reported on the feeding deterrence for spruce budworm to five solanum alkaloids. All five alkaloids, α-solanine, α-chaconine, solanidine, tomatine, and tomatidine, showed significant feeding deterrence at 10^{-3} *M*. At 10^{-4} *M*, insect feeding was reduced by several of these alkaloids; the alkaloids did not indicate wide differences in effectiveness. In a similar study using Colorado potato beetles,[197] the solanum glycoalkaloids elicited bursting activity in galeal and tarsal chemosensilla of adult beetles. The effect had an average latency of 6 to 12 s. A 20-s alkaloid treatment rendered galeal sensilla unresponsive to gamma-aminobutyric acid, normally an effective stimulant. A similar effect was observed on labeller chemosensilla of the blowfly. The possibility of a specific chemoreceptor in beetles to the glycoalkaloids was excluded.

Mitchell[198] suggested that taste sensitivity in insects can be modified by a variety of alkaloids. Solanine significantly inhibited the responses of the amino acid-sensitive cells to gamma-aminobutyric acid or L-alanine and of the sugar-sensitive cells to sucrose. He suggested that the mechanisms do not appear to be a result of a specific evolutionary modification of the sensilla, but are likely to be general.

α-Tomatine, a glycoalkaloid present in solanaceous plants, decreased larval survival, lowered pupal weights, extended the pupation period, and prolonged the period of adult emergence in *Ceratitis capitata* (Mediterranean fruit fly).[199] The LC$_{50}$ of this alkaloid to the larvae was estimated to be 52.7 ppm. Biochemical and toxicological aspects of the glycoalkaloids depicting the importance of the problem have been briefly reviewed by many workers.[189,200-204]

IX. CONTROL OF GLYCOALKALOIDS

A. PHYSICAL TREATMENTS

The use of various packaging,[205-210] colored film bags,[211,212] colored lights, and colored film filters[71,208,212] to protect tubers from light have been investigated. Certain pre- and postharvest procedures involving light, temperature, horticultural practices, handling, and storage conditions have been recommended. Curing of Kennebec, Norchip, and York tubers at 50°C prior to storage retarded the glycoalkaloid buildup 20 to 40%, compared to untreated (dark) or illuminated tubers. Controlled-atmosphere storage of tubers was explored.[213,214] Application of subatmospheric pressure (hypobaric) is a new approach.[215] Several workers who studied ionizing radiation in relation to greening and glycoalkaloid formation in illuminated potato tubers found that it was ineffective on the glycoalkaloids.[216-218] Mechanically damaged tubers and cut cubes of three cultivars of potato subjected to γ-irradiation produced significantly less glycoalkaloids than the control samples under identical storage conditions.[217] The inhibitory effect of γ-irradiation at 25 to 100 krd was 11 to 79% in Russet Burbank potatoes subjected to mechanical damage. At a dose of 200 krd of γ-irradiation, wound-induced glycoalkaloid formation was inhibited 81 to 92%. Similar results were reported on White Rose and Red Pontiac cultivars (Table 15). γ-Irradiation did not influence the existing glycoalkaloids. Vacuum packaging with polyethylene bags at 38 and 63.5 cmHg inhibited light-induced glycoalkaloid formation in three cultivars of potato tubers (Figure 6).[219] The inhibition of photo-induced greening and the glycoalkaloid synthesis in tubers submerged in water was explored by Wu and Salunkhe.[220] The use of sodium hypochlorites or circulation of freshwater was recommended to maintain the physical condition of the tubers. Anoxia water treatment up to 20 d prior to illumination was also found to be effective.[221] It is also possible to control glycoalkaloid synthesis in the aged potato slices by water soaking under controlled conditions.[88] Some of the above methods are of limited practicality because of the present marketing trends and storage practices.

B. CHEMICAL TREATMENTS

Various chemicals are known to prevent greening and/or glycoalkaloid synthesis in potato tubers.[28,29,72,222] Wilson and Frank[223] applied systemic pesticides as a banded side dressing 1 week after planting of three cultivars (Cobbler, Katahdin, and B5141-6); carbofuran-treated potato tubers contained significantly less glycoalkaloids than control tubers at harvest. However, in a greenhouse study, carbofuran applied during tuberization to Norland, Kennebec, and Abnaki cultivars significantly increased glycoalkaloid content of tubers. A sprout inhibitor, isopropyl-N-(3-chlorophenyl)-carbamate (CIPC), was tested by Wu and Salunkhe[224] on three cultivars of potato tubers as a possible inhibitor of wound-induced glycoalkaloid formation. Dipping of mechanically damaged Russet Burbank tubers and cubes into 100 and 1000 ppm CIPC emulsified water solution completely inhibited the wound-induced glycoalkaloids. A dip concentration at 1 and 10 ppm CIPC was 9 to 70% effective. Fumigation with 100 and 1000 mg CIPC per cubic meter inhibited wound-induced glycoalkaloid synthesis by 6 to 86% (Table 16). CIPC treatment, however, did not affect existing glycoalkaloids and photoinduced glycoalkaloid formation of normal whole potato tubers. The increase in glycoalkaloids during storage could be prevented by treating harvested tubers with the sprout inhibitor[225,226] or by spraying maleic hydrazide on the plants before harvest.[227]

The waxing of potatoes was once a fairly widespread marketing practice because it attracted consumer attention, but it appears to be rapidly declining in importance in the U.S. However, the discovery by Wu and Salunkhe[228] that hot paraffin wax effectively controls the photoinduced formation of chlorophyll and glycoalkaloids in potato tubers has created a new interest in this area. These authors treated Russet Burbank potatoes with paraffin wax at 60, 80, 100, 120, 140, and 160°C for 0.5 s and exposed them to fluorescent light (2152 lx) for 10 d at 16°C and 60% relative humidity. There was no inhibition at 100 and 120°C and almost complete inhibition at 140 and 160°C. Heating the tubers at 160°C in air for 3 to 5 min and subsequent exposure to light did not prevent chlorophyll and glycoalkaloid formation. Combined waxing and heating retarded the chlorophyll and glycoalkaloid formation. They concluded that this treatment is especially useful because paraffin wax does not create problems like most physicochemical treatments and can be easily removed by peeling the tubers before processing or cooking.

In a later study, potato tubers were dipped in corn oil at 22, 60, 100, and 160°C for 0.5 s and excess oil was removed with tissue paper.[229] Oiling at 22°C reduced chlorophyll levels by 93 to 100% and glycoalkaloid formation by 92 to 97%. At elevated temperatures

TABLE 15
Effects of Gamma-Irradiation on Wound-Induced Glycoalkaloid
Formation of Mechanically Damaged Tubers and Cut Cubes of
Three Cultivars

Type of material	γ-Irradiation (krad)	Days storage at 4°C			
		0	5	10	20
'Russet Burbank'					
Damaged whole tubers	Control	23.0	107.9**	161.5**	190.7**
	25		94.3**	145.0**	171.5**
	50		86.3**	132.4**	160.1**
	100		60.7**	74.5**	83.6**
	200		29.8**	39.6**	46.4**
Cut cubes	Control	1.4	19.6**	25.4**	28.1**
	25		17.2**	22.8**	25.2**
	50		11.5**	18.4**	22.5**
	100		5.3**	6.7**	7.2**
	200		2.9*	3.8*	4.1**
'White Rose'					
Damaged whole tubers	Control	26.1	89.6**	140.2**	157.4**
	25		77.5**	130.6**	142.2**
	50		63.5**	104.5**	128.7**
	100		51.9**	72.5**	83.5**
	200		33.8**	40.4**	45.1**
Cut cubes	Control	1.7	14.0**	19.3**	23.1**
	25		12.0**	17.1**	20.5**
	50		10.2**	16.2**	18.3**
	100		5.3**	6.0**	6.9**
	200		3.3*	4.2**	4.9**
'Red Pontiac'					
Damaged whole tubers	Control	27.5	110.3**	142.2**	188.4**
	25		94.5**	129.3**	168.0**
	50		85.4**	111.1**	160.3**
	100		59.6**	67.7**	89.4**
	200		36.9**	40.5**	43.2**
Cut cubes	Control	1.6	19.0**	24.6**	27.3**
	25		18.3**	20.0**	22.6**
	50		15.4**	17.9**	20.5**
	100		5.4**	7.1**	8.2**
	200		3.6**	3.9**	4.0**

Note: Single asterisk (*) indicates, significantly different from zero time sample at 0.05 level; double asterisk (**), significantly different from zero time sample at 0.01 level; total glycoalkaloid, mg/100 g dry weight.

From Wu, M. T. and Salunkhe, D. K., *Lebensm. Wiss. Technol.*, 10, 141, 1977. With permission.

such as 60, 100, and 160°C, the treatment completely inhibited both chlorophyll and gly-coalkaloid synthesis. Wu and Salunkhe[230] further reported that treatments with corn oil, peanut oil, olive oil, or mineral oil at 22°C were equally effective, but the tubers appeared oily. Moreover, the oils and fat might become rancid due to oxidation. To decrease the amount of oil used, corn oil was diluted with acetone. Treatment with one half, one fourth,

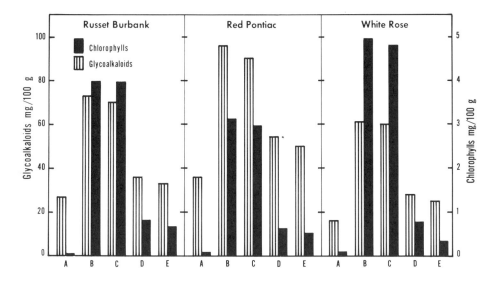

FIGURE 6. Effects of vacuum packing on chlorophyll and glycoalkaloid formation in peels of 'Russet Burbank', 'Red Pontiac', and 'White Rose' potatoes. After packaging, tubers were exposed to light (2152 lx) for 12 d at 16°C. (A) original, zero-time sample; (B) control; (C) packaging at atmospheric pressure with conventional brown polyethylene bags; (D) packaging at 38 cm of Hg with clear polyethylene bags; (E) packaging at 63.5 cm/Hg with clear polyethylene bags. (From Wu, M. T. and Salunkhe, D. K., *Can. Inst. Food Sci. Technol. J.*, 8, 185, 1975. With permission.)

and one eighth oil significantly and effectively inhibited the formation of chlorophyll and glycoalkaloids. Treatments with 1/16, 1/32, and 1/64 oil reduced chlorophyll levels by 95, 72, and 22%, and reduced glycoalkaloid content by 82, 49, and 28%, respectively. Levels in tubers treated with acetone alone or with 1/128 oil were similar to those in untreated control tubers. A concentration of one eighth corn oil and seven eighths acetone was the minimum effective dilution. They estimated that 1 g of oil in 50 g of acetone could treat 8000 g of potato tubers. The acetone treatment had no apparent harmful effect on the tubers. For practical application, acetone can be recovered by passing the treated tubers through a warm-air chamber and condensing the acetone in the warm air by a cooling coil system.

Wu and Salunkhe[231] recommended an alternative method, spraying potato tubers with a commercial spray lecithin such as Pam®, Mazola®, No Stick®, Cooking Ease®, and Griddle Mate®. These treatments significantly inhibited light-induced glycoalkaloid formation (89 to 98%) in potato tubers (Table 17). Additionally, coating tubers with lecithin (Centromix C® or Centrolex F®) or hydroxylated lecithin (Centrolene A®) inhibited glycoalkaloids when used at 5 to 20% concentration in petroleum ether.[232]

Jadhav and Salunkhe[222] reported that effectiveness of treatment with mineral oil increased until concentrations reached 10% (w/v in petroleum ether) and then remained almost constant as concentrations increased to 100%. Tubers treated with mineral oil up to 10% concentration appeared more attractive than untreated tubers or those treated with higher concentrations. At a concentration of 10% mineral oil, tubers did not turn green after exposure to light for 4 weeks, while glycoalkaloid synthesis at the end of the first, second, third, and fourth weeks was 93, 67, 49, and 65%. In general, oil treatment is a simple, effective, and inexpensive method of controlling greening and glycoalkaloid formation in potato tubers.

Sinden[104] immersed tubers in a 2 or 3% detergent solution at 21.1°C for 18 to 40 min and then rinsed them under tap water. On exposure to light (1291 lx) for 10 d, greening and light-induced glycoalkaloid synthesis were inhibited. Chlorophyll synthesis in Russet Burbank tubers was inhibited by 92% for the first 2 d after tubers were treated with 3% detergent solution for 30 min. The inhibitory effect decreased with extended exposure, but

TABLE 16
Effects of Dipping and Fumigation Treatment of CIPC on Wound-Induced Glycoalkaloid Formation of Mechanically Damaged Tubers and Cut Cubes of Three Cultivars

Type of material	CIPC treatment	Days storage at 4°C			
		0	5	10	20
	'Russet Burbank'				
	Dipping (ppm)				
Damaged whole tubers	Control	23.2	118.8**	163.0**	194.2**
	1		80.3**	118.4**	130.5**
	10		53.3**	71.6**	84.2**
	100		24.2	24.0	24.8
	1000		23.6	24.4	23.9
Cut cubes	Control	1.5	20.0**	26.3**	28.4**
	1		18.3**	20.1**	21.5**
	10		7.4**	8.9**	9.8**
	100		1.6	1.5	1.6
	1000		1.4	1.5	1.5
	Fumigation (mg CIPC/m³)				
Damaged whole tubers	10		104.9**	170.3**	190.0**
	100		84.5**	128.4**	133.6**
	1000		38.6**	47.8**	53.9**
Cut cubes	10		21.4**	25.8**	29.3**
	100		18.8**	20.6**	24.3**
	1000		4.0*	4.9**	5.4**
	'White Rose'				
	Dipping (ppm)				
Damaged whole tubers	Control	26.4	90.3**	142.6**	160.5**
	1		76.5**	104.8**	120.6**
	10		50.6**	63.8**	77.5**
	100		27.0	27.4	27.9
	1000		26.2	26.0	26.8
Cut cubes	Control	1.8	14.5**	20.8**	23.6**
	1		10.5**	16.3**	20.6**
	10		5.4**	5.9**	7.7**
	100		1.9	1.8	2.0
	1000		1.8	1.7	1.8
	Fumigation (mg CIPC/m³)				
Damaged whole tubers	10		91.4**	140.3**	158.8**
	100		80.3**	119.4**	130.8**
	1000		35.3**	39.6**	44.4**
Cut cubes	10		16.6**	21.8**	24.0**
	100		10.3**	14.5**	18.4**
	1000		5.3*	5.5*	6.0**

TABLE 16 (continued)
Effects of Dipping and Fumigation Treatment of CIPC on Wound-Induced Glycoalkaloid Formation of Mechanically Damaged Tubers and Cut Cubes of Three Cultivars

Type of material	CIPC treatment	Days storage at 4°C			
		0	5	10	20
'Red Pontiac'					
	Dipping (ppm)				
Damaged whole tubers	Control	27.3	115.4**	146.3**	186.2**
	1		77.6**	89.7**	114.9**
	10		54.9**	60.4**	65.3**
	100		28.3	29.2	29.0
	1000		27.0	27.6	28.0
Cut cubes	Control	1.6	19.3**	24.4**	27.8**
	1		13.3**	18.4**	22.5**
	10		5.3**	6.0**	6.4**
	100		1.8	1.8	1.9
	1000		1.7	1.9	1.8
	Fumigation (mg CIPC/m³)				
Damaged whole tubers	10		121.6**	140.9**	189.0*
	100		100.4**	113.3**	134.4**
	1000		39.4**	46.6**	48.7**
Cut cubes	10		20.6**	25.0**	28.3**
	100		13.1**	20.4**	22.6**
	1000		4.0*	4.8**	4.9**

Note: Single asterisk (*) indicates, significantly different from zero time sample at 0.05 level; double asterisk (**), significantly different from zero time sample at 0.01 level; total glycoalkaloids, mg/100 g dry weight.

From Wu, M. T. and Salunkhe, D. K., *J. Food Sci.,* 42, 622, 1977. With permission.

TABLE 17
Effects of Spray Lecithin Treatment on Light-Induced Chlorophyll and Glycoalkaloid Formations of Russet Burbank Potato Tubers after Exposure to 2152 lux Light Intensity for 14 d at 10°C and 60% RH

Treatment	Total chlorophylls (mg/100 g peels)	Total glycoalkaloids (mg/100 g peels)
Original sample (zero time)	0.14**	28.45**
Control (nontreated)	3.83	69.49
PAM®	0.41**	33.05**
Mazola No-Stick®	0.34**	31.93**
Cooking Ease®	0.26**	29.46**
Griddle Mate®	0.23**	30.38**

Note: ** Significantly different from control (nontreated) tubers at 0.01 level.

From Wu, M. T. and Salunkhe, D. K., *J. Food Sci.,* 42, 1413, 1977. With permission.

inhibition still exceeded 50% after 10 d. With Kennebec and Sebago cultivars, both of which green rapidly under light, treatment with 2% detergent (Joy®) for 20 min inhibited chlorophyll levels by 47 and 33%, respectively. The glycoalkaloid content of the fresh peels of Kennebec was 61% less than in untreated tubers. With the cultivar Green Mountain, which greens less rapidly, inhibition was only 14%.

Poapst et al.[233] reported that photoinduced greening of potatoes can be prevented by simply washing or spray rinsing the tubers with an aqueous solution of an edible surfactant with the brand name of Tween 85®. Applying the surfactant 0.04% of tuber weight prevented chlorophyll for 15 d or more, while a spray containing 4 to 5% Tween 85® formed an effective film on the surface of the most susceptible cultivars. Since the antigreening mechanism was related to the buildup of high carbon dioxide in the peel, the efficiency of several polysorbates (Tween® [s]) were evaluated.[234] Poapst et al.[235,236] suggested treatment with food-grade ethoxylated mono- and diglyceride as well as Tween® surfactants in combination with citric acid.

REFERENCES

1. **Zitnak, A.,** Occurrence of bitter potatoes in Ontario, Information Bulletin No. 257/81, University of Guelph, Ontario, Canada, 1, 1970.
2. **Bömer, A. and Mattis, H.,** High solanine content of potatoes, Z. *Unters. Nahr. Genussm. Gebrauchs-gegenstaende,* 47, 97, 1924.
3. **Bushway, A. A., Bushway, A. W., Belyea, P. R., and Bushway, R. J.,** The proximate composition and glycoalkaloid content of 3 potato meals, *Am. Potato J.,* 57, 167, 1980.
4. **Renwick, J. H.,** Hypothesis: anencephaly and spina bifida are preventable by avoidance of a specific but unidentified substance in certain potato tubers, *Br. J. Prev. Soc. Med.,* 26, 67, 1972.
5. **Kühn, R. and Löw, I.,** in *Origins of Resistance to Toxic Agents,* Sevag, M. G., Reid, R. D., and Reynolds, O. E., Eds., Academic Press, New York, 1955, 122.
6. **Kühn, R. and Low, I.,** New alkaloid glycosides in the leaves of *Solanum chacoense, Angew. Chem.,* 69, 236, 1957.
7. **Kühn, R. and Low, I.,** New (steroidal) alkaloids occurring in the leaves of *Solanum chacoense, Tagungsber. Dtsch. Akad. Landwirtschaftswiss. Berlin,* 27, 7, 1961.
8. **Zitnak, A.,** The occurrence and distribution of free alkaloid solanidine in Netted Gem potatoes, *Can. J. Biochem. Physiol.,* 39, 1257, 1961.
9. **Sinden, S. L. and Sanford, L. L.,** Origin and inheritance of solamarine glycoalkaloids in commercial potato cultivars, *Am. Potato J.,* 58, 305, 1981.
10. **Shih, M. and Kuć, J.,** Alpha-solamarine and beta-solamarine in Kennebec solanum-tuberosum leaves and aged tuber slices, *Phytochemistry,* 13, 997, 1974.
11. **Osman, S. F., Herb, S. F., Fitzpatrick, T. J., and Schmiediche, P.,** Glycoalkaloid composition of wild and cultivated tuber-bearing solanum species of potential value in potato breeding programs, *J. Agric. Food Chem.,* 26, 1246, 1978.
12. **Osman, S. F., Herb, S. F., Fitzpatrick, T. J., and Sinden, S. L.,** Commersonine, a new glycoalkaloid from 2 solanum species, *Phytochemistry,* 15, 1065, 1976.
13. **Jadhav, S. J., Sharma, R. P., and Salunkhe, D. K.,** Naturally occurring toxic alkaloids in foods, *CRC Crit. Rev. Toxicol.,* 9, 21, 1981.
14. **Wolf, M. J. and Duggar, B. M.,** Solanine in the potato and the effects of some factors on its synthesis and distribution, *Am. J. Bot.,* (Suppl. 20s), 27, 1940.
15. **Morgenstern, F.,** Uber den Solaningehalt der speiseund Futterkartoffeln und uber den Eenfluss der Bodenkultur auf die bildung von solanin in der Kartoffelpflanze, *Landwirtsch. Vers. Stn.,* 65, 301, 1907.
16. **Lampitt, L. H., Bushill, J. H., Rooke, H. S., and Jackson, E. M.,** Solanine, glycoside of the potato. II. Distribution in the potato plant, *J. Soc. Chem. Ind.,* 62, 48, 1943.
17. **Paseshnichenko, V. A.,** Content of solanine and chaconine in the potato during the vegetation period, *Biokhimyiya,* 22, 981, 1957.
18. **Wood, F. A. and Young, D. A.,** TGA in potatoes, *Agric. Canada,* 1533, 1974.
19. **Guseva, A. R., Borikhina, M. G., and Paseshnichenko, V. A.,** Utilization of acetate for the biosynthesis of chaconine and solanine in potato sprouts, *Biokhimiya,* 25, 282, 1960.
20. **Ahmed, S. S. and Muller, K.,** Effect of wound damages on the glycoalkaloid content in potato tubers and chips, *Lebensm. Wiss. Technol.,* 11, 144, 1978.
21. **Cadle, L. S., Stelzig, D. A., Harper, K. L., and Young, R. J.,** Thin-layer chromatographic system for identification and quantitation of potato-tuber glycoalkaloids, *J. Agric. Food Chem.,* 26, 1453, 1978.
22. **Fitzpatrick, T. J., Herb, S. F., Osman, S. F., and McDermott, J. A.,** Potato glycoalkaloids. Increase and variations of ratios in aged slices over prolonged storage, *Am. Potato J.,* 54, 539, 1977.
23. **Wolf, M. J. and Duggar, B. M.,** Estimation and physiological role of solanine in the potato, *J. Agric. Res.,* 73, 1, 1946.

24. **Hilton, R. J.,** Factors in relation to tuber quality in potatoes. II. Preliminary trial on bitterness in Netted Gem potatoes, *Sci. Agric.,* 31, 61, 1951.
25. **Reeve, R. M., Hautala, E., and Weaver, M. L.,** Anatomy and compositional variation within potatoes. II. Phenolics, enzymes, and other minor components, *Am. Potato J.,* 46, 374, 1969.
26. **Zitnak, A. and Johnston, G. R.,** Glycoalkaloid content of B5141-6 potatoes, *Am. Potato J.,* 47, 256, 1970.
27. **Roddick, J. G.,** Subcellular localization of steroidal glycoalkaloids in vegetative organs of *Lycopersicon esculentum* and *Solanum tuberosum, Phytochemistry,* 16, 805, 1977.
28. **Jeppsen, R. B., Salunkhe, D. K., and Jadhav, S. J.,** Formation and anatomical distribution of chlorophyll and solanine in potato tubers and their control by chemical and physical treatments, *33rd Annu. Inst. Food Technol. Meeting,* Miami Beach, FL, 1973, 153.
29. **Jeppsen, R. B., Wu, M. T., Salunkhe, D. K., and Jadhav, S. J.,** Some observations on occurrence of chlorophyll and solanine in potato-tubers and their control by N^6-benzyladenine, ethephon and filtered lights, *J. Food Sci.,* 39, 1059, 1974.
30. **Han, S. R.,** Ultrastructural Location of Solanidine in Potato Tubers, M.S. thesis, Utah State University, Logan, 1978.
31. **Han, S. R., Campbell, W. F., and Salunkhe, D. K.,** Ultrastructural localization of solanidine in potato tubers, *J. Food Biochem.,* 13, 377, 1989.
32. **Osman, S. F., Zacharius, R. M., Kalan, E. B., Fitzpatrick, T. J., and Krulick, S.,** Stress metabolites of the potato and other solanaceous plants, *J. Food Prot.,* 42, 502, 1979.
33. **Gregory, P., Sinden, S. L., Osman, S. F., Tingey, W. M., and Chessin, D. A.,** Glycoalkaloids of wild, tuber-bearing solanum species, *J. Agric. Food Chem.,* 29, 1212, 1981.
34. **Gregory, P.,** Glycoalkaloid composition of potatoes: diversity and biological implications, *Am. Potato J.,* 61, 115, 1984.
35. **Bushway, R. J., Bureau, J. L., and McGann, D. F.,** Alpha-chaconine and alpha-solanine content of potato peels and potato products, *J. Food Sci.,* 48, 84, 1983.
36. **Maga, J. A.,** Total and individual glycoalkaloid composition of stored potato slices, *J. Food Proc. Pres.,* 5, 23, 1981.
37. **Bushway, A. L., Bureau, J. L., Bergeron, D., Stickney, M. R., and Bushway, R. J.,** The nutrient and glycoalkaloid content of a new potato meal, *Am. Potato J.,* 62, 301, 1985.
38. **Parfitt, D. E., Peloquin, S. J., and Jorgensen, N. A.,** The nutritional value of pressed potato vine silage, *Am. Potato J.,* 59, 415, 1982.
39. **Coxon, D. T.,** The glycoalkaloid content of potato berries, *J. Sci. Food Agric.,* 32, 412, 1981.
40. **Clayton, R. B.,** Biosynthesis of sterols, steroids, and terpenoids. II. Phytosterols, terpenes and the physiologically active steroids, *Q. Rev. Chem. Soc.,* 19, 201, 1965.
41. **Heftmann, E.,** Biochemistry of plant steroids, *Annu. Rev. Plant Physiol.,* 14, 225, 1963.
42. **Heftmann, E.,** Biochemistry of steroidal saponins and glycoalkaloids, *Lloydia,* 30, 209, 1967.
43. **Heftmann, E.,** Biosynthesis of plant steroids, *Lloydia,* 31, 293, 1968.
44. **Heftmann, E. and Mosettig, E.,** *Biochemistry of Steroids,* Van Nostrand-Reinhold, Princeton, NJ, 1960.
45. **Willuhn, G.,** Biogenesis of pharmaceutically important plant steroids, *Pharm. Ztg.,* 110, 96, 1965.
46. **Guseva, A. R. and Paseshnichenko, V. A.,** A study of the biogenesis of potato glycoalkaloids by the method of labeled atoms, *Biokhimiya,* 23, 412, 1958.
47. **Guseva, A. R., Paseshnichenko, V. A., and Borikhina, M. G.,** Synthesis of radioactive mevalonic acid and its use in the study of the biosynthesis of steroid glycoalkaloids from *Solanum, Biokhimiya,* 26, 723, 1961.
48. **Wu, M. T. and Salunkhe, D. K.,** Difference between light-induced and wound-induced biosynthesis of alpha-solanine and alpha-chaconine in potato-tubers, *Biol. Plant.,* 20, 149, 1978.
49. **Johnson, D. F., Heftmann, E., and Houghland, G. V. G.,** Biosynthesis of sterols in *Solanum tuberosum, Arch. Biochem. Biophys.,* 104, 102, 1964.
50. **Hartmann, M. A. and Benveniste, P.,** Effect of aging on sterol-metabolism in potato-tuber slices, *Phytochemistry,* 13, 2667, 1974.
51. **Johnson, D. F., Bennett, R. D., and Heftmann, E.,** Cholesterol in higher plants, *Science,* 40, 198, 1963.
52. **Schreiber, K. and Osske, G.,** Sterols and triterpenoids. II. The isolation of cycloartenol from leaves of *Solanum tuberosum, Kulturpflanze,* 10, 372, 1962.
53. **Ripperger, H., Mortiz, W., and Schreiber, K.,** Zur Biosynthese von slanum-alkaloiden aus cycloartenol oder lanosterin, *Phytochemistry,* 10, 2699, 1971.
54. **Tschesche, R. and Hulpke, H.,** Zur Biosynthese von Steroid-Derivaten in Pflanzenreich. VIII. Biogenese von Solanidin aus chole sterin, *Z. Naturforsch. Teil B,* 22, 791, 1967.
55. **Canonica, L., Ronchetti, F., Russ, G., and Sportoletti, G.,** Fate of the 16 hydrogen atom of cholesterol in the biosynthesis of tomatidine and solanidine, *J. Chem. Soc. Chem. Commun.,* 8, 286, 1977.
56. **Guseva, A. R. and Paseshnichenko, V. A.,** A study of solasodin biosynthesis by the method of oxidative breakdown, *Biokhimiya,* 27, 853, 1962.

57. **Heftmann, E. and Weaver, M. L.**, 26-Hydroxycholesterol and cholest-4-en-3-one, first metabolites of cholesterol in potato plants, *Phytochemistry,* 13, 1801, 1974.
58. **Tschesche, R., Goossens, B., and Topfer, A.**, Biosynthesis of steroid derivatives in plants. 22. Introduction of nitrogen and common occurrence of 25 (R) steroid alkaloids and 25 (S) steroid alkaloids in solanaceae, *Phytochemistry,* 15, 1387, 1976.
59. **Kaneko, K., Tanaka, M. W., and Mitsuhashi, H.**, Origin of nitrogen in the biosynthesis of solanidine by *Veratrum grandiflorum, Phytochemistry,* 15, 1391, 1976.
60. **Jadhav, S. J., Salunkhe, D. K., Wyse, R. E., and Dalvi, R. R.**, Solanum alkaloids: biosynthesis and inhibition by chemicals, *J. Food Sci.,* 38, 453, 1973.
61. **Jadhav, S. J. and Salunkhe, D. K.**, Enzymatic glucosylation of solanidine, *J. Food Sci.,* 38, 1099, 1973.
62. **Osman, S. F. and Zacharius, R. M.**, Biosynthesis of potato glycoalkaloids, *Am. Potato J.,* 56, 475, 1979.
63. **Guseva, A. R. and Paseshnichenko, V. A.**, Enzymic splitting of potato glycoalkaloids by the method of labeled atoms, *Biokhimiya,* 22, 843, 1957.
64. **Swain, A. P., Fitzpatrick, T. J., Talley, E. A., Herb, S. F., and Osman, S. F.**, Enzymatic-hydrolysis of alpha-chaconine and alpha-solanine, *Phytochemistry,* 17, 800, 1978.
65. **Zacharius, R. M., Kalan, E. B., Osman, S. F., and Herb, S. F.**, Solanidine in potato tuber tissue disrupted by *Erwinia atroseptica* and by *Phytophthora infestans, Physiol. Plant Pathol.,* 6, 301, 1975.
66. **Holland, H. L. and Taylor, G. J.**, Transformations of steroids and the steroidal alkaloid, solanine by *Phytophthora infestans, Phytochemistry,* 18, 437, 1979.
67. **King, R. R. and McQueen, R. E.**, Transformations of potato glycoalkaloids by rumen microorganisms, *J. Agric. Food Chem.,* 29, 1101, 1981.
68. **Zitnak, A.**, The significance of glycoalkaloids in the potato plant, *Proc. Can. Soc. Hortic.,* 3, 81, 1964.
69. **Hardenburg, R. E.**, Greening of potatoes during marketing—a review, *Am. Potato J.,* 41, 215, 1964.
70. **Jadhav, S. J. and Salunkhe, D. K.**, Formation and control of chlorophyll and glycoalkaloids in tubers of *Solanum tuberosum* L. and evaluation on glycoalkaloids toxicity, *Adv. Food Res.,* 21, 307, 1975.
71. **Lilijemark, A. and Widoff, E.**, Greening and solanine development of white potato in fluorescent light, *Am. Potato J.,* 37, 379, 1960.
72. **Patil, B. C., Salunkhe, D. K., and Singh, B.**, Metabolism of solanine and chlorophyll in potato tubers as affected by light and specific chemicals, *J. Food Sci.,* 36, 474, 1971.
73. **Zitnak, A.**, Factors Influencing the Initiation and Rate of Solanine Synthesis in Tubers of *Solanum tuberosum* L., Ph.D. thesis, University of Alberta, Edmonton, 1955.
74. **Zitnak, A.**, The Influence of Certain Treatments upon Solanine Synthesis in Potatoes, M.S. thesis, University of Alberta, Edmonton, 1953.
75. **Zitnak, A.**, Photoinduction of glycoalkaloids in cured potatoes, *Am. Potato J.,* 58, 415, 1981.
76. **Conner, H. W.**, The effect of light on solanine synthesis in the potato tubers, *Plant Physiol.,* 12, 79, 1937.
77. **Wintgen, M.**, Uber die quantitative Bestimmung des cholesterins und der cholesterinester in einegen normalen and pathologischen nieren, *Z. Unters. Nahr. Genussm. Gebrauchsgegenstaende,* 12, 113, 1906.
78. **Bomer, A. and Mattis, H.**, High Solanine content of potatoes, *Z. Unters. Nahr. Genussm. Gebrauchsgegenstaende,* 45, 288, 1923.
79. **Gull, D. D. and Isenberg, F. M.**, Chlorophyll and solanine content and distribution in four varieties of potato tubers, *Proc. Am. Soc. Hortic. Sci.,* 75, 545, 1960.
80. **Sanford, L. L. and Sinden, S. L.**, Inheritance of potato glycoalkaloids, *Am. Potato J.,* 49, 209, 1972.
81. **Sinden, S. L. and Webb, R. E.**, Effect of variety and location on the glycoalkaloid content of potatoes, *Am. Potato J.,* 49, 334, 1972.
82. **Lepper, W.**, Solaningehalt von 58 Kartoffelsorten, *Z. Lebensm. Unters. Forsch.,* 89, 264, 1949.
83. **Sinden, S. L. and Deahl, K. L.**, Effect of glycoalkaloids and phenolics on potato flavor, *J. Food Sci.,* 41, 520, 1976.
84. **Tingey, W. M., Mackenzie, J. D., and Gregory, P.**, Total foliar glycoalkaloids and resistance of wild potato species to *Empoasca fabae* (Harris), *Am. Potato J.,* 55, 577, 1978.
85. **Hutchinson, A. and Hilton, R. J.**, Influence of certain cultural practices on the solanine content and tuber yields in Netted Gem potatoes, *Can. J. Agric. Sci.,* 35, 485, 1955.
86. **Braun, H.**, Uber solanin-anreicherungen in kartoffelknollen, *Beitr. Agrarwiss.,* 2, 61, 1948.
87. **Arutyunyan, L. A.**, The solanine content of potatoes, *Vopr. Pitan.,* 9, 30, 1940.
88. **Mondy, N. I. and Chandra, S.**, Reduction of glycoalkaloid synthesis in potato slices by water soaking, *HortScience,* 14, 173, 1979.
89. **Pallmann, H. and Schindler, K.**, Beeinflusst die Duengung den Solaningehalt der kartoffeln?, *Schweiz. Landwirtsch. Monatsch.,* 20, 21, 1942.
90. **Lepper, W.**, Beitrag zur Solaninfrage. Belichtung und Solaningehalt der kartoffein, *Z. Lebensm. Unters. Forsch.,* 86, 247, 1943.
91. **Mondy, N. I., Naylor, L. M., and Phillips, J. C.**, Total glycoalkaloid and mineral content of potatoes grown in soils amended with sewage sludge, *J. Agric. Food Chem.,* 32, 1256, 1984.

92. **Evans, D. and Mondy, N. I.,** Effect of magnesium fertilization on glycoalkaloid formation in potato tubers, *J. Agric. Food Chem.,* 32, 465, 1984.

93. **Mondy, N. I. and Ponnampalam, R.,** Effect of magnesium fertilizers on total glycoalkaloids and nitrate nitrogen in Katahdin potatoes, *J. Food Sci.,* 50, 535, 1985.

94. **Ponnampalam, R. and Mondy, N. I.,** Effect of foliar application of indoleacetic acid on the total glycoalkaloids and nitrate nitrogen content of potatoes, *J. Agric. Food Chem.,* 34, 686, 1986.

95. **Wilson, A. M., McGann, D. F., and Bushway, R. J.,** Effect of growth-location and length of storage on glycoalkaloid content of roadside-stand potatoes as stored by consumers, *J. Food Prot.,* 46, 119, 1983.

96. **Bushway, R. J., Bushway, A. A., and Wilson, A. M.,** α-Chaconine and α-solanine content of MH-30 treated Russet Burbank, Katahdin and Kennebec tubers stored for nine months at three different temperatures, *Am. Potato J.,* 58, 498, 1981.

97. **Haard, N. F. and Salunkhe, D. K.,** Stress metabolites in fruits and vegetables—introduction, *J. Food Prot.,* 42, 495, 1979.

98. **Wood, G. E.,** Stress metabolites of white potatoes, *Adv. Chem. Ser.,* 149, 369, 1976.

99. **McKee, R. K.,** Host-parasite relationships in the dry-rot disease of potatoes, *Ann. Appl. Biol.,* 43, 147, 1955.

100. **Allen, E. H. and Kuć, J.,** Alpha-solanine and alpha-chaconine as fungitoxic compounds in extracts of Irish potato tubers, *Phytopathology,* 58, 776, 1968.

101. **Locci, R. and Kuć, J.,** Steroid alkaloids as compounds produced by potato tubers under stress, *Phytopathology,* 57, 1272, 1967.

102. **Ishizaka, N. and Tomiyama, K.,** Effect of wounding or infection by *Phytophthora infestans* on the content of terpenoids in potato tubers, *Plant Cell Physiol.,* 13, 1053, 1972.

103. **Ozeretskovaskaya, O. L., Davylova, M. A., Vasyukov, N. I., and Metlitski, L. V.,** Glycoalkaloids in sound and injured potato-tuber, *Dokl. Akad. Nauk SSSR,* 196, 1470, 1971.

104. **Sinden, S. L.,** Effect of light and mechanical injury on the glycoalkaloid content of greening-resistant potato tubers, *Am. Potato J.,* 49, 368, 1972.

105. **Kuć, J.,** Phenolics, in *Normal and Diseased Fruits and Vegetables,* Runeckles, V. C., Ed., Plant Phenolics Group of North America, Norwood, MA, 1964, 63.

106. **Salunkhe, D. K., Wu, M. T., and Jadhav, S. J.,** Effects of light and temperature on formation of solanine in potato slices, *J. Food Sci.,* 37, 969, 1972.

107. **Jadhav, S. J., Wu, M. T., and Salunkhe, D. K.,** Glycoalkaloids of hollow heart and blackheart potato-tubers, *HortScience,* 15, 147, 1980.

108. **Wu, M. T. and Salunkhe, D. K.,** Changes in glycoalkaloid content following mechanical injuries to potato tubers, *J. Am. Soc. Hortic. Sci.,* 101, 329, 1976.

109. **Maga, J. A.,** Total and individual glycoalkaloid composition of stored potato slices, *J. Food Proc. Pres.,* 5, 23, 1981.

110. **Mondy, N. I., Leja, M., and Gosselin, B.,** Changes in total phenolic, total glycoalkaloid and ascorbic acid content of potatoes as a result of bruising, *J. Food Sci.,* 52, 631, 1987.

111. **Olssen, K.,** The influence of genotype on the effect of impact damage on the accumulation of glycoalkaloids in potato tubers, *Potato Res.,* 29, 1, 1986.

112. **Fitzpatrick, T. J., McDermott, J. A., and Osman, S. F.,** Evaluation of injured commercial potato samples for total glycoalkaloid content, *J. Food Sci.,* 43, 1617, 1978.

113. **Kuć, J.,** Steroid glycoalkaloids and related compounds as potato quality factors, *Am. Potato J.,* 61, 123, 1984.

114. **Sinden, S. L., Sanford, L. L., and Webb, R. E.,** Genetic and environmental control of potato glycoalkaloids, *Am. Potato J.,* 61, 141, 1984.

115. **Siegfried, R.,** Determination of solanine-chaconine ratio in potato sprouts and tubers, *Z. Lebensm. Unters. Forsch.,* 162, 253, 1976.

116. **Shih, M. and Kuć, J.,** Incorporation of C-14 from acetate and mevalonate into rishitin and steroid glycoalkaloids by potato-tuber slices inoculated with *Phytophthora infestans, Phytopathology,* 63, 826, 1973.

117. **Maas, M. R., Post, F. J., and Salunkhe, D. K.,** Production of steroid glycoalkaloids by *Phytophthora infestans in vitro, J. Food Safety,* 1, 107, 1977.

118. **Cheema, A. S. and Haard, N. F.,** Induction of rishitin and lubimin synthesis in potato-tuber slices by nonspecific elicitors—role of gene depression, *J. Food Prot.,* 42, 512, 1979.

119. **Bushway, R. J. and Ponnampalam, R.,** α-Chaconine and α-solanine content of potato products and their stability during several modes of cooking, *J. Agric. Food Chem.,* 29, 814, 1981.

120. **Ponnampalam, R. and Mondy, N. I.,** Effect of cooking on the total glycoalkaloid content of potatoes, *J. Agric. Food Chem.,* 31, 493, 1983.

121. **Maga, J. A.,** Glycoalkaloid stability during the extrusion of potato flakes, *J. Food Sci. Proc. Pres.,* 4, 291, 1980.

122. **Sizer, C. E., Maga, J. A., and Craven, C. J.,** Total glycoalkaloids in potatoes and potato chips, *J. Agric. Food Chem.,* 28, 578, 1980.
123. **Schwardt, E.,** Changes in glycoalkaloid content in industrial treatment processes for potatoes, *Veroeff. Arbeitgem Kartoffelforsch.,* 4, 48, 1982.
124. **Bushway, A. A., Bushway, A. W., Belyea, P. R., and Bushway, R. J.,** The proximate composition and glycoalkaloid content of three potato meals, *Am. Potato J.,* 57, 167, 1980.
125. **Jones, P. G. and Fenwick, G. R.,** The glycoalkaloid content of some edible solanaceous fruits and potato products, *J. Sci. Food Agric.,* 32, 419, 1981.
126. **Bushway, R. J., Bureau, J. L., and McGann, D. F.,** Alpha-chaconine and alpha-solanine content of potato peels and potato products, *J. Food Sci.,* 48, 784, 1983.
127. **Mondy, N. I. and Ponnampalam, R.,** Determination of total glycoalkaloids (TGA) in dehydrated potatoes, *J. Food Sci.,* 48, 612, 1983.
128. **Bushway, R. J., Burea, J. L., and Stickney, M. R.,** A new efficient method for extracting glycoalkaloids from dehydrated potatoes, *J. Agric. Food Chem.,* 33, 45, 1985.
129. **Davis, A. M. C. and Blincow, P. J.,** Glycoalkaloid content of potatoes and potato products, *J. Sci. Food Agric.,* 35, 553, 1984.
130. **Bushway, A. A., Bureau, J. L., Bergeron, D., Stickney, M. R., and Bushway, R. J.,** The nutrient and glycoalkaloid content of a new potato meal, *Am. Potato J.,* 62, 301, 1985.
131. **Schober, B.,** Physiological changes in the potato tuber after wounding and infection with *Phytophthora infestans, Potato Res.,* 14, 39, 1971.
132. **Sato, N., Katazawa, K., and Tomiyama, K.,** Role of rishitin in localizing invading hyphae of phyto-phthora-infestan sites at cut surfaces of potato-tubers, *Physiol. Plant Pathol.,* 1, 289, 1971.
133. **Deahl, K. L., Young, R. J., and Sinden, S. L.,** Study of relationship of late blight resistance to glycoalkaloid content in 15 potato clones, *Am. Potato J.,* 50, 248, 1973.
134. **Frank, J. A., Wilson, J. M., and Webb, R. E.,** Relationship between glycoalkaloids and disease resistance in potatoes, *Phytopathology,* 65, 1045, 1975.
135. **Roddick, J. G.,** Distribution of steroidal glycoalkaloids in reciprocal grafts of *solanum tuberosum* L. and *Lycopesicon esculentum* Mill, *Experientia,* 38, 460, 1982.
136. **Meyer, G.,** Ueber Vergiftungen durch Kartoffeln. I. Ueber den Gehalt der Kartoffeln an solanin und uber die Bildung desselben wahrend der keimung, *Arch. Exp. Pathol. Pharmakol.,* 36, 361, 1895.
137. **Willimott, S. G.,** An investigation of solanine poisoning, *Analyst (London),* 58, 431, 1933.
138. **Golubeva, S. N.,** Experiences in the diagnosis of food poisoning and treatment with solanine, *Vestn. Otorinolaringol.,* 28, 23, 1966.
139. **Satoh, T.,** Glycemic effects of solanine in rats, *Jpn. J. Pharmacol.,* 17, 652, 1967.
140. **Nishie, K., Gumbmann, H. R., and Keyl, A. C.,** Pharmacology of solanine, *Toxicol. Appl. Pharmacol.,* 19, 81, 1971.
141. **Orgell, W. H., Vaidya, K. A., and Dahm, P. A.,** Inhibition of human plasma cholinesterase *in vitro* by extracts of solanaceous plants, *Science,* 128, 1136, 1958.
142. **Pokrovskii, A. A.,** The effect of the alkaloids of the sprouting potato on cholinesterase, *Biokhimiya,* 21, 683, 1956.
143. **Harris, H. and Whittaker, M.,** Differential response of human serum cholinesterase types to an inhibitor in potato, *Nature (London),* 183, 1808, 1959.
144. **Harris, H. and Whittaker, M.,** Differential inhibition of the serum cholinesterase phenotypes by solanine and solanidine, *Ann. Hum. Genet.,* 26, 73, 1962.
145. **Orgell, W. H.,** Inhibition of human plasma cholinesterase *in vitro* by alkaloids, glycosides, and other substances, *Lloydia,* 26, 36, 1963.
146. **Zitnak, A.,** Cholinesterase inhibitors, *Science,* 131, 66, 1960.
147. **Patil, B. C., Sharma, R. P., Salunkhe, D. K., and Salunkhe, K.,** Evaluation of solanine toxicity, *Food Cosmet. Toxicol.,* 10, 395, 1972.
148. **Swinyard, C. A. and Chaube, S.,** Are potatoes teratogenic for experimental animals?, *Teratology,* 8, 349, 1973.
149. **Gull, D. D., Isenberg, F. M., and Bryan, H. H.,** Alkaloid toxicology of *Solanum tuberosum, HortScience,* 5, 316, 1970.
150. **Alozie, S. O., Sharma, R. P., and Salunkhe, D. K.,** Physiological disposition, subcellular distribution and tissue binding of α-chaconine (^{3}H), *J. Food Safety,* 1, 257, 1978.
151. **Alozie, S. O., Sharma, R. P., and Salunkhe, D. K.,** Inhibition of rat cholinesterase isoenzymes *in vitro* and *in vivo* by the potato alkaloid α-chaconine, *J. Food Biochem.,* 2, 259, 1978.
152. **Alozie, S. O., Sharma, R. P., and Salunkhe, D. K.,** Extraction of alpha-chaconine-H-3, a steroidal glycoalkaloid from *Solanum tuberosum* L. and its metabolites in hamsters, *Pharmacol. Res. Commun.,* 11, 483, 1979.
153. **Norred, W. P., Nishie, K., and Osman, S. F.,** Excretion, distribution and metabolic fate of (^{3}H-α-chaconine), *Res. Commun. Chem. Pathol. Pharmacol.,* 13, 161, 1976.

154. **Nishie, K., Norred, W. P., and Swain, A. P.,** Pharmacology and toxicology of chaconine and tomatine, *Res. Commun. Chem. Pathol. Pharmacol.,* 12, 657, 1975.

155. **Nishie, K., Fitzpatrick, T. J., Swain, A. P., and Keyl, A. C.,** Positive inotropic action of solanacae glycoalkaloids, *Res. Commun. Chem. Pathol. Pharmacol.,* 15, 601, 1976.

156. **Sharma, R. P., Willhite, C. C., Shupe, J. L., and Salunkhe, D. K.,** Acute toxicity and histopathological effects of certain glycoalkaloids and extracts of *Alternaria solani* on *Phytophthora infestans* in mice, *Toxicol. Lett.,* 3, 349, 1978.

157. **Aldous, C. N., Sharma, R. P., and Salunkhe, D. K.,** Effects of α-chaconine on brain biogenic amines, electroencephalogram, cardiac rates and respiratory response in rats, *J. Food Safety,* 2, 20, 1980.

158. **Daróczy, A. and Hernadi, F.,** Glucocorticoid-like effects of solanum alkaloids, *Acta Biochim. Biophys. Acad. Sci. Hung.,* 6, 327, 1971.

159. **Kupchan, S. M., Barboutis, S. J., Knox, J. R., and Lau Cam, C. A.,** β-Solamarine: tumor inhibitor isolated from *Solanum dulcamara, Science,* 150, 1827, 1965.

160. **Dalvi, R. R. and Peeples, A.,** *In vivo* effect of toxic alkaloids on drug metabolism, *J. Pharm. Pharmacol.,* 33, 51, 1981.

161. **Damon, S. R.,** *Food Infections and Food Intoxications,* Williams & Wilkins, Baltimore, 1928.

162. **Harris, F. W. and Cockburn, T.,** Alleged poisoning by potatoes, *Analyst (London),* 43, 133, 1918.

163. **Rothe, J. C.,** Illness following the eating of potatoes containing solanine, *Z. Hug.,* 88, 1, 1918.

164. **Griebel, C.,** Injurious potatoes rich in solanine, *Z. Unters. Nahr. Genussm. Gebrauchsgegenstaende,* 45, 175, 1923.

165. **Hansen, A. A.,** Two fatal cases of potato poisoning, *Science,* 61, 340, 1925.

166. **Wilson, G. S.,** A small outbreak of solanine poisoning, *Mon. Bull. Minist. Health Public Health Lab. Serv.,* 18, 207, 1959.

167. **McMillan, M. and Thompson, J. G.,** Outbreak of suspected solanine poisoning in schoolboys—examination of criteria of solanine poisoning, *Q. J. Med.,* 48, 227, 1979.

168. **Poswillo, D. E., Sopher, D., and Mitchell, S.,** Experimental induction of fetal malformation with ''blighted'' potato: a preliminary report, *Nature (London),* 239, 462, 1972.

169. **Poswillo, D. E., Sopher, D., Mitchell, S. J., Coxon, D. T., Curtis, R. F., and Price, K. R.,** Investigations into teratogenic potential of imperfect potatoes, *Teratology,* 8, 339, 1973.

170. **Wu, M. T. and Salunkhe, D. K.,** Toxicity of *Phytophthora infestans* and *Alternaria solani* to chick-embryos, *Experientia,* 34, 214, 1978.

171. **Jelinek, R., Kyzlink, V., and Blattny, C.,** Evaluation of embryotoxic effects of blighted potatoes on chicken embryos, *Teratology,* 14, 335, 1976.

172. **Mun, A. M., Barden, E. S., Wilson, J. M., and Hogan, J. M.,** Teratogenic effects in early chick-embryos of solanine and glycoalkaloids from potatoes infected with late-blight, *Phytophthora infestans, Teratology,* 11, 73, 1975.

173. **Chaube, S., Swinyard, C. A., and Daines, R. H.,** Failure to induce malformation in fetal rats by feeding blighted potatoes to their mothers, *Lancet,* 1, 329, 1973.

174. **Ruddick, J. A., Harwig, J., and Scott, P. M.,** Nonteratogenicity in rats of blighted potatoes and compounds contained in them, *Teratology,* 9, 165, 1974.

175. **Keeler, R. F., Douglas, D. R., and Stallknecht, G. F.,** Testing of blighted, aged, and control Russet Burbank potato-tuber preparation for ability to produce spina-bifida and anencephaly in rats, rabbits, hamsters, and mice, *Am. Potato J.,* 52, 125, 1975.

176. **Sharma, R. P., Willhite, C. C., Wu, M. T., and Salunkhe, D. K.,** Teratogenic potential of blighted potato concentrate in rabbits, hamsters, and miniature swine, *Teratology,* 18, 55, 1978.

177. **Keeler, R. F., Douglas, D. R., and Stallknecht, G. F.,** Failure of blighted Russet Burbank potatoes to produce congenital deformities in rats, *Proc. Soc. Exp. Biol. Med.,* 146, 284, 1974.

178. **Allen, J. R., Marlar, R. J., Chesney, C. F., Helgeson, J. P., Kelman, A., Weckel, K. G., Traisman, E., and White, J. W., Jr.,** Teratogenicity studies on late blighted potatoes in nonhuman primates (*Macaca mulatta* and *Saguinus labiatusi*), *Teratology,* 15, 17, 1977.

179. **Keeler, R. F., Young, S., Brown, D., Stallknecht, G. F., and Douglas, D.,** Congenital deformities produced in hamsters by potato sprouts, *Teratology,* 17, 327, 1978.

180. **Renwick, J. H., Claringbold, W. D. B., Earthy, M. E., Few, J. D., and McLean, A. C. S.,** Neural-tube defects produced in Syrian hamsters by potato glycoalkaloids, *Teratology,* 30, 371, 1984.

181. **Pierro, L. J., Haines, J. S., and Osman, S. F.,** Teratogenicity and toxicity of purified α-chaconine and α-solanine, *Teratology,* 15, 31, 1977.

182. **Harvey, M. H., Morris, B. A., McMillan, M., and Marks, V.,** Potato steroidal alkaloids and neural tube defects: serum concentrations fail to demonstrate a casual relation, *Hum. Toxicol.,* 5, 249, 1986.

183. **Keeler, R. F., Brown, D., Douglas, D. R., Stallknecht, G. F., and Young, S.,** Teratogenicity of the solanum alkaloid solasodine and of Kennebec potato sprouts in Hamsters, *Bull. Environ. Contamin. Toxicol.,* 15, 552, 1976.

184. **Keeler, R. F.,** Comparison of teratogenicity in rats of certain potato-type alkaloids and veratrum teratogen cyclopamine, *Lancet,* 1, 1187, 1973.
185. **Keeler, R. F., Young, S., and Brown, D.,** Spina-bifida, anencephaly and cranial bleb produced in hamsters by solanum alkaloid solasodine, *Res. Commun. Chem. Pathol. Pharmacol.,* 13, 723, 1976.
186. **Brown, D. and Keeler, R. F.,** Structure-activity relation to steroid teratogens. 3. Solanidan epimers, *J. Agric. Food Chem.,* 26, 566, 1978.
187. **Brown, D. and Keeler, R. F.,** Structure-activity relation of steroid teratogens. 1. Jervine ring-system, *J. Agric. Food Chem.,* 26, 561, 1978.
188. **Keeler, R. F.,** Cyclopamine and related alkaloid teratogens—their occurrence, structural relationship, and biological effects, *Lipids,* 13, 708, 1978.
189. **Kuć, J.,** Teratogenic constituents of potatoes, in *Recent Advances in Phytochemistry,* Runeckles, V. C., Ed., Plenum Press, New York, 1975, 139.
190. **Sinden, S. L., Goth, R. W., and O'Brien, M. J.,** Effect of potato alkaloids on growth of *Alternaria solani* and their possible role as resistance factors in potatoes, *Phytopathology,* 63, 303, 1973.
191. **McKee, R. K.,** Factors affecting the toxicity of solanine and related alkaloids to *Fusarium caeruleum, J. Gen. Microbiol.,* 20, 686, 1959.
192. **Paquin, R. and Lachance, R. A.,** Effects des glycoalcaloides de le pomme de terre sur la croissance de corynebacteriumsepedonicum, *Can. J. Microbiol.,* 10, 115, 1964.
193. **Paquin, R.,** Study on the role of glycoalkaloids in the resistance of potato to bacterial ring rot, *Am. Potato J.,* 43, 349, 1966.
194. **Sinden, S. L., Sanford, L. L., and Osman, S. F.,** Glycoalkaloids and resistance to the Colorado potato beetle, in *Solanum chacoense* bitt., *Am. Potato J.,* 56, 479, 1979.
195. **Dahlmann, D. L. and Hibbs, E. T.,** Responses of *Empoasca fabae* (Cicadellidae-Homoptera) to tomatine solanine leptine. 1. Tomatidine solanidine and deissidine, *Ann. Entomol. Soc. Am.,* 60, 732, 1967.
196. **Bentley, M. D., Leonard, D. E., and Bushway, R. J.,** Solanum alkaloids as larval feeding deterrents for spruce budworm, Choistoneura fumiferana (Lepidoptera: Tortricidae), *Ann. Entomol. Soc. Am.,* 77, 401, 1984.
197. **Mitchell, B. K. and Harrison, G. D.,** Effects of Solanum glycoalkaloids on chemosensilla of the Colorado potato beetle. A mechanism of feeding difference, *J. Chem. Ecol.,* 11, 73, 1983.
198. **Mitchell, B. K.,** Modification of taste sensitivity in insects by alkaloids derived from plants, *Pestic. Sci.,* 16, 535, 1985.
199. **Chan, H. T. and Tam, S. Y. T.,** Toxicity of α-tomatine to larvae of the Mediterranean fruit fly (Diptera: Tephrididae), *J. Econ. Entomol.,* 78, 305, 1985.
200. **Osman, S. F.,** Glycoalkaloids of the solanaceae, in *Recent Advances in Phytochemistry,* Vol. 14 (The Resource Potential in Phytochemistry), Swain, T. and Kleiman, R., Eds., Plenum Press, New York, 1979, 75.
201. **Tingey, W. M.,** Glycoalkaloids as pest resistance factors, *Am. Potato J.,* 61, 157, 1984.
202. **Osman, S. F.,** Glycoalkaloids in potatoes, *Food Chem.,* 11, 235, 1983.
203. **Morris, S. C. and Lee, T. H.,** The toxicity and teratogenicity of solanceae glycoalkaloids, particularly those of the potato (*Solanum tuberosum*): a review, *Food Technol. Aust.,* 36, 118, 1984.
204. **Kuć, J.,** Steroid glycoalkaloids and related compounds as potato quality factors, *Am. Potato J.,* 61, 123, 1984.
205. **Salunkhe, D. K. and Wu, M. T.,** Control of post-harvest glycoalkaloid formation in potato-tubers, *J. Food Proc.,* 42, 519, 1979.
206. **Hardenburg, R. E.,** Comparison of polyethylene with various other 10-pound consumer bags for Sebago, Katahdin, and Green Mountain potatoes, *Am. Potato J.,* 31, 29, 1954.
207. **Isenberg, F. M. and Gull, D. D.,** Potato greening under artificial light, *N.Y. Agric. Exp. Stn. Ithaca Bull.,* 1033, 1, 1959.
208. **Larsen, E. C.,** Investigations on cause and prevention of greening in potato tubers, *Idaho Agric. Exp. Stn. Res. Bull.,* 16, 1, 1949.
209. **Lutz, J. M., Findlen, H., and Ramsey, G. B.,** Quality of Red River Valley potatoes in various types of consumer packages, *Am. Potato J.,* 28, 589, 1951.
210. **Newman, L.,** Trends in merchandising quality potatoes, *8th Proc. Potato Ind. Conf.,* Nova Scotia, Canada, 1966, 1.
211. **Gull, D. D. and Isenberg, F. M.,** Lightburn and off-flavor development in potato tubers exposed to fluorescent lights, *Proc. Am. Soc. Hortic. Sci.,* 71, 446, 1958.
212. **Yamaguchi, M., Hughes, D. L., and Howard, F. D.,** Effect of color and intensity of fluorescent lights and application of chemicals and waxes on chlorophyll development of White Rose potatoes, *Am. Potato J.,* 37, 229, 1960.
213. **Forsyth, F. R. and Eaves, C. A.,** Greening of potatoes. CA cure, *Food Technol.,* 22, 48, 1968.
214. **Patil, B. C., Singh, B., and Salunkhe, D. K.,** Formation of chlorophyll and solanine in Irish potato (*Solanum tuberosum* L.) tubers and their control by gamma radiation and CO_2 enriched packaging, *Lebensm. Wiss. Technol.,* 4, 123, 1971.

215. **Jadhav, S. J., Patil, B. C., and Salunkhe, D. K.,** Controls potato greening. Storage at reduced pressure inhibits light-induced greening, *Food Eng.,* 45, 111, 1973.

216. **Schwimmer, S. and Weston, W. J.,** Chlorophyll formation in potato tubers as influenced by gamma irradiation and by chemicals, *Am. Potato J.,* 35, 534, 1958.

217. **Wu, M. T. and Salunkhe, D. K.,** Effect of gamma irradiation on wound induced glycoalkaloid formation in potato tubers, *Lebensm. Wiss. Technol.,* 10, 141, 1977.

218. **Ziegler, R., Schanderl, S. H., and Markakis, P.,** Gamma irradiation and enriched CO_2 atmosphere storage effects on the light-induced greening of potatoes, *J. Food Sci.,* 33, 533, 1968.

219. **Wu, M. T. and Salunkhe, D. K.,** Effects of vacuum packaging on light-induced greening and glycoalkaloid formation of potato-tubers, *Can. Inst. Food Sci. Technol. J.,* 8, 185, 1975.

220. **Wu, M. T. and Salunkhe, D. K.,** Influence of temporary anoxia by submerging in water on light-induced greening and glycoalkaloid formation of potato tubers, *J. Food Biochem.,* 1, 275, 1977.

221. **Wu, M. T. and Salunkhe, D. K.,** After effect of submersion in water on greening and glycoalkaloid formation of potato-tubers, *J. Food Sci.,* 43, 1330, 1978.

222. **Jadhav, S. J. and Salunkhe, D. K.,** Effects of certain chemicals on photoinduction of chlorophyll and glycoalkaloid synthesis and on sprouting of potato-tubers, *Can. Inst. Food Sci. Technol. J.,* 7, 178, 1974.

223. **Wilson, J. M. and Frank, J. S.,** Effect of systemic pesticides on total glycoalkaloid content of potato-tubers at harvest, *Am. Potato J.,* 52, 179, 1975.

224. **Wu, M. T. and Salunkhe, D. K.,** Inhibition of wound induced glycoalkaloid formation in potato-tubers (*Solanum tuberosum* L.) by isopropyl-n-(3-chlorophenyl)-carbamate, *J. Food Sci.,* 42, 622, 1977.

225. **Ahmed, S. S. and Müller, K.,** Effect of storage time, light and temperature on the solanine and α-chaconine content in potatoes with and without sprout suppressant treatment, *Potato Res.,* 24, 93, 1981.

226. **Mondy, N. I. and Ponnampalam, R.,** Effect of sprout inhibitor isopropyl-N-(3-chlorophenyl) carbamate on total glycoalkaloid content of potatoes, *J. Food Sci.,* 50, 258, 1985.

227. **Mondy, N. I., Tymiak, A., and Chandra, S.,** Inhibition of glycoalkaloid formation in potato tubers by sprout inhibitor, maleic hydrazide, *J. Food Sci.,* 43, 1033, 1978.

228. **Wu, M. T. and Salunkhe, D. K.,** Control of chlorophyll and solanine synthesis and sprouting of potato tubers by hot paraffin wax, *J. Food Sci.,* 37, 629, 1972.

229. **Wu, M. T. and Salunkhe, D. K.,** Inhibition of chlorophyll and solanine formation, and sprouting of potato tubers by oil dipping, *J. Am. Soc. Hortic. Sci.,* 97, 614, 1972.

230. **Wu, M. T. and Salunkhe, D. K.,** Control of chlorophyll and solanine formation in potato tubers by oil and diluted oil treatments, *HortScience,* 7, 466, 1972.

231. **Wu, M. T. and Salunkhe, D. K.,** Use of spray lecithin for control of greening and glycoalkaloid formation of potato tubers, *J. Food Sci.,* 42, 1413, 1977.

232. **Wu, M. T. and Salunkhe, D. K.,** Responses of lecithin and hydroxylated lecithin coated potato-tubers to light, *J. Agric. Food Chem.,* 26, 513, 1978.

233. **Poapst, P. A., Price, I., and Forsyth, F. R.,** Controlling post storage greening in table stock potatoes with ethoxylated monoglyceride and diglyceride surfactants and adjuvant, *Am. Potato J.,* 55, 35, 1978.

234. **Poapst, P. A. and Forsyth, F. R.,** The role of internally produced carbon dioxide in the prevention of greening in potato tubers, *Acta Hortic.,* 38, 277, 1974.

235. **Poapst, P. A. and Forsyth, F. R.,** Relative effectiveness of Tween surfactants when used to control greening in Kennebec potato tubers after cold storage, *Can. J. Plant Sci.,* 55, 337, 1975.

236. **Poapst, P. A., Price, I., and Forsyth, F. R.,** Prevention of post storage greening in table stock potato tubers by application of surfactants, and adjuvant, *J. Food Sci.,* 43, 900, 1978.

Chapter 9

PHYTOALEXINS

A. Kumar, S. J. Jadhav, and D. K. Salunkhe

TABLE OF CONTENTS

I. INTRODUCTION

Various stress conditions, such as temperature adversity, light exposure, mechanical injury, chemical insult, and attack by microorganisms, frequently cause formation and accumulation of unusual metabolites in plant tissues. These abnormal compounds produced by plants in response to various exogenous stimuli may generally be referred to as phytoalexins. At the end of the 19th or in the beginning of the current century, several scientists[1,2] began to think that plants, like animals, might have resistance to infection. In this direction, proteins were thought to be capable of binding pathogens,[3] while Meyer[4] regarded tannins as the antibodies of plants. Bernard[5] reported that the penetration of a particular type of fungus into the orchid embryo was confronted and finally disintegrated. Consequently, the embryo acquired resistance to infection by a fungus, which has generally a destructive nature. Subsequently, several experiments were carried out to reveal chemical interactions between fungus and orchid cells. These involved testing the response between intact vs. crushed or pieces of tuber with a developing mycelium in a culture medium. The results indicated that active cells in the intact tuber cuts interacted with fungal secretions and the released substances, which on diffusion through the culture medium inhibited fungal growth. These findings were supported by Nobecourt.[6] Thus, evidence for the production of defensive substances by plant tissues in response to infections by a fungus began to attract the attention of researchers.

II. CONCEPT OF PHYTOALEXINS

The phenomenon of producing the phytoalexins by plants under stress conditions was first proposed by Müller and Borger[7] in 1940. According to them, the phytoalexins were produced as a result of interaction between host and parasite. Müller and Borger[7] conducted systematic studies on the effect of hypersensitive reaction on the development of the race of *Phytophthora infestans,* a fungus found to be a parasite on the plant. The inhibition of the development of fungus was caused by a substance produced in the hypersensitive reaction and the substance was named phytoalexin. They pointed out that phytoalexins are formed or activated in the reacting host cells and might be considered as the end product of a "necrobiosis" released by the parasite. The defensive reaction is confined to the tissue infected by the fungus and its immediate neighborhood. The earlier definition of phytoalexins as chemical compounds produced as a result of invasion on living cells by a parasite was later modified by Müller[8] as "antibiotics which were produced as a result of the interaction of two metabolic systems, host and parasite, and which inhibit the growth of microorganisms pathogenic to plants".

Cruickshank[9] and Cruickshank and Perrin[10] presented a detailed description of properties of phytoalexins and their role in host-parasite interactions. In a new concept advanced by Kuć,[11] it was suggested that "the term phytoalexins should serve as an umbrella under which chemical compounds contributing to disease resistance can be classified whether they are formed in response to injury, physiological stimuli and the presence of infectious agents or are the products of such agents". Wood,[12] therefore, has accordingly used both the terms "phytoalexins" and "stress metabolites" interchangeably. Kiraly et al.[13] had also adopted a similar definition, according to which phytoalexins are the substances produced by plants after infection and adverse treatments and are responsible for resistance to infecting agents. However, Osman et al.[14] chose to classify phytoalexins as a special class of stress metabolites, i.e., compounds that are not normally found in healthy tissue, but accumulate in response to a disease situation and have a deleterious effect on the disease organism. Thus, all phytoalexins are stress metabolites, but not all stress metabolites are phytoalexins.

The phytoalexin concept has been summarized in several reviews.[15-22] Due to the confusion created by various definitions of phytoalexins, attempts were made by a large number of researchers to agree on a stringent definition. Recently, Paxton[23] formulated the definition of phytoalexins as low-molecular weight antimicrobial compounds that are both synthesized by and accumulated in plants after their exposure to microorganisms. Opinions advanced in recent years suggest that phytoalexins are part of the resistance mechanism of plants against pathogens.

III. OCCURRENCE IN PLANTS

It appears from Table 1 that the Solanaceous plants produce acetylenic, phenolic (phenyl-propanoid), and terpenoid stress metabolites. These compounds are toxic to a wide range of fungi and bacteria when accumulated to a substantial concentration. The utmost exploitation of the Solanaceous plants in phytoalexin research could become possible largely because of the ease of their isolation, identification, and assays. Diversity of chemical types within a plant family as well as within a species is revealed by the occurrence of phenolics, polyacetylenes, and terpenoids in tomato plants. In general, it may be concluded that the Solanaceous plants studied so far have produced mainly terpenoids in spite of some other types in certain species. The predominance of certain phytoalexins may vary depending upon the tissues used, the species of the infecting fungus, and the duration after inoculation or treatment. Numerous fungi are known for their ability to induce phytoalexin formation in the Solanaceous plants. However, it is evident that bacteria,[60] viruses,[46,53] and other factors also induce phytoalexin formation.

A. PHENOLICS, SUBERIN, AND LIGNIN

Reference to the phenolic compounds in potatoes was made as early as 1952 by Johnson and Schaal[57,61] and later on by Kuć.[62] Chlorogenic and caffeic acids, both phenolic compounds, are present in all parts of the potato. The levels of these phenolic compounds were found to increase after slicing.[63,64] The hydrolysis of chlorogenic acid, which is a major phenolic compound, produces caffeic acid. The hydrolysis is facilitated by certain fungi.[65] Johnson and Schaal[57,61] noted that the endogenous phenols in potato peels and those produced around wounds played an important role in the resistance of tubers to scab. A positive correlation between the content of chlorogenic acid and the resistance to potato scab was suggested.[61] The oxidation products of chlorogenic and caffeic acids formed after infection exhibited more fungitoxic activity than the parent compounds. Both the phenolic compounds are believed to be associated with the disease resistance of potatoes. Chlorogenic acid in the culture medium stimulated the growth of *P. infestans,* although it also undergoes transformation to quinic and caffeic acids. The former offers stimulating action on the growth of the organism, while the latter depresses it.[66] Certain experiments[67,68] have demonstrated that light imposes stress on the potato and that it serves as an inducer of phenolic synthesis.

The process of suberization and lignification associated with wound healing of potatoes have been implicated in disease resistance. Suberin, a polymer containing phenolics and long-chain decarboxylic, hydroxy, and perhaps epoxy fatty acids, forms a highly lipophilic coating within and on the surface of tissues and may act as a physical barrier to undesirable biological activities. The fatty acids liberated by the action of acylhydrolases may also act as phytoalexins, and the peroxides or epoxides produced by the enzyme lipoxygenases after injury or infection may have further toxic effects on the infectious agents.[69] The localized protection of plants would result from the covalent linkage of antimicrobial components to suberin polymers.

Lignification onto cell wall constituents or suberin would also lead to a possible barrier to infection.[70,71] The by-products such as oxidized phenolics and free radicals produced during lignification can be further toxic to infectious agents. It is increasingly evident that

TABLE 1
Occurrence of Stress Metabolites in Solanaceous Plants

Species	Stress metabolites	Cause and remarks
Pepper		
Capsicum frutescens	Phenolics	Fungi, bacteria, or chilling injury
	Capsidiol	Accounts for 33% of ether extract of diffusates of fruit injected with *Monilinia fructicola*[24]
		Bacterial elicitation[25]
		Accumulation in fruit and leaves[26,27]
		Fungitoxic, can be degraded by fungi[28]
		Also, rapidly metabolizes by healthy fruit[29,30]
		Synthesized via acetate-mevalonate pathway[31,32]
Jimsonweed		
Datura stramonium	Lubimin (6); Capsidiol; Oxylubi- min; 2,3-Dihydroxygermacrene	Main terpenoid stress metabolites isolated from immature capsules of jimsonweed[33,34]
Tomato		
Lycopersicon escu- lentum	Cinnamic acid derivatives	Infection of foliage with fungi and bacteria
	Tomatine (glycoalkaloid)	Fungistatic
		Perhaps a factor in resistance to disease caused by *Septoria lycopersici*,[35] *Pseudomonas solen- asearum*,[36] *Verticillium*[37] or *Fusarium* wilt[38,39]
	Rishitin(1)	Main sesquiterpenoid isolated from fruit, foli- age, and roots of plant[40-42]
	Falcarindol	The first two polyacetylenes accumulate in fruit
	Falcarinol; cis-Tetradeca-6-ene- 1,3-dyne-5,8 diol	and leaves, while the last accumulate only in fruit inoculated with *C. fulvum*[43]
Tobacco		
Nicotina tabacum	Clorogenic acid	Phenolics accumulation in response to tobacco
	Caffeoyl-,feruloyl-,coumaryl- quinic acid, and their esters of glucose; scopolin; rutin	mosaic virus (TMV) injection[44]
	Scopoletin	Aglycone of scopolin accumulation in response to *P. hyoscyami* f. sp. *tabacina*, the incitant of blue mold[45]
	Capsidiol	Sequiterpenoid accumulation in response to in- fection with tobacco necrosis virus (TNV),[46] *Peronospora tabacina*,[47] or *P. parasitica* var. *nicotianae*[48]
	Phytuberin(16)	Sesquiterpenoids accumulation in the response
	Rishitin(1)	to infection with *Pseudomonas lachrymans*,[49]
	Solavetivone(10)	*P. parasitica* var. nicotina,[50] and TMV,[51] re- spectively
	Capsidiol	Sesquiterpenoids accumulation in foliage inocu-
	Lubimin(6)	lated with nonpathogen *P. lachrymans*[52]
	Phytuberin(16)	
	Phytuberol(16a)	
	Rishitin(1)	
	Solavetivone(10)	
Nicotiana glutinosa	Glutinosone	Sesquiterpenoid accumulation in response to TMV infection[53]
Eggplant		
Solanum melongena	Lubimin(6)	Sesquiterpenoids isolated from eggplant fruits
	Aubergenone	inoculated with *M. fructicola* and other
	9-Oxonerolidol	fungi[54-56]
	9-Hydroxynerolidol	
	11-hydroxy-9, 11-dehydro- nerolidol	

TABLE 1 (continued)
Occurrence of Stress Metabolites in Solanaceous Plants

Species	Stress metabolites	Cause and remarks
Potato		
Solanum tuberosum	Phenolics	Fungitoxic[57]
	Suberin and lignin	Wound healing
	Glycoalkaloids	Fungitoxic[58]
	Noresesqui- and sesquiterpenoids	Accumulation after infection
	Sterols and n-octacosanol	Accumulation in aged potato slices[14,59]

lignification is important in the *P. infestans*-potato interaction[71,72] and also in disease resistance and immunization of some other plants.

B. NORSESQUI- AND SESQUITERPENOIDS

The norsesqui- and sesquiterpenoid phytoalexins in potatoes have attracted attention in recent years, mainly because of their accumulation after infection and in many instances because of their vigorous formation in greater magnitude in resistant infections. The physiological state of tuber and environment are of great importance in determining the amount and rate of synthesis of various phytoalexins. The major SSMs in potatoes are rishitin (1), lubimin (6), solvavetivone (10), phytuberin (16), phytuberol (16a), and anhydro-β-rotunol (11). Rishitin (1), lubimin (6), and solavetivone (10) usually account for at least 85% of the total accumulated SSMs. Although rishitin (1) is a predominant component, lubimin (6) or solavetivone (10) can become the major SSM under certain experimental conditions. The enhanced accumulation of phytuberin (16), phytuberol (16a), rishitin (1), and phytuberin (16) has also been reported.[73-75] The accumulation of nearly 20 more terpenoid stress metabolites has also been reported in infected tubers. The structures of all these compounds are presented in Figure 1. These compounds normally accumulate in small amounts.

IV. FORMATION AND INHIBITION OF PHYTOALEXINS

Phytoalexin accumulation depends not only upon a specific host-parasite relationship, but also upon various other factors. These factors are of a physical, chemical, and biological nature which may incite the injured cells to accumulate the antimicrobial agents.[76,77] The physical and chemical trauma may have different interactions with different tissues. It has, however, been demonstrated[76] that the amount of accumulation of stress metabolites in the case of host-parasite relation is higher than that induced by any physical or chemical trauma.

A. PHYSICAL FACTORS

Tomiyama[78] and Sato et al.[79] studied the influence of aging of potato slices before inoculation. Accumulation of rishitin (1) and glycoalkaloids were affected by aging in these experiments. It was shown that a period of 5 to 24 h between slicing and the surface inoculation reduced the time required for cell death and rishitin (1) accumulation in compatible interactions. After a 24-h period, rishitin (1) accumulation was reduced markedly.[80] Aging slices for 72 h made them highly resistant to *P. infestans,* though there was no rishitin (1) accumulation. It was felt that suberization, phenol accumulation, and oxidation at the slice surface may be responsible for the defense system of the potato.

No appreciable levels of terpenes have been noticed in the freshly harvested tubers if subjected to different kinds of stimuli listed in Table 2. However, the various cell-disruptive treatments induced accumulations of terpenes in tubers only after about 1 month of postharvest storage, which further increased during the storage. Bostock et al.[81] found that incompatible races of *P. infestans* elicited a hypersensitive response in potato slices from tubers stored

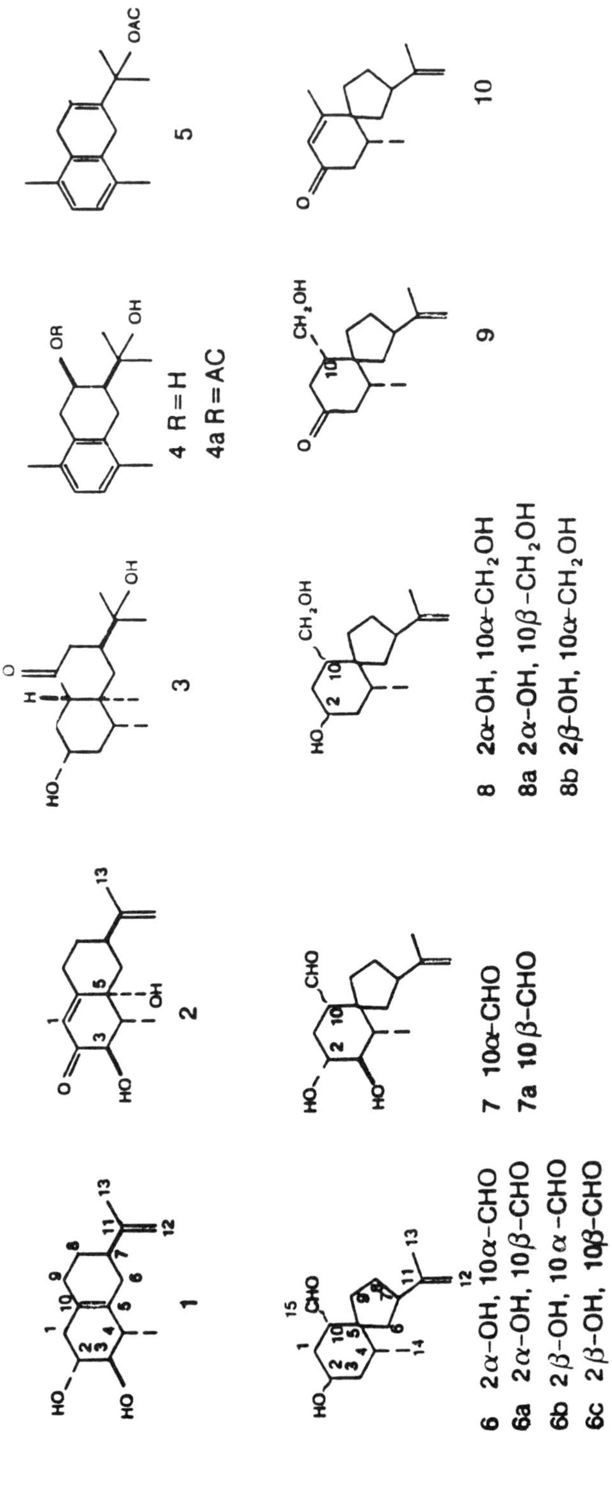
1
2
3
4 R = H
4a R = AC
5
6 2α-OH, 10α-CHO
6a 2α-OH, 10β-CHO
6b 2β-OH, 10α-CHO
6c 2β-OH, 10β-CHO
7 10α-CHO
7a 10β-CHO
8 2α-OH, 10α-CH2OH
8a 2α-OH, 10β-CH2OH
8b 2β-OH, 10α-CH2OH
9
10

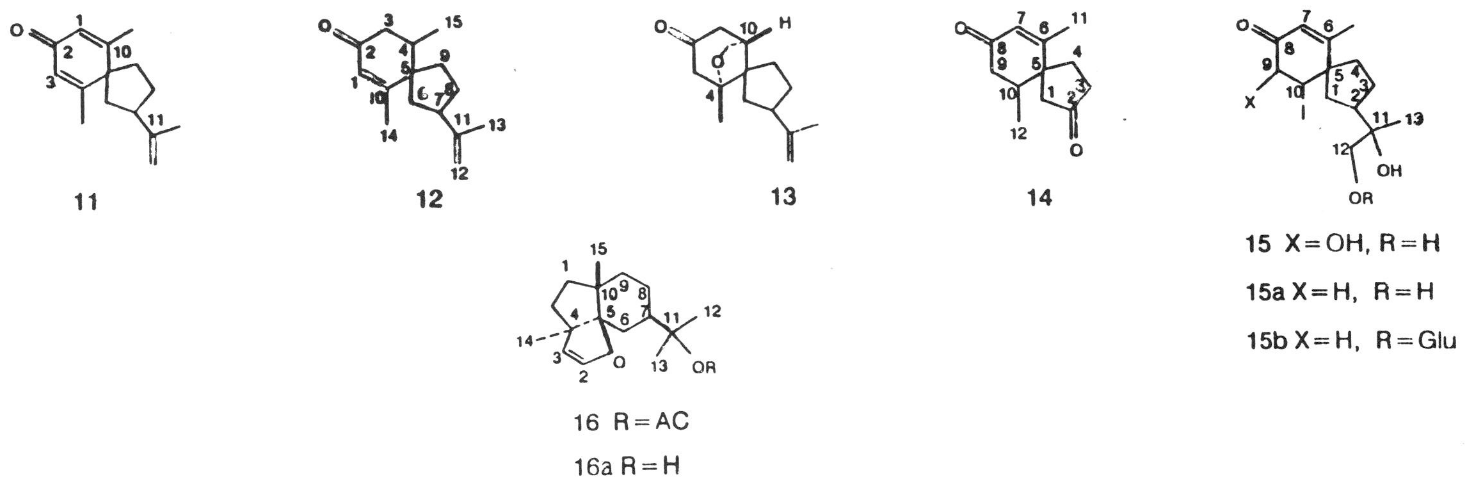

15 X = OH, R = H

15a X = H, R = H

15b X = H, R = Glu

16 R = AC

16a R = H

FIGURE 1. Molecular structures of terpenoid stress metabolites of potato tubers. (1) Rishitin; (2) oxyglutinosone; (3) rishitinone; (4) rishitinol; (4a) 8-*O*-acetyl rishitinol; (5) acetyldehydrorishitinol; (6) lubimin; (6a) 10-epilubimin; (6b) 2-epilubimin; (6c) 2,10-epilubimin; (7) oxylubimin (hydroxylubimin); (7a) epioxylubimin (epihydroxylubimin); (8) 15-dihydrolubimin; (8a) 15-dihydro-10-epilubimin; (8b) 15-dihydro-2-epilubimin; (9) isolubimin; (10) solavetivone (katahdinone); (11) anhydro-β-rotunol—spirovetiva-1(10), 3-11-trien-2-one; (12) spirovetiva-1(10), 11-dien-2-one; (13) cyclodehydroisolubimin; (14) 6,10-dimethylspiro[4,5]dec-6-en-2, 8-dione; (15) 2-(11,12-dihydroxy-11-methylethyl)-6,10-dimethyl-9-hydroxyspiro [4,5] dec-6-en-8-one; (15a) 2-(11,12-dihydroxy-11-methylethyl)-6,10-dimethyl-spiro [4,5] dec-6-en-8-one; (15b) 12-*O*-β-D-glucopyranoside of compound 15a; (16) phytuberin; (16a) phytuberol (desacetylphytuberin).

TABLE 2
Some Trauma Known to Elicit Stress
Metabolites in Plant Tissue

Physical
    Cut injury and bruising
    Chilling injury
    Ultraviolet light
    Visible light
    Low oxygen tension
Chemical
    Mercuric and other heavy-metal salts
    Sodium lauryl sulfate and other detergents
    Cysteine and various thiol reagents
    Ethylene, cyclic AMP
    DNA intercalating agents, basic proteins, and polyamines
    Fungal and insect extracts

From Cheema, A. S. and Haard, N. F., *J. Food Prot.*, 42, 512, 1979. With permission.

at 4°C. The response was characterized in part by the rapid accumulation of the fungitoxic sesquiterpenes from the stored tubers. Potato slices from unstored tubers harvested during July, August, and September accumulated low levels of rishitin (1) and lubimin (6) after inoculation with an incompatible race. The amount of sesquiterpenes accumulated in inoculated potato slices from tubers before cold storage was 10 to 20% of that accumulated in tubers from cold storage. These authors also noted that storage at 4°C generally increased the accumulation of terpenes in slices treated with a crude elicitor preparation from the fungus. Tubers stored at 4°C for 1 month after harvest, on treatment with mercuric acetate, accumulated approximately four times more terpenes than those stored at 25°C.[82] The behavior of 4°C-stored tubers was reversed by conditioning them at 25°C.

Henfling et al.[83] demonstrated that cold-stored tubers become susceptible to *P. infestans*, regardless of their R-gene designation, when treated with abscisic acid. The induced susceptibility was accompanied by a marked decline in sesquiterpenoid stress metabolite (SSM) accumulation. Immature tubers showed the typical compatible and incompatible reactions to appropriate races of *P. infestans*. Nevertheless, the incompatible reactions did not accumulate SSM more rapidly or in greater magnitude than the compatible interactions.[84,85]

Currier and Kuć[86] studied the effect of temperature and observed a remarkable reduction in browning and rishitin (1) accumulation at 19°C as compared to 25 to 30°C. It was also noted that no rishitin (1) accumulation took place at 14 or 37°C. The incubation at 25 or 30°C was found to increase the accumulation of fungitoxic glycoalkaloids, which are characteristics of wound response. Thus incubation at 25 and 30°C accentuates the wound response, but inhibits browning and rishitin (1) accumulation, which are associated with the hypersensitive response to incompatible races of *P. infestans*.

Controlled atmosphere storage conditions also induce the production of phytoalexins. Alves et al.,[87] in a very systematic study, monitored the levels of solavetivone (10), lubimin (6), rishitin (1), and phytuberin (16) in white potato tuber slices which were challenged with an extract of *P. infestans* and incubated under controlled atmospheres. A mixture of ethylene in air enhanced the stress metabolite production. This enhancement was amplified by higher partial pressures of oxygen. The stress metabolite production was inhibited by salicylhydroxamic acid. These studies suggested the involvement of a cyanide-resistant respiration in the production of potato stress metabolites which may serve as phytoalexins.

Another important elicitor is ultraviolet (UV) light which induces terpene production in potato tubers. Cheema and Haard[82] noted that potatoes stored at 4°C are more disposed to

TABLE 3
Examples of Heavy-Metal Salts
which Induce Rishitin (1) in Potato
Tuber Slices

Inducer (5 mM)	Rishitin(1) (48 h)[a]
HgCl$_2$	5.2
HgCH$_2$COOH	11.5
AgNO$_3$	3.6
CuCl$_2$	13.4
FeCl$_3$	N.D.[b]
FeCl$_2$	N.D.
NiCl$_2$	N.D.
CoCl$_2$	N.D.
MnCl$_2$	N.D.
Ce(NH$_4$)NO$_3$	N.D.
H$_2$O	N.D.

[a] μg/g fresh weight of discs (5 mm × 25 mm).
[b] N.D. = Not detectable.

From Cheema, A. S. and Haard, N. F., *J. Food Prot.*, 42, 512, 1979. With permission.

terpene induction by UV light than 25°C-stored tubers, and freshly harvested tubers are not responsive to this treatment. It has also been observed that the accumulation of rishitin (1) is a strong function of exposure time to light which in turn is dependent upon the potato disc thickness. No tissue browning and necrosis were noted on exposure to UV light when treated with heavy metals. A loss in cell turgor as evidenced by the flacid nature of discs was also observed after long exposure to UV light. Schmidt et al.[88] examined the phytoalexin accumulation elicited by *P. megasperma* in irradiated and/or heat-treated conditions. Rishitin (1), lubimin (6), and a very small quantity of phytuberol (16a) were noted to be present. The irradiated tubers accumulated higher concentrations of phytoalexins than the nonirradiated ones. On the samples containing the highest phytoalexin level, intensive growth of the fungus was observed. In the bioassay, an increased amount of phytoalexins showed a stronger fungistatic effect on *Cladosporium cueumerinium*. It was concluded by the above authors that phytoalexin accumulation was not clearly related to the mechanism of resistance to fungi.

B. CHEMICAL FACTORS

Cheema and Haard[82] noted that the accumulation of rishitin (1) was highly dependent upon the concentration of mercuric acetate. Optimal induction of rishitin (1), due to the level of mercuric acetate, was found to be related to tissue disc thickness. The lower concentrations of the chemical were enough to create response in thinner discs, while for thick disc tissues higher concentrations were required. The injury and depth to surface cells of discs were caused by the action of mercuric acetate. The retention of a healthy layer of cells beneath the necrotic zone can alter the rishitin (1) accumulation. Other heavy-metal salts and chemicals also play a vital role in inducing rishitin (1) formation. Examples of some heavy-metal salts which can induce rishitin (1) accumulation in potato tuber slices are listed in Table 3. In general, monitoring metal ion concentration and thickness of the disc are the major criteria to observe the effects of these chemicals. Trace quantities of rishitin (1) were observed in tuber discs treated with mercuric chloride after a 4-d incubation period.[89] Metlitskii et al.[90] studied the effects of iodoacetates, *p*-chloromercuribenzoate, *p*-benzoquinone, sodium fluoride, copper sulfate, chloramphenicol, fungicides, and many other chem-

TABLE 4
Induction of Terpenes in Potato Tuber Discs by Reagents which Interact with Sulfydryl Groups

Inducers (5 m*M*)	4°C Stored tubers		25°C Stored tubers	
	Rishitin[a](1)	Lubimin[a](6)	Rishitin[a](1)	Lubimin[a](6)
Mercuric acetate	9.7	0.8	2.1	0.4
Mercuric chloride	5.2	1.1	0.7	0.1
N-Methyl maleimide	0.6	0.3	N.D.[b]	N.D.
Iodoacetamide	4.1	2.3	0.2	N.D.
p-Chloromercuribenzoic acid (PCMB)	3.9	3.3	0.1	0.1
H₂O	N.D.	N.D.	N.D.	N.D.

[a]   μg Terpene per gram fresh weight 48 h after treatment (5 mm × 25 mm discs).
[b]   N.D. = Not detected.

From Cheema, A. S. and Haard, N. F., *Physiol. Plant Pathol.*, 13, 233, 1978. With permission.

icals on the induction of rishitin (1) and lubimin (6) and concluded that the fungicides may be useful in increasing the resistance of plants while acting as parasiticides.

The treatment of potato tuber discs with Blasticidin S[91] and Puromycin[82] alone were ineffective on the production of rishitin (1). However, inclusion of mercuric acetate or UV treatment induced some production of rishitin (1).[82] Testosterone was noted to have the greatest stimulatory activity on rishitin (1) among 13 steroids studied by Mustafa and D'yakov.[92,93] A study[94] on the influence of alkyl halides on the induction of rishitin (1) and browning in potato tissues revealed that the treatment with bromides produced more of rishitin (1) than of chlorides. Similarly, dihalogen alkyl halides induced a higher rishitin (1) content than single halogen alkyl halides. On the other hand, allylamine and allyl alcohol did not show any action on potato tuber tissues. Abscisic acid inhibited rishitin (1) and lubimin (6) accumulation in potato tuber slices. This consistent reduction in accumulation of rishitin (1) and lubimin (6) did not appear to be responsible for the conversion of incompatible to compatible interactions with *P. infestans*.[83]

The role of calcium was examined in the elicitation of rishitin (1) and lubimin (6) accumulation in potato tuber tissue.[95] The authors noted that Ca²⁺ and Sr²⁺ ions enhanced rishitin (1), but not lubimin (6) accumulation in the tuber tissues treated with arachidonic acid. Their studies suggest that the mobilization of calcium may play a central regulatory role in the expression of the phytoalexin accumulation following the elicitation in potato tissues. Murai et al.[96] conducted very exhaustive studies on hydrogen peroxide as a dynamic elicitor for phytoalexin production. They observed that the phytoalexin elicitation had some relationship with active oxygen species, generated from the interaction of diseased plants and air, and concluded that an inoculation of hydrogen peroxide induced phytoalexin production.

Cheema and Haard[82] designed an experiment to test the effect of various reagents, which interact with sulfhydryl groups, on terpene induction. The impact created by them on rishitin (1) and lubimin (6) in potato discs from two storage temperatures was studied (Table 4). Higher terpene concentrations were noted to accumulate on the discs from the cold-stored potatoes as compared to those stored at 25°C. All sulfhydryl chemicals were capable of inducing the accumulations. These authors noted that the levels of rishitin (1) and lubimin (6) were nearly equal following *p*-chloro mercury benzoate (PCMB) treatment, while rishitin (1) concentration was approximately 10 to 12 times higher than lubimin (6) in the mercuric acetate-treated discs. In another article, Cheema and Haard[77] have pointed out the role of various cell-disruptive agents such as detergents, metabolic inhibitors, and lysosomal enzyme

TABLE 5
Influence of (A) DNA Intercalating Compounds and (B) Protein Synthesis Inhibitors Functioning by Mechanisms other than DNA Intercalation, on Rishitin(1) and Lubimin(6) Accumulation

Compound	Conc	μg/g Fresh weight	
		Rishitin(1)	Lubimin(6)
Compound (A)			
Actinomycin D	25 μg/ml	52.5	30.1
Mitomycin C	100 μg/ml	6.6	N.D.
Distamycin A	25 μg/ml	12.6	5.4
Ethidium bromide	100 μg/ml	7.4	2.8
Compound (B)			
6-Methyl purine (blocks transcription by mechanism other than intercalation)	50 μm	8.2	3.7
Cycloheximide (blocks translation by interfering with peptide formation	10 μm	17.4	3.5
Puromycin (blocks translation by mimicking amino-acyl-tRNA	50 μm	12.4	20.6

Note: Tubers were previously stored at 4°C; terpenes were analyzed after 72 h of incubation at 20°C. All treatments were for 30 min.

From Cheema, A. S. and Haard, N. F., *J. Food Prot.,* 42, 512, 1979. With permission.

preparations in inducing the accumulation of phytoalexins which were somewhat effective. Again, cold-stored tubers showed a greater propensity for terpene accumulation than those at 25°C, and freshly harvested tubers were insensitive to nonspecific elicitors.[77] Ersek et al.[97,98] found inhibition of growth of *P. infestans* on potato tuber slices treated with chloramphenicol or streptomycin and concluded that hypersensitivity associated with phytoalexin production is a consequence and not the cause of host resistance to infection. Treatment of potato discs with actinomycin D alone is effective in promoting terpene accumulation when applied at low concentration. The time interval of rishitin (1) accumulation resulting after actinomycin D treatment continued to reach up to 96 h.[77,99] The concentration of rishitin (1) accumulating is comparable to that observed in *P. infestans*-infected potatoes when optimal concentrations of actinomycin D are applied to freshly prepared discs. The compounds which act by binding to DNA and those which affect protein synthesis by mechanisms other than DNA intercalation also induce terpenes in potato discs (Table 5). Cheema and Haard[77] stated that DNA binding is not a prerequisite for terpene induction. Both translational and transcriptional inhibitors were effective in the induction of rishitin (1) if applied for a short period immediately after slicing of potatoes. Cycloheximide was particularly effective at low concentration. The authors proposed a model indicating that terpene induction by various agents is dependent upon the availability of a DNA template, which contains necessary information for phytoalexin production. Thus, potatoes stored in the cold have higher template availability than those stored at a warm temperature. The translational and transcriptional inhibitors perturbing biosynthesis of phytoalexin repressor proteins at the respective level immediately after mechanical injury, without also inhibiting total protein synthesis, will result in expression of the phytoalexin genome. The theory is applicable to any agent which blocks repressor synthesis.

Potato tuber slices treated with Ethrel followed by inoculation with *Helminthosporium victoriae, H. carbonum,* or an incompatible race of *P. infestans* accumulated considerably more phytuberin (16) and phytuberol (16a) than did control slices treated with water and then with the respective agent.[100] The accumulation of the two terpenoids also was enhanced in slices treated with Ethrel and then with cell-free autoclaved sonicates of *Pythium aphan-idermatum* or *Achlya flagellata*. The changes in the accumulation of rishitin (1) and lubimin (6) in tissues accumulating a high level of phytuberin (16) and phytuberol (16a) were not consistently elicited by Ethrel. Moreover, Ethrel was ineffective in increasing the accumulation of these four terpenoids in noninoculated, inoculated with compatible races of *P. infestans*, or slices treated with sonicates of *H. carbonum*, *H. victoriae*, or *Neurospora crassa*. More accumulation of phytuberin (16) and phytuberol (16a) was noted in slices treated with Ethrel followed by a sonicate of *P. infestans* at 19°C than at 14 or 25°C, and little terpenoids were detected in tissues incubated at 30 and 36°C. Although ethylene alone is a nonelicitor of terpene accumulation, it markedly increased (100 μl/l) accumulation of phytuberin (16) and phytuberol (16a) in slices treated with cell-free sonicates of *P. infestans*. It appears that Ethrel does not alter the resistance or susceptibility of potato cultivars Kennebec or Russet Burbank to race 4 and 1234 of *P. infestans* or the resistance of both cultivars to *H. victoriae* and *H. carbonum*. Beczner[101] noticed that externally added ethylene suppressed the CO_2 evolution by the inoculated tubers, but did not influence that of healthy tubers. The resistance of inoculated tubers to *Erwinia atroseptica* was not significantly influenced by ethylene treatment. No considerable change was noticed in the concentration of rishitin (1) and lubimin (6) as a result of ethylene treatment. The phytuberin (16) production of infected tubers was suppressed by ethylene, but cannot be induced in healthy potato tubers by ethylene treatment only.

Production of phytoalexins has also been reported to be induced by fatty acids. In 1975, Ersek[102] observed that lipoid components obtained from mycelia of *P. infestans* caused slight browning on potato tuber slices. No fungal growth was observed on slices treated with the lipoid component and then challenged with *P. infestans*. The lipoid extract had no inhibitory effect on the germination of zoospores. The lipoid extract as well as the crude mycelial homogenate induced phytoalexin accumulation. Bostock[103] also identified polyunsaturated fatty acids from *P. infestans* as elicitors of hypersensitivity in potato tuber and studied related metabolism. Mycelial extracts from *P. infestans* caused necrosis and elicited the accumulation of antimicrobial stress metabolites in potato tubers.[104] They found that the most active elicitors of stress metabolites were eicosapentaenoic and arachidonic acids. Both the acids were found in either free or esterified form in all active fractions of the mycelial extracts. Bostock et al.,[105] later on, reported that the elicitor activity of the unsaturated fatty acids was enhanced by heat and base-stable factors in the mycelium. In a separate study, it was demonstrated that the synthesis of rishitin (1) from arachidonic acid was inhibited by salicylhydroxamic acid and also to some extent by tetraethylthiuram disulfide.[106] Similar studies relating to the synthesis of phytoalexins and phenols from fatty acids have been conducted.[107-109] It was noted by Maina et al.[107] that a high concentration of nonanoic acid promoted necrosis and accumulation of low levels of phytoalexins, but decreased levels of phenols, phenylalanine ammonia-lyase, and lignin. The addition of arachidonic acid to the fungicides like triadimenol and fenarimol elicited sequiterpenoid accumulation, although arachidonic acid was noted to shift the acetate-mevalonate pathway to sesquiterpenoid metabolism.[110] Application of pesticides induced formation of rishitin (1), which increased in response to inoculation with spores of *P. infestans*.[111]

Sonicates and boiled aqueous extracts of three races of *P. infestans* caused necrosis and accumulation of rishitin (1) and phytuberin (16) in tuber slices of potatoes having R genes for resistance and those susceptible to all known races of the fungus.[112] Autoclaved sonicates and cell wall preparations of oomycetes, consisting of glucan-cellulose as the main cell wall component, elicited the accumulation of SSM and browning of slices, but the similar preparations from other fungi and heat-treated bacteria were ineffective.[113] According to Henfling et al.,[114] cell walls of *P. infestans* contain an elicitor of SSM accumulation. Stoessl[115] noted the accumulation of SSM in tubers inoculated with mycelial suspensions of an alternaric acid. Two water-soluble glucans which inhibit the hypersensitive reaction of potato tuber tissue (Kennebec R_1) to *P. infestans* were isolated from mycelia and zoospores of the race 1234 (compatible) and the race 4 (incompatible) of the fungus.[116,117] The concept of resistance suppression as a mechanism for susceptibility was supported by the fact that appropriate glucans were produced extracellularly by germinating cytospores.[118] Further, Doke and Tomiyama[119] reported the suppression of the hypersensitive response of potato tuber protoplasts to hyphal wall constituents isolated from the fungus by water-soluble glucans of *P. infestans* origin. Doke et al.[116] suggested the possibility that protoplasmic membrane sites may serve as recognition sites for the elicitor. This was supported by Doke and Tomiyama,[120] indicating that crude elicitor from fungal walls of *P. infestans* was toxic to protoplasts of potato tuber tissues. It appears that the suppressor is a low-molecular weight polysaccharide, whereas the nonspecific elicitors are unsaturated fatty acids.[121]

C. BIOLOGICAL FACTORS

Potato phytoalexins have been studied with R gene resistance of potato to the late blight pathogen *P. infestans*. These studies are associated with the ability of compatible races of the fungus to suppress hypersensitive cell death, necrosis, and SSM accumulation.[122-125] Obviously, compatible races of *P. infestans* markedly reduced SSM accumulation caused by subsequent inoculation of tissue with incompatible races or treatment with elicitor preparation.[123,124] They further stated that treatment with the incompatible race is a prerequisite for the suppression of SSM accumulation by the compatible race. Kuć[16] noted that the criteria of determining a compatible or incompatible reaction of potato to *P. infestans* depends upon the interaction time.

The stress metabolites such as rishitin (1), phytuberin (16), phytuberol (16a), and solavetivone (10) were detected in different cultivars of potato tubers inoculated with *P. infestans, Fusarium avenaccum,* and *Phoma exigua* var. foneata.[126] Potatoes inoculated with nonpathogens were found to accumulate rishitin (1) and other terpenoid phytoalexins.[112,113,127,128] Thus, it became clear that the lack of a genetic potential to produce the terpenoid phytoalexins does not determine susceptibility of *Phytophthora infestans*. Cell-free preparations from all the races of *P. infestans* were found to elicit accumulation of sesquiterpenoid phytoalexins in all cultivars of potatoes, including those lacking R genes for resistance.[112,123,129,130] The differences in phytoalexin elicitation by *P. infestans* and *H. carbonum* were observed in potatoes.[131] Living spores and mycelia of *H. carbonum* and an incompatible race of *P. infestans* elicited the accumulation of rishitin (1) and lubimin (6) in potato tubers. The same inoculum from *H. carbonum* lost the ability to elicit rishitin (1) and lubimin (6) after heat, ethanol, or liquid nitrogen treatment. However, spores and mycelia of *P. infestans* retained eliciting activity after heat or ethanol treatment.

Horikawa et al.[132] studied accumulation and transformation of rishitin (1) and lubimin (6) in potato tuber tissue infected by an incompatible race of *P. infestans* and detected rishitin (1), lubimin (6), rishitinol (4), and eight unknown compounds. It appears that rishitin (1), synthesized in the healthy tissue around browned tissue in tubers infected by the compatible race, would be transformed to other compounds. This could be the reason for less accumulation of rishitin (1) in healthy tubers. Zacharius et al.[133] studied the effect of the R-3 gene in resistance of Wauseon potato to *P. infestans*. When challenged by the race 4, the tuber slices developed a penetrating necrosis and a high accumulation of the terpenoids, phytuberin (16) and solavetivone (10). It was noted that in the compatible interaction of Wauseon with the race 1, production of rishitin (1) was almost equal to that produced during the hypersensitive response of Kennebec (R_1) to race 4. Based on the study, these authors concluded that rishitin (1) was not the primary factor involved in the resistance of potato tubers to races of *P. infestans*.

Lisker and Kuć[134] demonstrated that the reaction of potato sprouts to compatible and incompatible races of *P. infestans* was different from that of tubers and leaves. Browning and accumulation of rishitin (1), lubimin (6), phytuberin (16), and phytuberol (16a) in sprouts of Kennebec, Russet Burbank, and Red Pontiac potatoes were induced by zoospores of the race 4 and 1234 of *P. infestans*. Regardless of the race, similar quantities of the terpenoids were accumulated in sprouts. Autoclaved, cell-free sonicates of some other pathogens also elicited browning and terpenoid accumulation. Tubers of cultivars lacking R genes for resistance and those from a cultivar with R_1 showed no difference in the extent of browning or amounts of terpenoid accumulation in sprouts when treated with spores. Kennebec sprouts treated with sonicates of *P. infestans* exhibited more intense browning and more accumulation of terpenoids at 19°C than at 14 or 25%C. However, the above phenomenon could not be detected at 30 and 36°C. Browning was first noticed near the growing point and then over the length of sprouts inoculated with zoospores of *P. infestans*. The sprouts treated with cell-free sonicates of the fungi showed no difference in the appearance of browning. Growth and sporulation of *P. infestans* was not apparent on inoculated sprouts.

Tissue culture callus of potato tubers lacking a resistance gene to fungus was observed to produce rishitin (1) and phytuberin (16), the quantities of which varied with the number of days.[135] Brindle et al.[136] noted the accumulation of lubimin (6), rishitin (1), and solavetivone (10) in potato suspension cultures with sporangia of either compatible or an incompatible race of the fungal pathogen *P. infestans*. The phytoalexins were accumulated in the cells and medium.

V. MECHANISM OF INFECTION AND PHYTOALEXIN PRODUCTION

It has been observed that the rate of formation of phytoalexins in natural infection is slower than that in an artificial one. The reason appears to lie in the fact that the spontaneous infections are not preceded by the metabolic activation resulting from cutting of tubers. Brishammer[137] has reported that in the case of a late blight infection, phytoalexins are not formed during spore germination, germ tube growth, or development of the appressorium. The elicitor acts upon the host cell membrane after the infection peg penetrates the tissues. Later on, a haustorium is developed by the fungus. Until this stage, *P. infestans* is at a biotrophic stage. If the phytoalexins are produced at this stage they act against the fungus, provided these phytoalexins have antimicrobial or antipathogenic activity. It has been demonstrated that pathogens developing haustoria are not very susceptible to the direct action of antimicrobial substances.[138] The acts of phytoalexins against the fungus can be of two types. First the direct, in which the phytoalexins act directly against the fungus at very low concentrations outside the host cell; and secondly the indirect, where the regulation of host cells takes place in such a way so that they are not easily infected by fungus. It has been observed through microscopical examinations that infected host cells die after 3 to 4 h, while rishitin (1) is formed 10 to 11 h after inoculation, provided the cut tubers had been allowed to age for 15 h.[26,139,140]

It is extremely difficult to suggest a single common mechanism for infection and subsequent phytoalexin production. However, if the inhibition takes place while the pathogen is in the necrotrophic stage in the host, a direct contact takes place between the host and the phytoalexins. On the other hand, phytoalexins may induce a disease-resistant mechanism in the cells as a cell activity regulator, if they inhibit pathogen development with their very low concentration when the host cells are still alive. In view of very minute quantities of phytoalexins and inadequate analytical techniques, the present day knowledge regarding this mechanism is insufficient.

VI. EFFECTS OF PHYTOALEXINS

Rishitin (1) and other potato phytoalexins when sprayed on leaves did not reduce the occurrence of late blight, whereas capsidiol sprayed on tomato leaves controlled development of *P. infestans*.[141] Ishizaka et al.[142] observed that rishitin (1) application to the surfaces of cut tubers and leaves inhibited spore formation by *P. infestans*. The formation of rishitin (1) and lubimin (6) is one of the factors in controlling the vertical resistance of potato to *P. infestans*.[143] The intensity of necrosis correlated in all 20 potato cultivars with the contents of these two compounds. The rate of rotting was inversely proportional to the rishitin (1) concentration.[144-146] Evidence implicating phytoalexins as potential inhibitors of fungal glucanases was presented by Hohl et al.[147] Potato leaf epidermal strips incubated with rishitin (1) in 1% Evans blue rapidly accumulated the stain within the cells and there was evidence of rapid effect on several cell membranes. Rishitin (1) reduced the rate of respiration to a little extent, whereas phaseolin first increased and then decreased the rate of respiration. Also, liposome permeability was influenced more by rishitin (1) than by phaseolin.[148]

VII. TOXICOLOGY

Numerous edible plants are capable of producing stress compounds when infected by certain microorganisms. This is of vital importance due to a possible harm to human health posed by these compounds. Potatoes are known to produce more than 25 terpenoid stress metabolites, and their number is increasing by virtue of their structural modifications resulting from biochemical or biological interactions. However, information is lacking on the toxic effects of these compounds which may be consumed by humans in low doses for long durations. ROTECTS[149] did not treat terpenes of current interest or similar compounds in the international index as toxic compounds. Wood[150] conducted a very systematic study in order to examine the effects of terpenes from infected potatoes on fertile chick eggs. No toxic or teratogenic effects were noticed in the embryos. However, no birth defects appeared when marmosets were fed a mixture of potato concentrate rejected by industry and tubers infected with *E. carotovora* during early pregnancy.[151] Several authors showed no teratogenic or embryolethal effects resulting in experimental animals from blighted or infected tubers[152-156] or exposed to either rishitin (1) or phytuberin (16).[157] Terpenes, if consumed by mammals, are probably converted to glycosides and are removed from bodies through the excretion process.[158] Some reports are available regarding toxic effects on plants. Since the quantities of phytoalexins produced are low, it should be pointed out that more precise analytical techniques for examining the toxicological effects should be developed.

REFERENCES

1. **Ward, H. M.,** Recent researches on the parasitism of fungi, *Ann. Bot.,* 19, 1, 1905.
2. **Stakman, E. C.,** Relation between *Puccinia graminis* and plants highly resistant to its attack, *J. Agric. Res.,* 4, 193, 1915.
3. **Chester, K. S.,** The problem of acquired physiological immunity in plants, *Q. Rev. Biol.,* 8, 129 & 275, 1933.
4. **Meyer, G.,** Zellphysiologische und Anatomische Untersuchungen uber die Reaktion der Kartoffdknolle auf den Angriff der *Phytophthora infestans* bei Sorten Verschiedener Resistenz, *Arb. Biol. Reichsanst. Land Forstwirtsch. Berlin-Dahlem,* 23, 97, 1939.
5. **Bernard, N.,** Sur la fonction fungicide des bulbes d'Ophrydees, *Ann. Sci. Nat. Bot. Biol. Veg.,* 14, 221, 1911.
6. **Nobecourt, P.,** Sur la Production d'anticrop par les tubercules des Ophrydees, *C.R. Acad. Sci.,* 177, 1057, 1923.

7. **Müller, K. O. and Borger, H.,** Experimentelle Untersuchungen die *Phytophthora-Resistans* der Kartoffel, *Arb. Biol. Reichsanst. Land Forstwirtsch. Berlin-Dahlem,* 23, 189, 1940.

8. **Müller, K. O.,** Einige einfache Versuche zum Nachweis von phytoalexinen, *Phytopathol. Z.,* 27, 237, 1956.

9. **Cruickshank, I. A. M.,** Phytoalexins, *Annu. Rev. Phytopathol.,* 1, 351, 1963.

10. **Cruickshank, I. A. M. and Perrin, D. R.,** Pathological function of phenolic compounds in plants, in *Biochemistry of Phenolic Compounds,* Harborne, J. B., Ed., Academic Press, New York, 1964, 511.

11. **Kuć, J.,** Phytoalexins, *Annu. Rev. Phytopathol.,* 10, 207, 1972.

12. **Wood, G. E.,** Stress metabolites of potato—a growing concern, *J. Food Prot.,* 42, 496, 1979.

13. **Kiraly, Z., Barna, B., and Ersek, T.,** Hypersensitivity as a consequence, not a cause, of plant resistance to infection, *Nature (England),* 239, 456, 1972.

14. **Osman, S. F., Zacharius, R. M., Kalan, E. B., Fitzpatrick, T. J., and Krulick, S.,** Stress metabolites of the potato and other solanaceous plants, *J. Food Prot.,* 42, 502, 1979.

15. **Grezelinska, A.,** Fitoaleksyny, *Postepy Biochem.,* 22, 53, 1976.

16. **Kuć, J.,** Phytoalexins, in *Physiological Plant Pathology,* Heitefuss, R. and Williams, P. H., Eds., Springer-Verlag, New York, 1976.

17. **Kuć, J., Currier, W. W., and Shih, M. J.,** Terpenoid phytoalexins, *Annu. Proc. Phytochem. Soc.,* 13 (Biochem. Aspects Plants Parasite Related Proc. Symp.), 225, 1975.

18. **Kuć, J., Henfling, J., Geras, N., Doke, N., and Lexington, K. Y.,** Control of terpenoid metabolism in the potato *Phytophthora infestans* interaction, *J. Food Prot.,* 42, 508, 1979.

19. **Kuć, J., Tjamos, E., and Bostock, R.,** Metabolic regulation of terpenoid accumulation and disease resistance in potato, in *Isopentenoids in Plants, Biochemistry and Function,* Nes, W. D. and Tsai, L. S., Eds., Marcel Dekkar, New York, 1984, 103.

20. **Ingham, J. L.,** Phytoalexins and other natural products as factors in plant disease resistance, *Bot. Rev.,* 38, 343, 1972.

21. **Uritani, I.,** Abnormal substances produced in fungus contaminated food stuffs, *J. Assoc. Off. Anal. Chem.,* 50, 105, 1967.

22. **Tibor, E.,** Fitoalexinek, *Novenytermeles,* 24, 359, 1975.

23. **Paxton, J. D.,** Phytoalexins—a working redefinition, *Phytopathol. Z.,* 101, 106, 1981.

24. **Stoessl, A., Unwin, C. H., and Ward, E. W. B.,** Post-infectional inhibitors from plants. I. Capsidiol, an antifungal compound from *Capsicum frutescens, Phytopathol. Z.,* 74, 141, 1972.

25. **Ward, E. W. B., Unwin, C. H., and Stoessl, A.,** Post-infectional inhibitors from plants. VI. Capsidiol production in pepper fruit infected with bacteria, *Phytopathology,* 63, 1537, 1973.

26. **Jones, D., Graham, W., and Ward, E. W. B.,** Ultrastructural changes in pepper cells in an incompatible interaction with *Phytophthora infestans, Phytopathology,* 65, 1274, 1975.

27. **Ward, E. W. B.,** Capsidiol production in pepper leaves in incompatible interactions with fungi, *Phytopathology,* 66, 175, 1976.

28. **Stoessl, A., Unwin, C. H., and Ward, E. W. B.,** Post-infectional fungus inhibitors from plants: fungal oxidation of capsidiol in pepper fruit, *Phytopathology,* 63, 1225, 1973.

29. **Ward, E. W. B., Stoessl, A., and Stothers, J. B.,** Metabolism of sesquiterpenoid phytoalexins capsidiol and rishitin to their 13-hydroxy derivatives by plant cells, *Phytochemistry,* 16, 2024, 1977.

30. **Stoessl, A., Robinson, J. R., Rock, G. L., and Ward, E. W. B.,** Metabolism of capsidiol by sweet pepper tissue: some possible implications for phytoalexin studies, *Phytopathology,* 67, 64, 1977.

31. **Baker, F. C. and Brooks, C. J. W.,** Biosynthesis of the sesquiterpenoid capsidiol in sweet pepper fruit inoculated with fungal spores, *Phytochemistry,* 15, 689, 1976.

32. **Stoessl, A., Ward, E. W. B., and Stothers, J. B.,** Biosynthetic relationships of sesquiterpenoidal stress compounds from the Solanaceae, in *Host Plant Resistance to Pests,* Hedin, P. A., Ed., American Chemical Society, Washington, D.C., 1977, 61.

33. **Ward, E. W. B., Unwin, C. H., Rock, G. L., and Stoessl, A.,** Postinfectional inhibitors from plants. XXIII. Sesquiterpenoid phytoalexins from fruit capsules of *Datura stramonium, Can. J. Bot.,* 54, 25, 1976.

34. **Birnbaum, G. I., Huber, C. P., Post, M. L., and Stothers, J. B.,** Sesquiterpenoid stress compounds of *Datura stramonium:* biosynthesis of the three major metabolites (1, 2^{13} C)acetate and the X-ray structure of 3-hydroxylubimin, *J. Chem. Soc. Chem. Comm.,* 330, 1976.

35. **Arneson, P. and Durbin, R.,** The sensitivity of fungi to α-tomatine, *Phytopathology,* 58, 536, 1968.

36. **Mohanakumaran, N., Gilbert, J., and Buddenhagen, I.,** Relationship between tomatine and bacterial wilt resistance in tomato, *Phytopathology,* 59, 14, 1969.

37. **Tjamos, E. C. and Smith, I. M.,** The role of phytoalexins in resistance of tomato to *Verticillium* wilt, *Physiol. Plant Pathol.,* 4, 249, 1974.

38. **McCance, D. and Drysdale, R.,** Production of tomatine and rishitin in tomato plants inoculated with *Fusarium oxysporum* F. sp. *lycopersici, Physiol. Plant Pathol.,* 7, 221, 1975.

39. **Langcake, P., Drysdale, R., and Smith, H.,** Post-infectional production of an inhibitor of *Fusarium oxysporum* f. sp. *lycopersici* by tomato plants, *Physiol. Plant Pathol.,* 2, 17, 1972.

40. **deWit, P. J. G. M. and Flach, W.,** Differential accumulation of phytoalexins in tomato leaves but not in fruits after inoculation with *Cladosporium fulvum, Physiol. Plant Pathol.,* 15, 257, 1979.

41. **Huston, R. A. and Smith, I. M.,** Phytoalexins and tyloses in tomato cultivars infected with *Fusarium oxysporium* f. sp. *lycopersici* or *Verticillium albo- atrum, Physiol. Plant Pathol.,* 17, 245, 1980.

42. **Grzelinska, A. and Sieakowska, J.,** Isolation of rishitin from tomato plants, *Phytopathol. Z.,* 91, 320, 1978.

43. **deWit, P. J. G. M. and Kodde, E.,** Induction of polyacetylenic phytoalexins in *Lycopersicon esculentum* after innoculation with *Cladosporium fulvum, Physiol. Plant Pathol.,* 18, 143, 1981.

44. **Tanguy, J. and Martin, C.,** Phenolic compounds and the hypersensitivity reaction in *Nicotiana tabacum* infected with tobacco mosaic virus, *Phytochemistry,* 11, 19, 1972.

45. **Reuveni, M. and Cohen, Y.,** Growth retardation and changes in phenolic compounds, with special reference to scopoletin, in mildewed and ethylene-treated tobacco plants, *Physiol. Plant Pathol.,* 12, 179, 1978.

46. **Bailey, J. A., Burden, R. S., and Vincent, G. G.,** Capsidiol: an antifungal compound produced in *Nicotiana tabacum* and *N. clevelandii* following infection with tobacco necrosis virus, *Phytochemistry,* 14, 597, 1975.

47. **Cruickshank, I. A. M., Perrin, D. R., and Mandryk, M.,** Capsidiol produced by tobacco infected with *Peronospora hyoseyami* f. sp. *tabacina, Annu. Rep. Div. Plant Ind. CSIRO Australia,* 1975, 65.

48. **Helgeson, J. P., Budde, A. D., and Haberlach, G. T.,** Capsidiol: a phytoalexin produced by tobacco callus tissues, *Plant Physiol.,* 61, Suppl. 53, 1978.

49. **Hammerschmidt, R. and Kuć, J.,** Isolation and identification of phytuberin from *Nicotiana tabacum* previously infiltrated with an incompatible bacterium, *Phytochemistry,* 18, 874, 1979.

50. **Budde, A. D. and Helgeson, J. P.,** Phytoalexins in tobacco callus tissues challenged by zoospores of *Phytophthora parasitica* var *nicotiana, Phytopathology,* 71, 206, 1981.

51. **Uegaki, R., Fujimoro, T., Kubo, S., and Kato, K.,** Sesquiterpenoid stress compounds from *Nicotiana* species, *Phytochemistry,* 20, 1567, 1981.

52. **Guedes, M. E. M., Kuć, J., Hammerschmidt, R., and Bostock, R.,** Accumulation of six sesquiterpenoid phytoalexins in tobacco infiltrated with *Pseudomonas lachrymans, Phytochemistry,* 21, 2987, 1982.

53. **Burden, R. S., Bailey, J. A., and Vincent, G. G.,** Glutinosone, a new antifungal sesquiterpene from *Nicotiana glutinosa* infected with tobacco mosaic virus, *Phytochemistry,* 14, 221, 1975.

54. **Ward, E. W. B., Unwin, C. H., Hill, J., and Stoessl, A.,** Sesquiterpenoid phytoalexins from fruits of eggplants, *Phytopathology,* 65, 859, 1975.

55. **Stoessl, A., Stothers, J. B., and Ward, E. W. B.,** The structure of some stress metabolites from *Solanum melongena, Can. J. Chem.,* 53, 3351, 1975.

56. **Murai, K., Abiko, A., Ono, M., Katsui, V., and Masamune, T.,** Structure revision and biogenetic relationship of aubergenone, a sesquiterpenoid phytoalexin of egg-plants, *Chem. Lett.,* 1209, 1978.

57. **Johnson, G. and Schaal, L.,** Relation of chlorogenic acid to scab resistance in potatoes, *Science,* 155, 627, 1952.

58. **McKee, R.,** Factors affecting the toxicity of solanine and related alkaloids to *Fusarium caeruleum, J. Gen. Microbiol.,* 20, 686, 1959.

59. **Hartmann, M. A. and Benveniste, P.,** Effect of ageing on sterol metabolism in potato tuber slices, *Phytochemistry,* 13, 2667, 1974.

60. **Lyon, G. D.,** Occurrence of rishitin and phytuberin in potato tubers inoculated with *Erwinia carotovora* Var. *atroseptica, Physiol. Plant Pathol.,* 2, 411, 1972.

61. **Johnson, G. and Schaal, L.,** The inhibitory effect of phenolic compounds on growth of streptomyces seabies as related to the mechanism of scab resistance, *Phytopathology,* 45, 626, 1955.

62. **Kuć, J.,** A biochemical study of the resistance of potato tuber to attack by various fungi, *Phytopathology,* 47, 676, 1957.

63. **Kuć, J., Henze, R. E., Ullstrup, A. J., and Quackenbush, F. W.,** Chlorogenic and caffeic acids as fungistatic agents produced by potatoes in response to inoculation with *Helminthosporium carbonum, J. Am. Chem. Soc.,* 78, 3123, 1956.

64. **Sakuma, T. and Tomiyama, K.,** The role of phenolic compounds in the resistance of potato tuber tissue to infection by *P. infestans, Ann. Phytopathol. Soc. Jpn.,* 33, 48, 1967.

65. **Lee, S. and Le Tourneau, D.,** Chlorogenic acid content and *Verticillium* wilt resistance of potatoes, *Phytopathology,* 48, 268, 1958.

66. **Sokolova, V. E.,** Toxicity of chlorogenic acid and its derivatives caffeic and quinic acids for *Phytophthora infestans* fungus, *Izv. Akad. Nauk SSSR, Ser. Biol.,* 28, 707, 1963.

67. **Zucker, M.,** Influence of light on synthesis of protein and chlorogenic acid in potato tuber tissue, *Plant Physiol.,* 38, 575, 1963.

68. **Zucker, M.,** Induction of phenylalanine deaminase by light and its relation to chlorogenic acid synthesis in potato tuber tissue, *Plant Physiol.,* 40, 779, 1965.

69. **Galliard, T.,** Degradation of plant lipid by hydrolytic and oxidative enzymes, in *Recent Advances in Chemistry and Biochemistry of Plant Lipids,* Galliard, T. and Mercer, E., Eds., Academic Press, New York, 1975, 319.

70. **Kolattukudy, P.,** Biochemistry of cutin, suberin, and waxes. The lipid barriers on plants, in *Recent Advances in the Chemistry and Biochemistry of Plant Lipids,* Galliard, T. and Mercer, E., Eds., Academic Press, New York, 1975, 203.

71. **Henderson, S. J. and Friend, J.,** Increase in PAL and lignin-like compounds as race-specific responses of potato tubers to *Phytophthora infestans, Phytopathol. Z.,* 94, 323, 1979.

72. **Friend, J.,** Lignification in infected tissue, in *Biochemical Aspects of Plant-Parasite Interactions,* Friend, J. and Threlfall, D., Eds., Academic Press, New York, 1976, 291.

73. **Katsui, N., Murai, A., Takasugi, M., Imaizumi, K., and Masamune, T.,** The structure of rishitin, a new anti-fungal compound from diseased potato tubers, *J. Chem. Soc. Chem. Comm.,* 43, 1968.

74. **Price, K. R., Howard, B., and Coxon, D. T.,** Stress metabolite production in potato tubers infected by *Phytophthora infestans, Fusarium avenaceum* and *Phoma exigua, Physiol. Plant Pathol.,* 9, 189, 1976.

75. **Lyon, G., Lund, B., Bayliss, C., and Wyatt, G.,** Resistance of potato tubers to *Erwinia carotovora* and formation of rishitin and phytuberin in infected tissue, *Physiol. Plant Pathol.,* 6, 43, 1975.

76. **Haard, N. F. and Cody, M.,** Stress metabolites in postharvest fruits and vegetables—role of ethylene, in *Post-harvest Biology and Biotechnology,* Hultin, H. O. and Milner, M., Eds., Food and Nutrition Press, Westport, CT, 1978, 111.

77. **Cheema, A. S. and Haard, N. F.,** Induction of rishitin and lubimin synthesis in potato tuber slices by non-specific elicitor-role of gene derepression, *J. Food Prot.,* 42, 512, 1979.

78. **Tomiyama, K.,** Some factors affecting the death of hypersensitive potato plant cells infected by *Phytophthora infestans, Phytopathol. Z.,* 39, 134, 1960.

79. **Sato, N., Kitazowa, K., and Tomiyama, K.,** The role of rishitin in localizing the invading hyphae of *Phytophthora infestans* in infection sites at the cut surfaces of potato tubers, *Physiol. Plant Pathol.,* 1, 289, 1971.

80. **Shih, M. J., Kuć, J., and Williams, E. B.,** Suppression of steroid glycoalkaloid accumulation as related to rishitin accumulation in potato tubers, *Phytopathology,* 63, 821, 1973.

81. **Bostock, R. M., Nuckles, E., Henfling, J. W. D. M., and Kuć, J.,** Effect of potato tuber age and storage on sesquiterpenoid stress metabolite accumulation, steroid glycoalkaloid accumulation, and response to abscisic acid and arachidonic acids, *Phytopathology,* 73, 435, 1983.

82. **Cheema, A. S. and Haard, N. F.,** Induction of rishitin and lubimin in potato tuber discs by non-specific elicitors and the influence of storage conditions, *Physiol. Plant Pathol.,* 13, 233, 1978.

83. **Henfling, J. W. D. M., Bostock, R., and Kuć, J.,** Effect of abscisic acid on rishitin and lubimin accumulation and resistance to *Phytophthora infestans* and *Cladosporium cucumerinum* in potato tuber tissue slices, *Phytopathology,* 70, 1074, 1980.

84. **Henfling, J. W. D. M.,** Aspects of the Elicitation and Accumulation of Terpene Phytoalexins in Potato-*Phytophthora infestans* Interaction, Ph.D. thesis, University of Kentucky, Lexington, 1979.

85. **Bostock, R. M., Henfling, J. W. D. M., and Kuć, J.,** Lack of correlation between resistance and the accumulation of sesquiterpene stress metabolites in potatoes inoculated with *Phytophthora infestans* during the growing season, *Phytopathology,* 73, 441, 1983.

86. **Currier, W. W. and Kuć, J.,** Effect of temperature on rishitin and steroid glycoalkaloid accumulation in potato tuber, *Phytopathology,* 65, 1194, 1975.

87. **Alves, L. M., Heisler, E. G., Kissinger, J. C., Patterson, J. M., III, and Kalan, E. B.,** Effects of controlled atmospheres on production of sesquiterpenoid stress metabolites by white potato tuber, *Plant Physiol.,* 63, 359, 1979.

88. **Schmidt, K., Langerak, D. I., and Van Duren, M.,** Effects of irradiation and/or heat treatment on the phytoalexin accumulation in potato tubers, *Z. Lebensm. Unters. Forsch.,* 180, 369, 1985.

89. **Tomiyama, K. and Fukaya, M.,** Accumulation of rishitin in dead tuber tissue following treatment with $HgCl_2$, *Ann. Phytopathol. Soc. Jpn.,* 41, 418, 1975.

90. **Metlitskii, L. V., D'yakov, Yu, T., Ozeretskovskaya, O. L., Yurganova, L. A., Chalova, L. I., and Vasyukova, N. I.,** Induction of potato phytoalexins, *Izv. Akad. Nauk SSSR. Ser. Biol.,* 3, 399, 1971.

91. **Doke, N., Nakae, Y., and Tomiyama, K.,** Effect of blasticidin S on production of rishitin in potato tuber tissue infected by an incompatible race of *Phytophthora infestans, Phytopathol. Z.,* 87, 337, 1976.

92. **Mustafa, M. and D'yakov, Yu. T.,** Effects of chemicals on the interactions between *Phytophthora infestans* (Mont.) d By. and its host plants. Effects of steroids, *Mikol. Fitopathol.,* 3, 241, 1978.

93. **Mustafa, M. and D'yakov, Yu, T.,** Effects of chemicals on the interactions of *Phytophthora infestans* (Mont.) d By with its host plants. II. Effects of fungicides and antibiotics, *Mikol. Fitopathol.,* 13, 33, 1979.

94. **Komai, K. and Sato, S.,** Studies on the browning in potato tubers caused by nematocides. V. Induction of rishitin formation and browning in potato tuber tissues by alkyl halides, *Nippon Nogei Kagaku Kaishi,* 50, 357, 1976.

95. **Zook, M. N., Jaffery, S., and Kuć, J.,** A role for calcium in the elicitation of rishitin and lubimin accumulation in potato tuber tissues, *Plant Physiol.,* 84, 520, 1987.

96. **Murai, A., Yoshizawa, Y., Sato, K., Hasegawa, T., Masamune, T., Sato, N., and Tamura, M.,** *Tennen Yaki Kagobutsu Toronkai Koen Yoshinshu,* 29, 193, 1987.

97. **Ersek, T., Barna, B., and Kiraly, Z.,** Hypersensitivity and the resistance of potato tuber tissues to *Phytophthora infestans, Acta Phytopathol.,* 8, 3, 1973.

98. **Ersek, T., Kiraly, Z., and Dobrovolszky, A.,** Lack of correlation between tissue necrosis and phytoalexins accumulation in tubers of potato cultivars, *J. Food Safety,* 1, 77, 1977.

99. **Cheema, A. S. and Haard, N. F.,** Actionomycin D induction of terpene stress metabolites in potato tuber slices, *J. Food Biochem.,* 2, 277, 1979.

100. **Henfling, J. W. D. M., Lisker, N., and Kuć, J.,** Effect of ethylene on phytuberin and phytuberol accumulation in potato tuber slices, *Phytopathology,* 68, 857, 1978.

101. **Beczner, J.,** Effect of ethylene on potato tubers inoculated with *E. carotovora* var. *Atroseptica, Acta Phytopathol. Acad. Sci. Hung.,* 11, 235, 1976.

102. **Ersek, T.,** A lipoid component from *Phytophthora infestans* inducing resistance and phytoalexin accumulation in potato tubers, *Curr. Top. Plant Pathol.,* (Proc. Symp.), 1977, 73.

103. **Bostock, R. M.,** The Identification of Polyunsaturated Fatty Acids from *Phytophthora infestans* as Elicitors of Hypersensitivity in Potato Tuber and Studies of Related Terpenoid Metabolism, Dissertation, University of Kentucky, Lexington, 1981.

104. **Bostock, R. M., Kuć, J., and Laine, R. A.,** Ecosapentaenoic and arachidonic acids from *Phytophthora infestans* elicit fungitoxic sesquiterpenes in the potatoes, *Science,* 212, 67, 1981.

105. **Bostock, R. M., Laine, R. A., and Kuć, J.,** Factors affecting the elicitation of sesquiterpenoid phytoalexin accumulation by eicosapentaenoic and arachidonic acids in potato, *Plant Physiol.,* 70, 1417, 1982.

106. **Stelzig, D. A., Allen, R. D., and Bhatia, S. R.,** Inhibition of phytoalexin synthesis in arachidonic acid-stress potato tissue by inhibitors of lipoxygenase and cyanide resistant respiration, *Plant Physiol.,* 72, 746, 1983.

107. **Maina, G., Allen, R. D., Bhatia, S. R., and Stelizig, D. A.,** Phenol metabolism, phytoalexins and respiration in potato tuber tissue treated with fatty acid, *Plant Physiol.,* 76, 735, 1984.

108. **Bloch, C. B., De Wit, P. J. G. M., and Kuć, J.,** Elicitation of phytoalexins by arachidonic and eicosapentaenoic acids: a host survey, *Physiol. Plant Pathol.,* 25, 199, 1984.

109. **Davis, D. A. and Currier, W. W.,** The effect of the phytoalexin elicitors, arachidonic and eicosapentaenoic acids and other unsaturated fatty acids on potato tuber protoplasts, *Physiol. Mol. Plant Pathol.,* 28, 431, 1986.

110. **Tjamos, E. C., Nuckles, E., and Kuć, J.,** Regulation of steroid glycoalkaloid and sesquiterpenoid stress metabolite accumulation in potato tubers by inhibitors of steroid synthesis and phytoharmones in combinations with arachidonic acid, *NATO ASI Ser., Ser. H., (Biol. Mol. Biol. Plant-Pathol. Interact.),* 1, 197, 1986.

111. **Mustafa, M. and D'yakov, Yu. T.,** Effect of pesticides on the vulnerability of potatoes and tomatoes to late blight, *Khim. Sel'sk. Khoz.,* 17, 56, 1979.

112. **Varns, J. L., Currier, W. W., and Kuć, J.,** Specificity of rishitin and phytuberin accumulation by potato, *Phytopathology,* 61, 968, 1971.

113. **Lisker, N. and Kuć, J.,** Elicitors of terpenoid accumulation in potato tuber slices, *Phytopathology,* 67, 1356, 1977.

114. **Henfling, J. W. D. M., Bostock, R. M., and Kuć, J.,** Cell wall of *Phytophthora infestans* contain an elicitor of terpene accumulation in potato tubers, *Phytopathology,* 70, 772, 1980.

115. **Stoessl, A.,** Stress compound formation in the interaction of potato tubers with *Alternaria solani* and *A. solani* metabolites, *Physiol. Plant Pathol.,* 20, 263, 1982.

116. **Doke, N., Garas, N. A., and Kuć, J.,** Partial characterization and aspects of the mode of action of a hypersensitive inhibiting factor (HIF) isolated from *Phytophthora infestans, Physiol. Plant Pathol.,* 15, 127, 1979.

117. **Garas, N. A., Doke, N., and Kuć, J.,** Suppression of the hypersensitive reactions in potato tubers by mycelial components from *Phytophthora infestans, Physiol. Plant Pathol.,* 15, 117, 1979.

118. **Doke, N., Garas, N., and Kuć, J.,** Effect of host hypersensitivity of suppressors related during the germination of cytospores of *Phytophthora infestans, Phytopathology,* 70, 35, 1980.

119. **Doke, N. and Tomiyama, K.,** Suppression of the hypersensitive response of potato tuber protoplasts to hyphal wall component by water soluble glucans isolated from *Phytophthora infestans, Physiol. Plant Pathol.,* 16, 177, 1980.

120. **Doke, N. and Tomiyama, K.,** Effect of hyphal wall components from *Phytophthora infestans* on protoplasts of potato tuber tissues, *Physiol. Plant Pathol.,* 16, 169, 1980.

121. **Garas, N. A. and Kuć, J.,** Lectin from potato lyses zoospores of *Phytophthora infestans* and precipitates the elicitor of terpenoid accumulation in potato, *Physiol. Plant Pathol.,* 18, 277, 1981.

122. **Tomiyama, K.,** Double infection by an incompatible race of *Phytophthora infestans* of potato cell which has previously been infected by a compatible race, *Ann. Phytopathol. Soc. Jpn.,* 32, 181, 1966.

123. **Varns, J. and Kuć, J.,** Suppression of rishitin and phytuberin accumulation and hypersensitive response in potato by compatible races of *Phytophthora infestans, Phytopathology,* 61, 178, 1971.

124. **Varns, J. and Kuć, J.,** Suppression of the resistance response as an active mechanism for susceptibility in the potato-*Phytophthora infestans,* in *Phytotoxins in Plant Diseases,* Wood, R. K. S. and Graniti, A., Eds., Academic Press, New York, 1972, 465.

125. **Doke, N.,** Prevention of hypersensitive reaction of potato cells to infection with an incompatible race of *Phytophthora infestans* by constituents of the zoospores, *Physiol. Plant Pathol.,* 7, 1, 1975.

126. **Price, K. R., Howard, B., and Coxon, D. T.,** Stress metabolite production in potato tubers infected by *Phytophthora infestans, Fusarium avenaceum* and *Phoma exigua, Physiol. Plant Pathol.,* 9, 189, 1976.

127. **Tomiyama, K., Sakuma, T., Ishizaka, N., Sato, N., Takasugi, M., and Katsui, T.,** A new antifungal substance isolated from potato tuber tissue infected by pathogens, *Phytopathology,* 58, 115, 1968.

128. **Stoessl, A., Stothers, J. B., and Ward, E. W. B.,** Biosynthetic studies of stress metabolites from potato: incorporation of sodium acetate-$^{13}C_2$ into 10 sesquiterpenes, *Can. J. Chem.,* 56, 645, 1978.

129. **Sato, N., Tomiyama, K., Katsui, N., and Masamune, T.,** Isolation of rishitin from tubers of interspecific potato varieties containing different late blight resistance genes, *Ann. Phytopathol. Soc. Jpn.,* 34, 140, 1968.

130. **Varns, J., Kuć, J., and Williams, E.,** Terpenoid accumulation as a biochemical response of the potato tuber to *Phytophthora infestans, Phytopathology,* 61, 174, 1971.

131. **Zook, M. N. and Kuć, J.,** Differences in phytoalexin elicitation by *Phytophthora infestans* and *Helminthosporium carbonum* in potato, *Phytopathology,* 77, 1217, 1987.

132. **Horikawa, T., Tomiyama, K., and Doke, N.,** Accumulation and transformation of rishitin and lubimin in potato tuber tissue infected by an incompatible race of *Phytophthora infestans, Phytopathology,* 66, 1186, 1976.

133. **Zacharius, R. M., Osman, S. F., Heisler, E. G., and Kissinger, J. C.,** Effect of the R-3 gene in resistance of the Wauseon potato tuber to *Phytophthora infestans, Phytopathology,* 66, 964, 1976.

134. **Lisker, N. and Kuć, J.,** Terpenoid accumulation and browning in potato sprouts inoculated with *Phytophthora infestans, Phytopathology,* 68, 1284, 1978.

135. **Ersek, T. and Sziraki, I.,** Production of sesquiterpene phytoalexins in tissue culture callus of potato tubers, *Phytopathol. Z.,* 97, 364, 1980.

136. **Brindle, P. A., Kuhn, P. J., and Threlfall, D. R.,** Accumulation of phytoalexins in potato cell suspension cultures, *Phytochemistry,* 22, 2719, 1983.

137. **Brishammar, S.,** Critical aspects of phytoalexins in potato, *J. Agric. Sci. Fin.,* 59, 217, 1987.

138. **Schöber, B.,** Phytoalexine in Knollen resistenter und anfalliger Kartoffelsorten nach Infection mit *Phytophthora infestans* (Mont.) de Bary, *Potato Res.,* 23, 435, 1980.

139. **Sato, N., Kitazawa, K., and Tomiyama, K.,** The role of rishitin in localizing the invading hyphae of *Phytophthora infestans* in infection sites at the cut surfaces of potato tubers, *Physiol. Plant Pathol.,* 1, 289, 1971.

140. **Jones, D. R., Unwin, C. H., and Ward, E. W. B.,** The significance of capsidiol induction in pepper fruit during an incompatible interaction with *Phytophthora infestans, Phytopathology,* 65, 1286, 1975.

141. **Ward, E. W. B., Unwin, C. H., and Stoessl, A.,** Experimental control of late blight of tomatoes with capsidiol, the phytoalexin from peppers, *Phytopathology,* 65, 168, 1975.

142. **Ishizaka, N., Sato, N., and Tomiyama, K.,** Attempt to test a possible role of rishitin to protect potato plant against the late blight, *Hokkaido Nogyo Shinkenjo Iho,* 99, 62, 1971.

143. **Metlitskii, L. V., Ozeretskovskaya, O. L., Vasyukova, N. I., Davydova, M. A., Saveleva, O. N., and Dyakov, Yu. T.,** Role of phytoalexins in the vertical resistance of the potato to *Phytophthora infestans, Mikol. Fitopathol.,* 8, 42, 1974.

144. **Lyon, G. D.,** Comparisons between phytoalexin concentrations and the extent of rotting of potato tubers inoculated with *Erwinia carotovora* sub. sp. *atroseptica, E. Carotovora* sub. sp. *carotovora* or *E. chrysanthemi, Phytopathol. Z.,* 11, 236, 1984.

145. **Lyon, G. D. and Bayliss, C. E.,** Effect of rishitin on *Erwinia carotovora* var. *atroseptica* and other bacteria, *Physiol. Plant Pathol.,* 6, 177, 1975.

146. **Lyon, G. D.,** Rishitin and phytuberin in relation to bacterial soft rots of potato tubers, *Ann. Appl. Biol.,* 85, 166, 1977.

147. **Hohl, H. R., Stoessl, P., and Hoechler, H.,** Papilla formation and partial inhibition of fungal glucanases by phytoalexins in the *Phytophthora infestans-Solanum tuberosum* system, *Ann. Phytopathol.,* 12, 353, 1980.

148. **Lyon, G. D.,** Evidence that the toxic effect of rishitin may be due to membrane damage, *J. Exp. Bot.,* 31, 957, 1980.

149. **Lewis, R. J., Sr. and Tatken, R. L., Eds.,** *Registry of Toxic Effects of Chemical Substances (ROTECS),* Cincinnati, OH, 1980.

150. **Wood, G. E.,** Stress metabolites of white potatoes, in *Mycotoxins and Other Fungal Related Food Problems,* Rodricks, J. V., Ed., American Chemical Society, Washington, D.C., 1976, 369.

151. **Poswillo, D. E., Sopher, D., Mitchell, S. J., Coxon, D. T., Curtis, R. F., and Price, K. R.,** Investigation into the teratogenic potential of imperfect potatoes, *Teratology,* 8, 339, 1973.

152. **Chaube, S., Swinyard, C. A., and Daines, R. H.,** Failure to induce malformations in fatal rats by feeding blighted potatoes to their mothers, *Lancet,* 1, 329, 1973.

153. **Keeler, R. F., Douglas, D. R., and Stallknecht, G. E.,** Failure of blighted Russet Burbank potatoes to produce congenital deformities on rats, *Proc. Soc. Exp. Biol. Med.,* 146, 284, 1974.

154. **Keeler, R. F., Dougals, D. R., and Stallknecht, G. F.,** The listing of blighted, aged, and control Russet Burbank potato tuber preparation for ability to produce spina bifida and anencephaly in rats, rabbits, hamsters and mice, *Am. Potato J.,* 52, 125, 1975.

155. **Ruddick, J. A., Harwig, J., and Scott, P. M.,** Nonteratogenicity in rats of blighted potatoes and compounds contained in them, *Teratology,* 9, 165, 1974.

156. **Sharma, R. P., Willhite, C. C., Wu, M. T., and Salunkhe, D. K.,** Teratogenic potential of blighted potato concentrate in rabbits, hamsters and miniature swine, *Teratology,* 18, 55, 1978.

157. **Neudecker, C. and Schober, B.,** Futterungsversuche zum Teratogenen Potential Phytoalexin haltiger Kartoffeln, *EAPR Abstr.,* Conf. Papers., 372, 1984.

158. **Ishag, K. E. A.,** Mono- and Sesquiterpene Alcohol Glucosides—Synthesis and Characterization, Ph.D. dissertation, Mathematisch-Naturwissenschaftlichen Fakultat der Universitat der Saarlands, Saarbrucken., West Germany, 1984.

Chapter 10

PRESENT OUTLOOK AND FUTURE PROSPECTS

S. S. Dhumal, D. K. Salunkhe, and S. S. Kadam

TABLE OF CONTENTS

I. INTRODUCTION

The potato (*Solanum tuberosum* L.) has spread around the world during the last 500 years and adapted to a wide variety of environments and an equally diverse range of human taste and preference. The potato crop is grown in about 80% of the countries of the world. It is second only to maize (corn) in terms of the number of producer countries and fourth after wheat, maize, and rice in global tonnage. The potato crop is increasing in most industrial nations. Since 1950, it has increased by about 80% in Australia and New Zealand and by more than 40% in the U.S. and Canada; it has remained constant in Eastern Europe and the U.S.S.R. The only region where potato production has fallen slightly is Western Europe. This is because of farm operations, income growth, and increased potato prices. In some tropical developing countries, it is a common vegetable while elsewhere consumption ranges from 1 to 100 kg per person annually. This chapter addresses the diversity in potato consumption patterns around the world with future prospects.

II. CULTIVATION OF POTATOES

The potato is one of the major food crops grown in a wide variety of soils and climatic conditions.[1,2] The dry matter and protein production of potatoes per unit area exceeds that of cereals (Table 1).[3] Because of increasing yield per unit area of land, total potato production has been increasing even though the area of land planted to potatoes is decreasing (Figures 1, 2, and 3).[4] However, fresh-potato consumption is decreasing (Figure 4),[4] and utilization of potatoes for processed products is increasing (Figure 5),[4] especially frozen products (Figure 6).[4] On the contrary, it has increased significantly in the tropical developing countries.[5] During the past 30 years, potato production and utilization have changed considerably around the world. No longer is the potato crop suitable only for cool climates. The potato has moved down the hills into warm climates, and into rice fields during the dry season. In fact, the potato fits very well into cereal-based farming systems; as the countries of Asia and Africa become self-sufficient in rice, the potato is becoming a priority alternative crop.[2] The potato has already demonstrated its flexibility to fit into farming in a few tropical areas where it has become important as a staple food. Still to be investigated is the role of potatoes in the farming systems of the rest of the tropics where it is still mainly a luxury vegetable. According to FAO statistics, the potato has had the greatest increase in production in the third-world countries of any major world food crop during the past 10 years. These statistics certainly attest to the acceptability of potatoes as a food and the increasing awareness of productive capabilities and the food value of potatoes in developing countries. India and China, examples of countries which are traditionally cereal (rice and wheat)-consuming nations, have found ready acceptability of potatoes.

The major limiting factors for the use of potatoes as a food in tropical climates is the availability of both true and certified seeds. Most tropical countries are presently relying on costly importations from northern-latitude countries as a base for potato production. Seed costs are from 50 to 75% of the total production costs under such conditions. The systemic study on the interrelationship of the climate to pest (vector) avoidance has helped in the realization of low-cost, healthy seed production in India. In the developing countries producing potatoes, the extension and/or modification of the Indian seed production system will be useful to identify suitable areas for potato seed production. Selection should be where climatological factors allow the required period for potato production as well as avoidance of vector populations. This will provide a solution to the production of low-cost, high-quality tuber seed and reduce or eliminate dependence on seed imports. New seed technology is permitting countries in the tropics to establish their own seed programs independent of expensive importations. The use of true potato seed instead of tubers is providing a way for

TABLE 1
Approximate Yields (1987) of Dry Matter and Crude Protein
from Potatoes and Cereals in the FAO Regions

Region	Dry matter (tonnes/ha)		Crude protein (tonnes/ha)	
	Potatoes	Cereals	Potatoes	Cereals
World	14.6	2.6	1.46	0.36
Developed countries	17.7	3.1	1.77	0.43
North America	14.5	4.1	1.45	0.56
Western Europe	11.1	4.3	1.11	0.61
Oceania	14.7	1.5	1.47	0.21
Other developed	19.9	2.7	1.99	0.38
Developing countries	14.5	2.2	1.45	0.32
Africa	5.7	1.1	0.57	0.16
Latin America	7.2	2.0	0.72	0.28
Near East	19.6	1.6	1.96	0.23
Far East	8.2	1.9	0.82	0.22
Other developing	4.7	2.0	0.47	0.28
Centrally planned economies	17.3	2.9	1.73	0.40
Asian CPE	17.3	3.8	1.73	0.55
European U.S.S.R.	17.4	2.3	1.74	0.32

From FAO, Food and Agricultural Organization of the U.N., 1988.

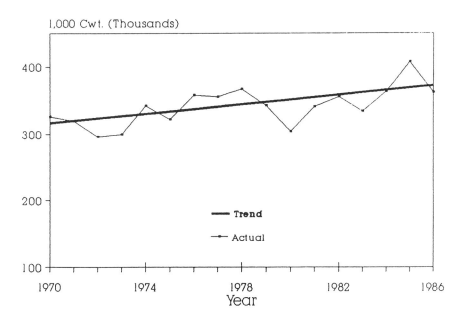

FIGURE 1.   Potato production. (From *Agricultural Statistics,* U.S. Department of Agriculture, Washington, D.C., 1988.)

good, clean planting material to reach small growers in tropical climates. Although the potato is a high-investment crop, returns are very high in relation to the investment. Several reports from developing countries in the tropics indicate that the potato produces a higher return to the grower per unit area than any other crop. Obviously, part of this is due to the high cost of potatoes in the market and the high demand for this commodity known mainly as a luxury vegetable in many tropical countries.

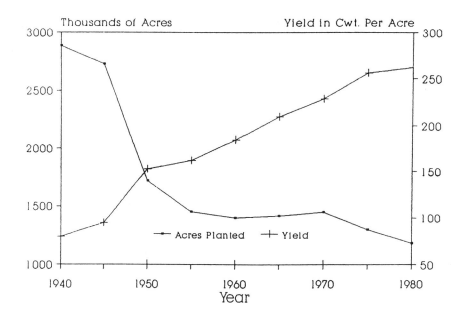

FIGURE 2.   Trends in white potato acreage planted in the U.S. and in yields per acre, 1938 to 1979. (From *Agricultural Statistics*, U.S. Department of Agriculture, Washington, D.C., 1988.)

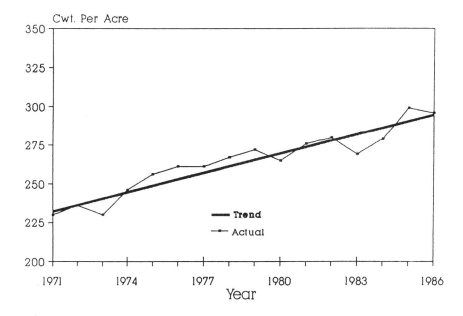

FIGURE 3.   Potato yield. (From *Agricultural Statistics*, U.S. Department of Agriculture, Washington, D.C., 1988.)

III. POTATO CONSUMPTION

Figures on potato consumption are usually derived from food balance sheets as the residual after net exports, nonfood uses (livestock feed, industry, seed), and waste have been subtracted from total production. All the components of these calculations are estimates, and the compounding of errors make consumption estimates highly suspect. Household surveys that have focused especially on the issue of potato consumption indicate that, in

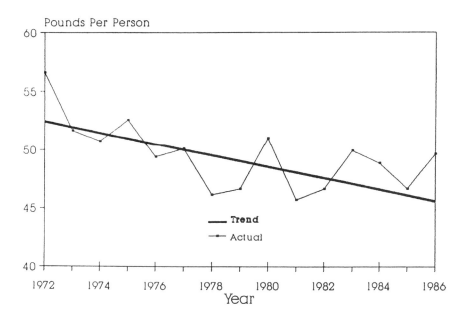

FIGURE 4. Fresh potatoes, per capita consumption and trend. (From *Agricultural Statistics*, U.S. Department of Agriculture, Washington, D.C., 1988.)

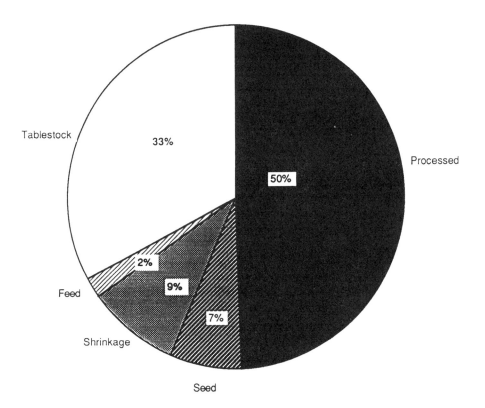

FIGURE 5. Utilization of potatoes in 1986 (U.S.). (From *Agricultural Statistics*, U.S. Department of Agriculture, Washington, D.C., 1988.)

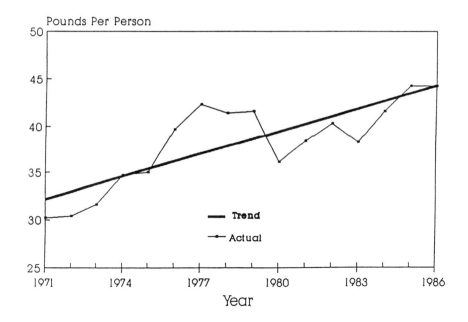

FIGURE 6.   Frozen potatoes, per capita consumption and trend. (From *Agricultural Statistics*, U.S. Department of Agriculture, Washington, D.C., 1988.)

TABLE 2
Potato Consumption in
Guatemala (1965 to 1967)

Location	kg/person/year
Urban	13.82
Rural	6.0
Average	8.32

Courtesy of CIP, Lima, Peru, 1988.

general, the food balance sheets substantially underestimate potato consumption. In addition, as national averages, potato consumption estimates mask important differences in consumption among groups within a country.

Within rural potato-growing areas, producers eat more potatoes than nonproducers. Both of these groups eat less potatoes than urban people in other areas, because potatoes marketed outside of the producing areas are usually consumed in large towns and cities (Table 2).[7] High transport costs and the low purchasing power of rural consumers discourage market agents from shipping potatoes from producing areas to rural deficit areas.

In towns and cities, the effect of household income on potato consumption depends upon the relative price of potatoes. In the few developing areas where potatoes are relatively cheap, such as in the Andes, potato consumption is highest among poor households. In the more common situation where potatoes are relatively expensive, as in most of Africa and Asia, potato consumption is highest among the wealthy. The implication is that as income levels increase in the future, so will potato consumption in most places. In many places where potatoes are still little known or eaten occasionally as a vegetable, they are likely to become an important vegetable in the diet (Table 3).

TABLE 3
Potato Consumption and Their Frequency

Role	Frequency	Rate (kg/year)	Beliefs
Staple	5 + Meals per week	60—200	Potato = food, meal incomplete without potatoes
Complementary as vegetable	2 to 4 Meals per week	10—60	Potato = vegetable
Luxury or special food	1 to 2 Meals per week	1—10	Potato = rich man's food

IV. NUTRITIVE VALUE OF POTATOES

The potato contributes not only energy, but also substantial amounts of high-quality protein and essential vitamins, minerals, and trace elements to the diet. The energy and protein contents of fresh potatoes are much lower than those of cereals, but the differences are narrowed when these foods are cooked. The biological value of potato protein is better than that of most other vegetable sources and is comparable to that of cow's milk. Its high lysine content makes potato protein a valuable compliment to cereal-based diets that are generally limiting in this amino acid. The potato is well balanced in the sense that the protein-to-calorie ratio is higher than root crops, most cereals, and plantains. If, however, one satisfies one's entire protein requirement with potatoes, one could still be short of calories. This illustrates the error of considering the potato primarily as a source of starch or calories. The potato is comparable to other common vegetables in vitamin content and is especially rich in vitamin C. As little as 200 g of boiled potato provides an adult's recommended daily allowance of vitamin C. Mineral content of a potato is strongly influenced by the soil in which it is grown. Normally, the potato is a moderately good source of iron, a good source of phosphorus and magnesium, and an excellent source of potassium. These nutritional facts indicate that contribution of the potato to the diet is not principally energy, but rather protein, vitamins, and minerals.

V. QUALITY OF POTATOES

Quality of potato tubers for processing depends upon genetic and environmental factors.[8] The successful production of potatoes of high quality is fundamentally influenced by the individual as well as the cumulative effects of these factors. Though these are well-recognized factors, their effects on quality need further exploration and definition. The particular factors such as selection of variety, date of planting, fertilizers, herbicides and pesticides, fungicides, vine-killing treatments, soil structure and texture, irrigations, locations, and their cumulative effects on processed products need periodic investigations. Likewise, storage temperature, duration and subsequent conditioning, specific gravity of the tubers, and several other factors influence the final quality (color, flavor, texture, and nutritive value) of the processed products.

Sowokinos[9] identified several research areas that need exploration before recombination of DNA technology can be efficiently utilized to improve potato quality. Future development of superior variety types utilizing genetic engineering techniques depends greatly upon the scientific contributions made in five research areas: (1) identification of potato genes of agricultural importance, (2) isolation and characterization of the structure and regulatory properties of these genes and their products, (3) identification of vectors that are compatible with potato cells that can be sliced with the desired genomic segment, (4) development of methods to effectively introduce the recombinant vector in potato cells, and (5) to identify regenerated plants that express the desired trait. A few of the possible benefits that may ultimately evolve from such research in potatoes are[9] (1) introduction of resistant genes to prevent attack of pests; (2) increase in plant products; (3) increase in plant resistance to salinity, acidity, heat- and cold-stress conditions; (4) insertion and expression of nitrogen fixation genes; and (5) increase in photosynthetic efficiency, carbon assimilation, and yield.

VI. STORAGE OF POTATOES

In developing countries, enormous losses result from spoilage, contamination, attack by microorganisms, insects, birds, rodents, and deterioration in storage. The postharvest losses of potatoes ranging from 5 to 40%[10] for all devleoping countries, 25 to 30% for Peru and Venezuela,[11] and 25% on a global basis[12] have been reported. If these losses could be reduced through research and development, there would be a significant improvement in the world food supply. Hence, appropriate and efficient postharvest technology and marketing are critical to the entire production and use system of the potato due to its bulkiness and perishability. Potatoes are available for consumption only during a part of the year due to inadequate storage. Unlike in temperate regions, in tropics and subtropics the potato crop is harvested at the time when summer sets in, and temperatures are high during postharvest handling and marketing, leading to deterioration in storability and consumer quality. It is therefore necessary that priorities be developed for making low-cost, renewable energy-dependent storage and processing systems available. Research is required to develop low-cost alternatives, particularly the application to potato storage of lower-cost cooling systems such as evaporative cooling in hot and dry areas.[13] At the technical level, the problem of seed tuber storage may be more readily overcome as it has been shown that seed tubers stored in inexpensive diffused light can be maintained in acceptable physiological conditions for 4 to 6 months at average storage temperatures of 20 to 25°C, provided that important pests and diseases, particularly potato tuber moth, are controlled.[13] In undertaking the technological research on storage technologies for hot regions, emphasis should simultaneously be placed on discovering the client's needs as a means to reduce the transfer gap.

VII. PROCESSING OF POTATOES

Dehydration or sun drying is the simplest and lowest-cost method of processing and preservation.[13] Wherever dried or simply cooked and dried products are acceptable, either directly for food or for secondary processing into needed products, this type of processing should be encouraged. This can ideally be used to upgrade the value of otherwise low-grade tubers. By the simple methods, tubers can be converted to stable items with long storage life and which can readily be transported. The development of alternative and somewhat more sophisticated systems may be required for regions where solar drying is not feasible in the available processing season and where large quantities of potatoes are produced in a short period which cannot be absorbed by the fresh market, stored, or processed by the above simple means.[13] This commonly occurs, for example, when immature tubers are harvested out of paddy fields prior to rice planting. Such processing methods should, if possible, also be relatively low in capital and energy requirements to enable them to be used economically for short periods of the year. Alternatively, the methods used should be applicable to other crops to enable the plants to be used for a greater portion of the year.

In the developing world, the growth of the fast-food industry in recent years has markedly changed the way potatoes are consumed and has called for changes in the varieties produced and how these are stored.[2] The rapid expansion in potato processing which took place some time ago in the developed world is now taking place in third-world countries. While potato production has stabilized in the developed world, in the developing world it is increasing faster than that of any other major world food crop. Many developing countries have just started to rely on the potato as a major food source. This trend can be expected to continue as scarcity of land resources requires a change from extensive production of certain food crops to intensive production with priority consideration given to calories and high-quality protein per unit area and per unit time. Thus, there is the need for new innovations in potato processing in order to quickly convert the potato to a nonperishable, culturally acceptable, highly nutritional food product.

Since 1960, consumption of potato products has been significantly increased due to research and developments in convenience and instant processed products from potato tubers such as potato chips, French fried potatoes, potato flakes, hash brown potatoes, potato granules, diced potatoes, potato flour, potato starch, canned potatoes, prepeeled potatoes, dehydrated instant mashed potatoes, restructured potato chips, potato soup, shoestring potatoes, potato pancake mixes, potato nuts, potato puffs, and potato salad. There has been a significant increase in per capita consumption of processed potatoes in the last 3 decades. Frozen products account for more than half of all potatoes that are processed (Figure 7).[6] In the U.S., the percent of potatoes utilized for production of frozen products has increased since 1959 (Figure 8).[4] Similarly, dehydrated products and chips are the other major products of the potato processing industry. The future of potato processing and instant and convenience products is great, particularly for frozen products, potato chips, restructured frozen French fries, and mashed potato granules. Factors that will influence the future of potato processing are demographics consumer attitudes and needs, new technologies, raw materials, environmental issues, and global issues. The processing of potatoes for French fries, chips, starch, and other products generates both solid and liquid wastes. It is essential that processing losses of potatoes be minimized by improving the existing processing practices. Also, the in-plant management should be oriented towards achieving maximum reuse of process water. The use of wastes for value-added products and energy recovery through waste treatment should be exploited for technical advantages and economic gains.

VIII. GLYCOALKALOIDS AND PHYTOALEXINS OF POTATOES

The role of potato glycoalkaloids as an associated agent in human teratology is far from being fully resolved. Although most animal studies indicate little possibility that eating potatoes, even with moderately high glycoalkaloid contents, can lead to human malformation, the contribution of some of the infectious agents in potatoes cannot be completely eliminated. Some of the organisms can lead to high glycoalkaloid contents in potatoes and can indeed synthesize such toxic chemicals even in a cell-free medium. Additionally, there is need to investigate other factors, such as nutritional status and composition of other food, that may alter the toxic or teratologic risk in people in certain geographical locations. There has been a considerable amount of interest in devising inhibitors of glycoalkaloid synthesis. Although there has been some success with both physical and chemical agents in achieving this goal, the practical value of such treatments is limited. However, before any chemical treatment is employed on a large scale, its influence on other processes in tubers, as well as food safety, should be fully assessed.

The induction of the terpenoid phytoalexins accumulation in response to infections is generally believed to be a part of the resistance mechanism against pathogens. Phytoalexin research has been centered around some major compounds such as rishitin, lubimin, phytuberin, and phytuberol. However, the occurrence of a wide range of terpenoids in potatoes would require further clarifications in order to provide an overall assessment of the role of individual compounds in disease resistance. The emergence of several new compounds, particularly glycoside and terpenes, should stimulate increased interest in extending chemical, biochemical, and biological investigations on these compounds. The elucidation of convincing relationships between three-dimensional chemical structure and biological activity

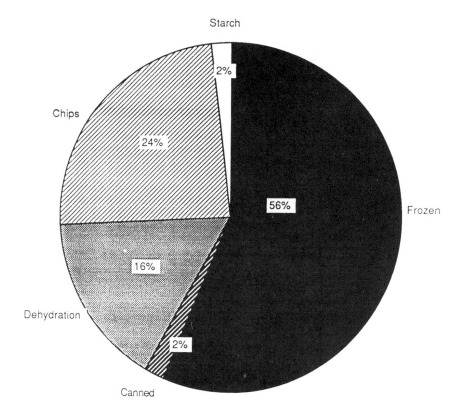

FIGURE 7.   Utilization of potatoes in 1986 (U.S.). (From Mazza, G., Personal communication, 1989.)

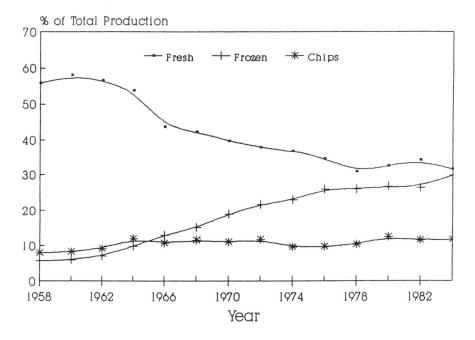

FIGURE 8.   U.S. percent utilization of potatoes. (From *Agricultural Statistics,* U.S. Department of Agriculture, Washington, D.C., 1988.)

TABLE 4
New Technologies Related to Potato Production

Recovery of minor components	Price ($)
Chlorogenic acid (0.05—0.2%)	20/g
Glycoalkaloids (0.01—0.07%)	
Chaconine	2,433/g
Solanine	2,838/g
Lectins (0.01%)	
Agglutinin	17,950/g or 8.1 M/lb
Enzymes	
Apyrase	4.30—21.0 per 100 units
Vanillin	57/kg
Anthocyanins	5—50/kg
Leptine	*
Potatin	**

* Research on Leptine from potato leaves is still in experimental stage. Leptine has potential in the biological control of Colorado potato beetle.

** Potato protein research is presently in progress.

From Mazza, G., personal communication, 1989.

may become useful in establishing definite trends in phytoalexin research. However, these and other chemicals are very costly as chemical reagents and medicinal uses. Their success depends upon development of new technologies (Table 4).[6]

IX. GLOBAL IMPACT OF POTATOES

On a global basis, the potato is a major world food crop occupying fourth place in importance, falling just after three cereals, rice, wheat, and corn. However, most of the importance of the potato as a food exists in the colder regions of the world, either in the highland tropics or the northern and southern latitude temperate climates.[2] The genetic potential which exists with the potato is being utilized for the tropical regions where a major portion of the world population exists with insufficient diet and an inadequate staple food supply. The major cereal crops have been highly exploited for many years. The potato resources have only recently been developed which would permit a similar kind of genetic exploitation as has taken place with cereals. Technology is being developed which will give exciting new approaches to production, storage, and processing for tropical regions of the world. Potential potato consumption in developing countries need to be increased, especially in Latin America, Asia, North Africa, the Near East, and sub-Sahara regions of the world (Figure 9).[14] National capabilities for research and transfer of technology in agriculture are rapidly being developed in a number of tropical countries. Sound national agricultural capabilities in research and transfer of technology are essential to the development of the potato for tropical climates.[2] The short- and long-term research goals are given in Table 5.[6] With coordination of research efforts and a concentration on the priority factors, the potato will take its proper place as the world food crop for the future.

X. SUCCESSFUL POTATO CULTIVATION PROGRAM IN DEVELOPING COUNTRIES

The backlog of scientific information on potatoes in developed countries, together with practical oriented research promoted by the International Potato Center, Lima, Peru, and

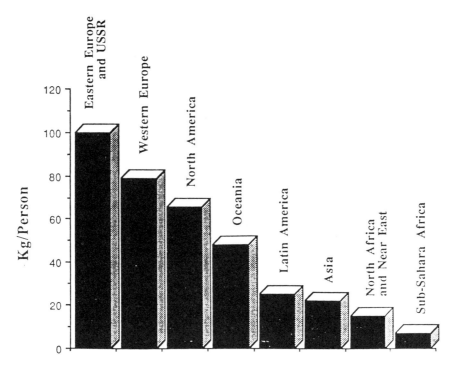

FIGURE 9.   Per capita consumption of potatoes by region. (From FAO Year Book, 1984.)

TABLE 5
Short- and Long-Term Research Goals

Short term (1 to 5 years)

No major changes in developed world
Encourage potato production in developing countries

Long term (5 to 20 years)

Limited expansion of national/North American market
Expansion of export markets
Development of new uses for potatoes
New technologies to reduce costs of production and improve product quality
Expansion of potato cultivation in tropical countries

From Mazza, G., personal communication, 1989.

regional networks, has created many opportunities for high returns to national potato programs in developing countries. But not all programs are successful. The optimal size and
structure of a potato program depends upon the level of development, the institutional
framework, and the personnel, physical, and financial resources of the country available for
potato improvement. Programs with foreign funding that attempt to minimize bureaucratic
problems by working outside of normal administrative channels generally collapse when the
external support is withdrawn. Therefore, strong local commitment and integration into the
existing agricultural research and development system are essential.

A national program needs a solid research base to be able to utilize and adapt the results
of research conducted at the international agricultural centers and other institutions around
the world. Most national programs should have some expertise in four key areas: seed

systems, varieties, pest management, and storage. The specific priorities within each of these areas should result from a careful examination of local production problems and resources of the program.

Detailed information on the main constraints to potato production and use is needed in order to set appropriate goals for technology, research, and extension. A number of programs that have generated an impressive amount of new information and indigenous technology have had negligible impact on production or social welfare because they worked on topics that had little practical importance or produced technologies that were impractical for the typical farmer. Interaction among researchers, extension workers, and their clients (farmers, market agents, consumers) is essential for identifying and solving problems. Communication can be facilitated by interdisciplinary team research that involves both technologists and social scientist. Farmers are usually eager to participate in research on their farms, as long as it has a practical short-term payoff. Once appropriate priorities have been set, the effectiveness of research depends upon the analytical capabilities of researchers and the information base from which they draw. Both can be enhanced through contact with other scientists and access to scientific literature generated by other institutions. The International Potato Center at Lima, Peru, can help national programs tap various information, training, and technical assistance sources.

National programs that need financial resources and resident foreign scientific personnel can benefit from direct associations with bilateral agencies. Increasingly, these agencies and the International Potato Center at Lima, Peru, are working together to eliminate duplication, improve coordination, and provide continuity to potato improvement programs.

XI. FUTURE RESEARCH NEEDS

The research needed may vary from basic research to the utilization of research already done, but several fields deserve particular attention. One is the number and complexity of problems involved in pure-seed production that could be available at a reasonable price for farmers in developing tropical countries of the world for successful potato cultivation. A multidisciplinary approach, as indicated in this chapter, should be employed for successful production of both yield and quality of potatoes which can be available to all consumers at a reasonable price. The magnitude of the problems should be viewed in the context of the potential of a better life for many people if the problems are indeed resolved. The potential for utilization of the potato as food, feed, and industrial products should be followed as indicated in the previous chapters. Increasing utilization of fresh tubers as food will depend mainly upon their availability to consumers year-round at a reasonable price. Likewise, increased utilization of potatoes will need the attention of economists and social scientists to remove any stigma consumers have, since this may be a new crop in many rice-growing regions of developing countries, and because increased utilization of potatoes does not depend only on the resolution of scientific problems. Utilization of processed potato products will depend upon the purchasing power of consumers. However, if large quantities of potatoes are produced and if they are available to consumers at a reasonable price, it will become a prosperous industry. Economic and marketing studies will be needed to determine if processing is feasible. The use of cull (odd size and smaller) potatoes can be used as cattle feed. As economies develop, it is certain that the demand for animal feed can be made from processing waste.

Research on various established processed products must be done to develop other products as per the taste of various cultures. Successful utilization of potatoes depends upon industrial uses of starch and other edible food items and utilization of solanine, chaconine, and leptine in medicine and biological sciences.

What is needed is an organized system for rapid and easy exchange of information.

Symposia and workshops have been arranged and successful progress has been made by the International Potato Center at Lima, Peru. International programs adequately funded by each nation to provide complete and up-to-date information would certainly increase the likelihood of developing the potato crop to fulfill its potential in both developed and developing nations.

REFERENCES

1. NAS, *Population and Food: Crucial Issues,* National Academy of Sciences, Washington, D.C., 1975.
2. **Swaminathan, M. S. and Sawyer, R. L.,** The potential of the potato as a world food, in *Proc. Intl. Congress on Research for the Potato in the Year 2000,* International Potato Center, Lima, Peru, 1983, 3.
3. Food and Agriculture Year Book, U.N., Rome, 1988.
4. Agricultural Statistics, U.S. Department of Agriculture, Washington, D.C., 1988.
5. **Hooker, W. J., Ed.,** *Compendium of Potato Diseases,* American Phytopathological Society, St. Paul, MN, 1986, 1.
6. **Mazza, G.,** Personal communication, 1989.
7. Annual Report of International Potato Center, Lima, Peru, 1988.
8. **Ugent, D.,** The potato, *Science,* 170, 1161, 1971.
9. **Sowokinos, J. R.,** Recombinant DNA technology and improvement of potato quality, in *Proc. Intl. Congress on Research for the Potato in the Year 2000,* International Potato Center, Lima, Peru, 1983.
10. FAO, Analysis of FAO survey of post-harvest crop losses for developing countries, Food and Agriculture Organization of the United Nations, 1977.
11. National Research Council, World Food and Nutrition Study: National Research Council, Commission on International Relations, National Academy of Sciences, Washington, D.C., 1977.
12. **Booth, R. H. and Shaw, R. L.,** *Principles of Potato Storage,* International Potato Center, Lima, Peru, 1981, 105.
13. **Burton, W. G. and Booth, R. N.,** Postharvest technology for developing country tropical climate, in *Proc. Intl. Congress on Research for the Potato in the Year 2000,* International Potato Center, Lima, Peru, 1983, 40.
14. Food and Agriculture Organization Year Book, U.N., Rome, 1984.

INDEX

Houma, 54
Human teratology, 277
Humidity, 70
Hydration, 163
Hydrogen bonding, 165
Hydrogen peroxide, 256
Hydrolysis, 160, 213
Hydroxyalkyl starch, 164, 167—168
Hylum, 117
Hypobaric pressure, 232
Hypobaric storage, 79

I

Idaho Russet, 204
Impact damage, 26, 29—31
Income, 274, 275
Incubation, 214
India, 270
Indoleacetic acid, 96
Industrial waste, 138
Infection, 260
Inhibited starches, see Cross-linked starch
Insecticides, 51, 59
Insects, 56—59, 232
Insulation, 101, 102
Integument, 12—14, 26
International Potato Center, 279
Interveinal mosaic, 53
Invertase, 19, 123
Irish, 4
Irish Cobbler, 38
Iron, 5, 12, 21, 123, 137
Irradiation, 12, 215
γ-Irradiation
    glycoalkaloids and, 233, 234
    greening and, 84—85, 88
    loss control and, 96—99
    sprouting and, 88, 89
Irrigation, 27, 41, 116, 195—196
Isolubimin, 252
Isopropyl-N-carbamate, 233, 236—237
Isopropyl-N-chlorophenyl carbamate, 85, 95
Isopropyl-N-phenylcarbamate, 95

J

Juice, 181, 182

K

Kallikrein, 22
Katahdin
    chilling injury and, 91
    disease resistance, 47, 54
    glycoalkaloids, 215
    peeling, 126, 127
    production, 38
    sprouting, 89
    storage, 82
Kennebec

browning, 260
density and yield, 62
disease resistance, 54
glycoalkaloids, 204, 207
production, 39
sprouting, 89
storage, 75—77, 82
temperature, 119
Ketoheptose, 15
Kinetin, 96
Kufri cultivars, 39, 40

L

Lactic acid, 139
Lactobacillus spp., 181
Lady beetle, 57
Lamellae, 117
Lanosterol, 212
Late blight, 47—48, 56, 221, 259—261
Leaflets, 10—11
Leaf rolling mosaic, see Interveinal mosaic
Leaf scar, 12
Leakage, 102
Leaves, 10—11
Lectin, 23
Lembi, 126, 127
Lenape, 26, 204
Lenticel, 12, 13
Leptine, 221, 232
Leptinine, 204, 205, 221
Leptinotarsa decemlineata, 56
Lesion nematodes, 59
Letine, 204, 205
Light
    greening and, 79, 82—84
    injury, 219
    intensity, 214—215
    production and, 40—41
    storage and, 100, 101
Lignification, 249, 251
Lignin, 249, 251
Lime, 176, 177
Linuron, 43
Lipids, 19, 73
Liquefaction, 184
Lophenol, 212
Losses
    causation, 90—92
    control, 92—99
    processing, 276, 277
    storage, 89—90
Low-browning, 121
Low-temperature storage, see Storage, low-temperature
Lubimin, 250—252, 254—256, 258, 261
Lysine, 18, 275

M

Macrosiphum euphorbiae, 56